Load Testing of Bridges

Structures and Infrastructures Series

ISSN 1747–7735

Book Series Editor:

Dan M. Frangopol

Professor of Civil Engineering and
The Fazlur R. Khan Endowed Chair of Structural Engineering and Architecture
Department of Civil and Environmental Engineering
Center for Advanced Technology for Large Structural Systems (ATLSS Center)
Lehigh University
Bethlehem, PA, USA

Volume 12

Load Testing of Bridges

Current Practice and Diagnostic Load Testing

Editor

Eva O.L. Lantsoght

Politécnico, Universidad San Francisco de Quito, Quito, Ecuador
Concrete Structures, Department of Engineering Structures, Civil Engineering and Geosciences, Delft University of Technology, Delft, The Netherlands

CRC Press is an imprint of the
Taylor & Francis Group, an **informa** business

A BALKEMA BOOK

Colophon

Book Series Editor:
Dan M. Frangopol

Volume Editor:
Eva O. L. Lantsoght

CRC Press/Balkema is an imprint of the Taylor & Francis Group, an informa business

© 2019 Taylor & Francis Group, London, UK

Typeset by Apex CoVantage, LLC

All rights reserved. No part of this publication or the information contained herein may be reproduced, stored in a retrieval system, or transmitted in any form or by any means, electronic, mechanical, by photocopying, recording or otherwise, without written prior permission from the publishers.

Although all care is taken to ensure integrity and the quality of this publication and the information herein, no responsibility is assumed by the publishers nor the author for any damage to the property or persons as a result of operation or use of this publication and/or the information contained herein.

Library of Congress Cataloging-in-Publication Data

Names: Lantsoght, Eva O. L., editor.
Title: Load testing of bridges : current practice and diagnostic load testing / editor: Eva O.L. Lantsoght.
Description: Leiden : CRC Press/Balkema, [2019] | Series: Structures and infrastructures series, ISSN 1747-7735 ; volumes 12-13 | Includes bibliographical references and index.
Identifiers: LCCN 2019016183 (print) | LCCN 2019018556 (ebook) | ISBN 9780429265426 (ebook) | ISBN 9780367210823 (volume 1 : hbk) | ISBN 9780429265426 (volume 1 : e-Book) | ISBN 9780367210830 (volume 2 : hbk) | ISBN 9780429265969 (volume 2 : e-Book) | ISBN 9781138091986 (set of 2 volumes : hbk)
Subjects: LCSH: Bridges—Testing. | Bridges—Live loads. | Dynamic testing.
Classification: LCC TG305 (ebook) | LCC TG305 .L63 2019 (print) | DDC 624.2/52—dc23
LC record available at https://lccn.loc.gov/2019016183

Published by: CRC Press/Balkema
 Schipholweg 107C, 2316 XC, Leiden
 e-mail: Pub.NL@taylorandfrancis.com
 www.crcpress.com – www.taylorandfrancis.com

ISBN: 978-0-367-21082-3 Volume 12 (Hbk)
ISBN: 978-0-429-26542-6 Volume 12 (e-Book)
ISBN: 978-0-367-21083-0 Volume 13 (Hbk)
ISBN: 978-0-429-26596-9 Volume 13 (e-Book)
ISBN: 978-1-138-09198-6 Set of 2 Volumes (Hbk)
Structures and Infrastructures Series: ISSN 1747–7735
Volume 12

Printed and bound in Great Britain by
TJ International Ltd, Padstow, Cornwall

DOI: https://doi.org/10.1201/9780429265426

Table of Contents

Editorial	xiii
About the Book Series Editor	xv
Preface	xix
About the Editor	xxvii
Author Data	xxix
Contributors List	xxxiii
List of Tables	xxxv
List of Fgures	xxxvii

Part I Background to Bridge Load Testing   **I**

Chapter 1   Introduction   3
       Eva O. L. Lantsoght

1.1 Background	3
1.2 Scope of application	4
1.3 Aim of this book	5
1.4 Outline of this book	6

Chapter 2   History of Load Testing of Bridges   9
       Mohamed K. ElBatanouny, Gregor Schacht
       and Guido Bolle

2.1 Introduction	9
2.2 Bridge load testing in Europe	11
2.3 Bridge load testing in North America	20
2.4 The potential of load testing for the evaluation of existing structures	24
2.5 Summary and conclusions	25
References	25

Chapter 3   Current Codes and Guidelines   29
       Eva O. L. Lantsoght

3.1 Introduction	29
3.2 German guidelines	30
3.2.1 General	30

	3.2.2 Safety philosophy and target proof load	31
	3.2.3 Stop criteria	32
3.3	British guidelines	34
	3.3.1 General	34
	3.3.2 Preparation and application of loading	35
	3.3.3 Evaluation of the load test	36
3.4	Irish guidelines	36
	3.4.1 General	36
	3.4.2 Recommendations for applied loading	37
	3.4.3 Evaluation of the load test	37
3.5	Guidelines in the United States	37
	3.5.1 Bridges: *Manual for Bridge Rating through Load Testing*	37
	3.5.1.1 General	37
	3.5.1.2 Preparation of load tests	39
	3.5.1.3 Execution of load tests	40
	3.5.1.4 Determination of the rating factor after a diagnostic load test	41
	3.5.1.5 Determination of the rating factor after a proof load test	42
	3.5.2 Buildings	43
	3.5.2.1 ACI 437.1: "Load tests of concrete structures: methods, magnitude, protocols, and acceptance criteria"	43
	3.5.2.2 New buildings: ACI 318-14	48
	3.5.2.3 Existing buildings: ACI 437.2M-13	49
3.6	French guidelines	53
	3.6.1 General	53
	3.6.2 Recommendations for load application	53
	3.6.3 Evaluation of the load test	54
3.7	Czech Republic and Slovakia	54
	3.7.1 General requirements	54
	3.7.2 Acceptance criteria	54
	3.7.3 Dynamic load tests	56
3.8	Spanish guidelines	57
	3.8.1 General considerations	57
	3.8.2 Loading requirements	58
	3.8.3 Stop and acceptance criteria for static load tests	59
	3.8.4 Acceptance criteria for dynamic load tests	62
3.9	Other countries	64
	3.9.1 Italy	64
	3.9.2 Switzerland	64
	3.9.3 Poland	65
	3.9.4 Hungary	65
3.10	Current developments	67
3.11	Discussion	67
3.12	Summary	68
	References	68

Table of Contents   VII

Part II Preparation, Execution, and Post-Processing of Load Tests on Bridges   **71**

Chapter 4   General Considerations   73
      Eva O. L. Lantsoght and Jacob W. Schmidt
4.1 Initial considerations   73
      4.1.1 Introductory remarks   73
      4.1.2 Load test types and their goals   73
      4.1.3 Type of bridge structure or element   74
      4.1.4 Structural inspections, background codes, and literature   74
4.2 Types of load tests, and which type of load test to select   78
      4.2.1 Diagnostic load tests   78
      4.2.2 Proof load tests   79
      4.2.3 Failure tests   81
4.3 When to load test a bridge, and when not to load test   82
4.4 Structure type considerations   84
      4.4.1 Steel bridges   84
      4.4.2 Reinforced concrete bridges   84
      4.4.3 Prestressed concrete bridges   85
      4.4.4 Masonry bridges   86
      4.4.5 Timber bridges   87
4.5 Safety requirements during load testing   87
      4.5.1 General considerations   87
      4.5.2 Safety of personnel and traveling public   88
      4.5.3 Structural safety   90
4.6 Summary and conclusions   90
      References   91

Chapter 5   Preparation of Load Tests   97
      Eva O. L. Lantsoght and Jacob W. Schmidt
5.1 Introduction   97
5.2 Determination of test objectives   98
5.3 Bridge inspection   99
      5.3.1 Inspection results   99
      5.3.2 Limitations of testing site   101
5.4 Preliminary calculations and development of finite element model   104
      5.4.1 Development of finite element model   104
      5.4.2 Assessment calculations   106
      5.4.3 Estimation of bridge behavior during load test   108
      5.4.4 Shear capacity considerations   110
5.5 Planning and preparation of load test   111
      5.5.1 Planning   111
      5.5.2 Personnel requirements   113
      5.5.3 Loading requirements   113
      5.5.4 Traffic control and safety   117
      5.5.5 Measurements and sensor plan   119

VIII Table of Contents

5.6 Summary and conclusions	125
References	127

Chapter 6 General Considerations for the Execution of Load Tests 129
Eva O. L. Lantsoght and Jacob W. Schmidt

6.1 Introduction	129
6.2 Loading equipment	130
6.3 Measurement equipment	132
6.3.1 Measurement requirements	132
6.3.2 Data acquisition and visualization equipment	133
6.3.3 Sensors	135
6.3.4 Interpretation of measurements during load test	136
6.4 Practical aspects of execution	137
6.4.1 Communication	137
6.4.2 Safety	137
6.5 Summary and conclusions	138
References	140

Chapter 7 Post-Processing and Bridge Assessment 141
Eva O. L. Lantsoght and Jacob W. Schmidt

7.1 Introduction	141
7.2 Post-processing of measurement data	142
7.2.1 Applied load	142
7.2.2 Verification of measurement data	142
7.2.3 Correction for support deformations	143
7.2.4 Correction for influence of temperature and humidity	144
7.2.5 Reporting of measurements	144
7.3 Updating finite element model with measurement data	147
7.4 Bridge assessment	148
7.5 Formulation of recommendations for maintenance or operation	149
7.6 Recommendations for reporting of load tests	149
7.7 Summary and conclusions	150
References	151

Part III Diagnostic Load Testing of Bridges **153**

Chapter 8 Methodology for Diagnostic Load Testing 155
Eva O. L. Lantsoght, Jonathan Bonifaz, Telmo A. Sanchez and Devin K. Harris

8.1 Introduction	155
8.2 Preparation of diagnostic load tests	157
8.2.1 New bridge diagnostic testing	157
8.2.2 Existing bridge diagnostic testing	161
8.3 Procedures for the execution of diagnostic load testing	162
8.3.1 Loading methods	162
8.3.2 Monitoring bridge behavior during test	163

8.4	Processing diagnostic load testing results	164
	8.4.1 On-site validation and review of test data	164
	8.4.2 Processing and reporting test data	166
	8.4.3 Verification of structural responses for new bridges	166
	8.4.4 Calibration of analytical model for existing bridges	167
8.5	Evaluation of diagnostic load testing results	168
	8.5.1 Evaluation of results for new bridges	168
	8.5.2 Improved assessment for existing bridges	171
8.6	Summary and conclusions	172
	References	172
	Appendix: Determination of Experimental Rating Factor According to Barker	176

Chapter 9   Example Field Test to Load Rate a Prestressed Concrete Bridge   181
Eli S. Hernandez and John J. Myers

9.1	Introduction	181
9.2	Sample bridge description	182
9.3	Bridge instrumentation plan	183
	9.3.1 Installation of embedded sensors	183
	9.3.2 Data acquisition by non-contact and remote equipment	184
	9.3.2.1 Automated total station (ATS)	185
	9.3.2.2 Remote sensing vibrometer (RSV-150)	186
9.4	Diagnostic load test program	186
	9.4.1 Static load test	187
	9.4.2 Dynamic load test	187
9.5	Test results	187
	9.5.1 Static load tests	187
	9.5.1.1 Vertical deflection	187
	9.5.1.2 Lateral distribution factor (deflection measurements)	191
	9.5.1.3 Girders' longitudinal strain	191
	9.5.1.4 Lateral distribution factor (strain measurements)	193
	9.5.2 Dynamic load tests	193
9.6	Girder distribution factors	195
9.7	Load rating of Bridge A7957 by field load testing	197
9.8	Recommendations	199
9.9	Summary	199
	References	200

Chapter 10   Example Load Test: Diagnostic Testing of a Concrete Bridge
with a Large Skew Angle   201
Mauricio Diaz Arancibia and Pinar Okumus

10.1	Summary	201
10.2	Characteristics of the bridge tested	202
10.3	Goals of load testing	202
10.4	Preliminary analytical model	203
10.5	Coordination of the load test	204
10.6	Instrumentation plan	205

X Table of Contents

	10.6.1 Sensor types and application methods	205
	10.6.2 Sensor locations	208
10.7	Data acquisition	209
10.8	Loading	209
	10.8.1 Load type and magnitude	209
	10.8.2 Load configurations and locations	210
10.9	Planning and scheduling	211
10.10	Redundancy and repeatability	211
10.11	Results	212
	10.11.1 Preliminary evaluation of results	212
	10.11.2 Shear strain influence lines and shear distribution	212
	10.11.3 Bending strain influence lines and moment distribution	213
	10.11.4 Deck strains under short-term loading	214
10.12	Conclusions and recommendations	214
	Ackowledgements	216
	References	216

Chapter 11	Diagnostic Load Testing of Bridges – Background and Examples of Application	217
	Piotr Olaszek and Joan R. Casas	
11.1	Background	217
	11.1.1 Definition	217
	11.1.2 Objectives	218
	11.1.3 Planning and execution	218
	11.1.4 Results and safety assessment	220
	11.1.4.1 Static tests	221
	11.1.4.2 Dynamic tests	222
11.2	Examples of diagnostic load testing	223
	11.2.1 Static load testing	223
	11.2.1.1 The estimation of the elastic and permanent values	223
	11.2.1.2 Examples of application to different types of bridges	224
	11.2.2 Dynamic load testing	234
	11.2.2.1 Extrapolation of values for quasi-static speed	234
	11.2.2.2 Extrapolation of values under higher speed	236
	11.2.2.3 Examples of dynamic testing	236
11.3	Conclusions and recommendations for practice	246
	References	247

Chapter 12	Field Testing of Pedestrian Bridges	249
	Darius Bačinskas, Ronaldas Jakubovskis and Arturas Kilikevičius	
12.1	Introduction	249
	12.1.1 Types of the tests	252
	12.1.2 Objectives of the tests	253
12.2	Preparation for testing	254
	12.2.1 General guidelines	254
	12.2.2 Preliminary inspection of the footbridge before the tests	255

12.2.3	The test program	257
12.2.4	Loading of the bridge	258
	12.2.4.1 Static tests	258
	12.2.4.2 Free vibration tests	262
	12.2.4.3 Forced and ambient vibration tests	263

12.3 Organization of the tests — 264
    12.3.1 General requirements — 264
    12.3.2 Measuring techniques and equipment — 265
    12.3.3 Execution of the tests — 269
        12.3.3.1 Static tests — 269
        12.3.3.2 Dynamic tests — 271
12.4 Analysis of test results — 271
    12.4.1 General guidelines — 272
    12.4.2 Methods for identification of static and dynamic parameters
        of the bridge — 272
        12.4.2.1 Methods for identification of static parameters
            of the bridge — 272
        12.4.2.2 Methods for identification of dynamic parameters
            of the bridge — 274
    12.4.3 Presentation of results — 278
12.5 Theoretical modeling of tested bridge — 278
    12.5.1 Introduction — 279
    12.5.2 Modeling techniques — 279
    12.5.3 Comparison of experimental and theoretical results — 280
    12.5.4 Model updating — 283
    12.5.5 Code requirements for serviceability of footbridges — 284
    12.5.6 Evaluation of footbridge condition based on test results — 286
12.6 Concluding remarks — 286
    Acknowledgments — 287
    References — 287

Author Index — 291
Subject Index — 293
Structures and Infrastructures Series — 301

Editorial

Welcome to the book series *Structures and Infrastructures*.

Our knowledge to model, analyze, design, maintain, manage, and predict the life-cycle performance of structures and infrastructures is continually growing. However, the complexity of these systems continues to increase and an integrated approach is necessary to understand the effect of technological, environmental, economic, social, and political interactions on the life-cycle performance of engineering structures and infrastructures. In order to accomplish this, methods have to be developed to systematically analyze structure and infrastructure systems, and models have to be formulated for evaluating and comparing the risks and benefits associated with various alternatives. We must maximize the life-cycle benefits of these systems to serve the needs of our society by selecting the best balance of the safety, economy, and sustainability requirements despite imperfect information and knowledge.

In recognition of the need for such methods and models, the aim of this book series is to present research, developments, and applications written by experts on the most advanced technologies for analyzing, predicting, and optimizing the performance of structures and infrastructures such as buildings, bridges, dams, underground construction, offshore platforms, pipelines, naval vessels, ocean structures, nuclear power plants, and also airplanes, aerospace, and automotive structures.

The scope of this book series covers the entire spectrum of structures and infrastructures. Thus it includes, but is not restricted to, mathematical modeling, computer and experimental methods, practical applications in the areas of assessment and evaluation, construction and design for durability, decision-making, deterioration modeling and aging, failure analysis, field testing, structural health monitoring, financial planning, inspection and diagnostics, life-cycle analysis and prediction, loads, maintenance strategies, management systems, nondestructive testing, optimization of maintenance and management, specifications and codes, structural safety and reliability, system analysis, time-dependent performance, rehabilitation, repair, replacement, reliability and risk management, service life prediction, strengthening, and whole life costing.

This book series is intended for an audience of researchers, practitioners, and students worldwide with a background in civil, aerospace, mechanical, marine, and automotive engineering, as well as people working in infrastructure maintenance, monitoring, management, and cost analysis of structures and infrastructures. Some volumes are monographs defining the current state of the art and/or practice in the field, and some are textbooks to be used in undergraduate (mostly seniors), graduate, and postgraduate courses. This book series is affiliated to *Structure and Infrastructure Engineering*

DOI: https://doi.org/10.1201/9780429265426

(www.tandfonline.com/toc/nsie20/current), an international peer-reviewed journal which is included in the Science Citation Index.

It is now up to you, authors, editors, and readers, to make *Structures and Infrastructures* a success.

Dan M. Frangopol
Book Series Editor

About the Book Series Editor

Dan M. Frangopol is the first holder of the Fazlur R. Khan Endowed Chair of Structural Engineering and Architecture at Lehigh University, Bethlehem, Pennsylvania, USA, and a Professor in the Department of Civil and Environmental Engineering at Lehigh University. He is also an Emeritus Professor of Civil Engineering at the University of Colorado at Boulder, USA, where he taught for more than two decades (1983–2006). Before joining the University of Colorado, he worked for four years (1979–1983) in structural design with A. Lipski Consulting Engineers in Brussels, Belgium. In 1976, he received his doctorate in Applied Sciences from the University of Liège, Belgium, and holds four honorary doctorates (Doctor Honoris Causa) from the Technical University of Civil Engineering in Bucharest, Romania (2001); the University of Liège, Belgium (2008); the Gheorghe Asachi Technical University of Iași, Romania (2014); and the Polytechnic University of Milan (Politecnico di Milano), Milan, Italy (2016).

Dr. Frangopol is an Honorary Professor at 13 universities (Hong Kong Polytechnic, Tongji, Southeast, Tianjin, Dalian, Hunan, Chang'an, Beijing Jiaotong, Chongqing Jiaotong, Shenyang Jianzhu, Royal Melbourne Institute of Technology (RMIT), Changsha University of Science and Technology, and Harbin Institute of Technology) and a Visiting Chair Professor at the National Taiwan University of Science and Technology. He is a Distinguished Member of the American Society of Civil Engineers (ASCE), Foreign Member of the Academia Europaea (Academy of Europe, London), Foreign Member of the Royal Academy of Belgium for Science and the Arts, Honorary Member of the Romanian Academy, Honorary Member of the Romanian Academy of Technical Sciences, Inaugural Fellow of both the Structural Engineering Institute and the Engineering Mechanics Institute of ASCE, Fellow of the American Concrete Institute (ACI), Fellow of the International Association for Bridge and Structural Engineering (IABSE), Fellow of the International Society for Health Monitoring of Intelligent Infrastructures (ISHMII), and Fellow of the Japan Society for the Promotion of Science (JSPS). He is the President of the International Association for Bridge Maintenance and Safety (IABMAS), Honorary Member of the Portuguese Association for Bridge Maintenance and Safety (IABMAS-Portugal Group), Honorary Member of the IABMAS-China Group, Honorary Member of the IABMAS-Australia Group, Honorary Member of the IABMAS-Japan Group, Honorary President of the IABMAS-Italy Group, Honorary President of the IABMAS-Brazil Group, Honorary

DOI: https://doi.org/10.1201/9780429265426

President of the IABMAS-Chile Group, Honorary President of the IABMAS-Turkey Group, and Honorary President of the IABMAS-Korea Group.

He is the initiator and organizer of the Fazlur R. Khan Distinguished Lecture Series (www.lehigh.edu/frkseries) at Lehigh University. He is an experienced researcher and consultant to industry and government agencies, both nationally and abroad. His main research interests are in the development and application of probabilistic concepts and methods to civil and marine engineering, including structural reliability; life-cycle cost analysis; probability-based assessment, design, and multi-criteria life-cycle optimization of structures and infrastructure systems; structural health monitoring; life-cycle performance maintenance and management of structures and distributed infrastructure under extreme events (earthquakes, tsunamis, hurricanes, and floods); risk-based assessment and decision making; multi-hazard risk mitigation; infrastructure sustainability and resilience to disasters; climate change adaptation; and probabilistic mechanics.

According to ASCE (2010), "Dan M. Frangopol is a preeminent authority in bridge safety and maintenance management, structural systems reliability, and life-cycle civil engineering. His contributions have defined much of the practice around design specifications, management methods, and optimization approaches. From the maintenance of deteriorated structures and the development of system redundancy factors to assessing the performance of long-span structures, Dr. Frangopol's research has not only saved time and money, but very likely also saved lives."

Dr. Frangopol's work has been funded by NSF, FHWA, NASA, ONR, WES, AFOSR, ARDEC, and numerous other agencies. He is the Founding President of the International Association for Bridge Maintenance and Safety (IABMAS, www.iabmas.org) and the International Association for Life-Cycle Civil Engineering (IALCCE, www.ialcce.org), and is Past Director of the Consortium on Advanced Life-Cycle Engineering for Sustainable Civil Environments (COALESCE). He is the Past Vice-President of the International Association for Structural Safety and Reliability (IASSAR), Past Vice-President of the Engineering Mechanics Institute (EMI) of ASCE, and Past Member of its Board of Governors. He is also the Founding Chair of the ASCE-SEI Technical Council on Life-Cycle Performance Safety, Reliability and Risk of Structural Systems and of the IASSAR Technical Committee on Life-Cycle Performance, Cost and Optimization. He has held numerous leadership positions in national and international professional societies including Chair of the Technical Activities Division of the 20,000+ members of the Structural Engineering Institute (SEI) of ASCE, Chair of Executive Board of IASSAR, Vice-President of the International Society for Health Monitoring of Intelligent Infrastructures (ISHMII), and Chair of IABSE Working Commission 1 on Structural Performance, Safety, and Analysis.

Dr. Frangopol is the recipient of several prestigious awards including the George W. Housner Medal, the 2016 ASCE OPAL Award for Lifetime Accomplishments in Education, the 2016 ASCE Alfredo Ang Award, the 2016 ASCE-Lehigh Valley Section Civil Engineer of the Year Award, the 2015 ASCE Noble Prize, the 2014 ASCE James R. Croes Medal, the 2012 IALCCE Fazlur R. Khan Life-Cycle Civil Engineering Medal, the 2012 ASCE Arthur M. Wellington Prize, the 2012 IABMAS Senior Research Prize, the 2008 IALCCE Senior Award, the 2007 ASCE Ernest Howard Award, the 2006 IABSE OPAC Award, the 2006 Elsevier Munro Prize, the 2006 T. Y. Lin Medal, the 2005 ASCE Nathan M. Newmark Medal, the 2004 Kajima Research Award, the 2003

ASCE Moisseiff Award, the 2002 and 2016 JSPS Fellowship Award for Research in Japan, the 2001 ASCE J. James R. Croes Medal, the 2001 IASSAR Research Prize, the 1998, 2004 and 2019 ASCE State-of-the-Art of Civil Engineering Award, and the 1996 Distinguished Probabilistic Methods Educator Award of the Society of Automotive Engineers (SAE). Among several awards he has received at the University of Colorado, he is the recipient of the 2004 Boulder Faculty Assembly Excellence in Research Scholarly and Creative Work Award, the 1999 College of Engineering and Applied Science's Research Award, the 2003 Clarence L. Eckel Faculty Prize for Excellence, and the 1987 Teaching Award. He is also the recipient of the Lehigh University's 2013 Eleanor and Joseph F. Libsch Research Award and of the Lehigh University's 2016 Hillman Award for Excellence in Graduate Education. He has given plenary keynote lectures at numerous major conferences held in Asia, Australia, Europe, North America, South America, and Africa.

Dr. Frangopol is the Founding Editor in Chief of *Structure and Infrastructure Engineering* (Taylor & Francis, www.tandfonline.com/toc/nsie20/current), an international peer-reviewed journal. This journal is dedicated to recent advances in maintenance, management, and life-cycle performance of a wide range of structures and infrastructures. He is the author or co-author of 3 books, 50 book chapters, more than 380 articles in refereed journals, and over 600 papers in conference proceedings. He is also the editor or co-editor of 48 books published by ASCE, Balkema, CIMNE, CRC Press, Elsevier, McGraw-Hill, Taylor & Francis, and Thomas Telford, and is an editorial board member of several international journals. Additionally, he has chaired and organized several national and international structural engineering conferences and workshops.

Dr. Frangopol has supervised 45 Ph.D. and 55 M.Sc. students. Many of his former students are professors at major universities in the United States, Asia, Europe, and South America, and several are prominent in professional practice and research laboratories.

For additional information on Dr. Frangopol's activities, please visit www.lehigh.edu/~dmf206/.

Preface

Load testing of bridges is a practice as old as bridge engineering. From the early days, when a load test prior to opening had to convince the traveling public of the safety of the bridge, engineers and the general public have seen the value of load testing. The purpose of this work, divided across two volumes, is to study load testing from different perspectives. As such, this work deals with the practical aspects related to load testing and the current scientific developments related to load testing of bridges, and it offers an international perspective on the topic of load testing of bridges. You can find general recommendations, advice, and best practices in these books along with detailed case studies, so that this work can be a guide for the engineer preparing a load test. You can also find open research questions, topics for future research, and the latest research findings related to load testing in these books, so that this work can be a guide for researchers who want to identify interesting topics to study.

The work is divided in eight parts across two volumes (Volumes 12 and 13). Part I of Volume 12 gives a background to bridge load testing, including the historical perspectives and currently governing codes and guidelines. The background is discussed from an international perspective, outlining the history of load testing in North America and Europe, and summarizing the current codes and guidelines for load testing from Germany, the United States, the United Kingdom, Ireland, Poland, Hungary, Spain, the Czech Republic and Slovakia, Italy, Switzerland, and France.

Part II of Volume 12 deals with general practical aspects of load testing, which are valid for the different types of load tests (diagnostic and proof load tests). These practical aspects cover the entire stage of a load testing project: from preparation, to execution, to post-processing and reporting about the load test. This first topic in this part contains general considerations for each load testing project: which type of load test is suitable for the project, and whether load testing is the best option for the considered bridge. If the decision is made to load test a bridge, the next step is to prepare this load test. Therefore, the next topic discusses the elements of the preparation of a load test: inspection, preliminary calculations, and planning of the project. Then, general aspects of the execution of the load test are discussed, with regard to loading equipment, measurement equipment, and practical aspects of communication and safety. The last topic of this part deals with post-processing of load testing data and reporting of load tests.

Part III of Volume 12 focuses on diagnostic load testing of bridges. It discusses the specific aspects of diagnostic load testing during the preparation, execution, and post-processing of a diagnostic load test and the general methodology for diagnostic load tests. Chapters describing detailed examples of diagnostic load tests from North America

DOI: https://doi.org/10.1201/9780429265426

and Europe are included. One chapter that deals with the particularities of testing pedestrian bridges is included as well.

Part I of Volume 13 focuses on proof load testing of bridges. It discusses the specific aspects of proof load testing during the preparation, execution, and post-processing of a proof load test. Important topics in this part are the interpretation of the measurements in real time during the experiment and the determination of the target proof load. Since high loads are applied in proof load tests, the risk of collapse or permanent damage to the structure exists. For this reason, careful instrumentation and monitoring of the bridge during the proof load test is of the utmost importance. Criteria based on the structural response that can be used to evaluate when further loading may not be permitted are discussed. Chapters describing examples of proof load test from North America and Europe are included.

Part II of Volume 13 describes how the practice of load testing can also be applied to buildings. Although the main focus of this book is on bridges, the same principles can be used for load testing of buildings. Buildings can require a load test prior to opening when there are doubts about the adequate performance of the building, or during the service life of the building when there are doubts regarding the capacity as a result of material degradation or deterioration, or when the use and associated live load of the building is to be changed.

Part III of Volume 13 discusses novel ideas regarding measurement techniques used for load testing. Methods using non-contact sensors such as photography- and video-based measurement techniques are discussed. With acoustic emission measurements, signals of microcracking and distress can be observed before cracks are visible. Fiber optics are explored and applied as a new measurements technique. The topic of measurements through radar interferometry is also discussed. These chapters contain the background of these novel measurement techniques and recommendations for practice, and identify topics for further research and improvement of these measurement techniques.

Part IV of Volume 13 discusses load testing in the framework of reliability-based decision-making and in the framework of a bridge management program. The first topic of this part deals with the topic of updating the reliability index of a bridge after load testing, and discusses the required proof load magnitude for this purpose, as well as systems reliability considerations. The second topic deals with the effect of load test results on the estimation of the remaining service life of a tested bridge, taking into account the effects of degradation. Finally, two chapters from the perspective of the bridge owner discuss how load testing fits in the framework of a bridge management program.

Part V of Volume 13 brings all information from the previous parts in Volumes 12 and 13 together and discusses the current state-of-the art on load testing. An overview of open questions for research is given, along with generally practical recommendations for load testing.

Besides the general structure in eight parts (three parts in Volume 12 and five parts in Volume 13), this work contains 25 chapters (12 chapters in Volume 12 and 13 chapters in Volume 13) written by a collective of international experts on the topic of load testing. Chapter 1 of Volume 12, by Lantsoght, introduces the general topic of load testing of structures. In this chapter, the reader can find a short background to the topic of load testing of structures, the scope of this work, the aim of this work, and a short discussion of the structure and outline of the two volumes making up this work.

Chapter 2 of Volume 12, by ElBatanouny, Schacht, and Bolle, provides an overview of the historical development and current practices of bridge load tests in Europe and the United States. The use of load tests to prove the capacity of structures is as old as mankind and plays an important role in the historical development of reinforced concrete design and construction. Historically, load tests provided a proof that a structure can carry a certain load and, therefore, they were likely used to convince the people of the bearing capability of bridges. With the development of static calculations and acceptable design rules, load tests became unnecessary for new structures. However, currently, strength evaluation of existing structures is becoming more and more important, and the advantages of experimental assessments are used. Chapter 2 of Volume 12 describes the development of the technology of load testing and shows the European and American way of practice.

In Chapter 3 of Volume 12, Lantsoght reviews the existing codes and guidelines for load testing of structures. A summary of the main requirements of each existing code is provided, with a focus on the determination of the required load and measurements. The requirements for load testing of bridges and buildings are revised, for new and existing structures. An international perspective is given, revising the practice from Germany, the United Kingdom, Ireland, the United States, France, the Czech Republic, Slovakia, Spain, Italy, Switzerland, Poland, and Hungary. The chapter concludes with a short overview of the current developments and a discussion of the differences between the available codes and guidelines.

In Chapter 4 of Volume 12, Lantsoght and Schmidt discuss aspects that should be considered prior to every load test. The first questions that need to be answered are "Is this bridge suitable for load testing?" and "If so, what are the goals of the load test?" To answer these questions, information must be gathered, and preliminary calculations should be carried out. In order to evaluate if a bridge is suitable for load testing, different types of testing are shown, the topic of when to load test a bridge is discussed, and structure type considerations are debated. Finally, some safety precautions that should be fulfilled during a load test are discussed.

Subsequently, Lantsoght and Schmidt discuss in Chapter 5 of Volume 12 the aspects related to the preparation of load tests, regardless of the chosen type of load test. After determination of the test objectives, the first step should be a technical inspection of the bridge and bridge site. With this information, the preparatory calculations (assessment for existing bridges and expected behavior during the test) can be carried out. Once the analytical results are available, the practical aspects of testing can be prepared: planning, required personnel, method for applying the load, considerations regarding traffic control and safety, and the development of the sensor and data acquisition plan. It is good practice to summarize all preparatory aspects in a preparation report and to provide this information to the client/owner as well as to all parties involved with the load test.

Then in Chapter 6 of Volume 12, Lantsoght and Schmidt discuss the aspects related to the execution of load tests, regardless of the chosen type of load test. The main elements required for the execution of the load test are the equipment for applying the load and the equipment for measuring and displaying the structural responses (if required). This chapter reviews the commonly used equipment for applying the loading and discusses all aspects related to the measurements. The next topic is the practical aspects related to the execution. This topic deals with communication on site during the load test and important safety aspects.

Finally, in Chapter 7 of Volume 12, Lantsoght and Schmidt discuss the aspects related to processing the results of a load test after the test. The way in which the data are processed depends on the goals of the test. As such, the report that summarizes the preparation, execution, and post-processing of the load test should clearly state the goal of the load test, how the test addressed this goal, and what can be concluded. Typical elements of the post-processing stage include discussing the applied load, the measured structural responses, and then evaluating the bridge based on the results of the load test.

Chapter 8 of Volume 12, by Lantsoght, Bonifaz, Sanchez, and Harris, deals with the methodology for diagnostic load testing. All aspects of diagnostic load testing that are shared with other load testing methods have been discussed in Part II of Volume 12. In this chapter, the particularities of diagnostic load testing of new and existing bridges are discussed. These elements include loading procedures, monitoring behavior during the test, reviewing test data, calibrating analytical models, and evaluating the test results.

In Chapter 9 of Volume 12, Hernandez and Myers illustrate an example diagnostic load test. This chapter introduces an example diagnostic load test conducted on the superstructure of Bridge A7957, built in Missouri, USA, to illustrate how experimental in-situ parameters can be included in the estimation of a bridge load rating. The experimental load rating was less conservative than the analytical load rating.

In Chapter 10 of Volume 12, Diaz Arancibia and Okumus illustrate an example diagnostic load test. They argue that bridge load testing is widely used for assessing bridge structural behavior and may be preferred over other means, since it is capable of capturing the actual response of structures. However, load testing is complex and requires careful consideration of several activities that precede its execution. They describe the planning, coordination, scheduling, execution, and data analysis of a load test, using an example of a highly skewed, prestressed concrete girder/reinforced concrete deck bridge under service loads. The load test allowed the evaluation of the effects of high skew angles and mixed pier support fixity arrangements on girder load distribution behavior and deck performance.

In Chapter 11 of Volume 12, Olaszek and Casas present principles and justification of diagnostic load tests of bridges as performed from the European point of view. Normally, diagnostic tests serve to verify and adjust the predictions of an analytical model. However, as presented in this chapter, the results of a diagnostic load test in a bridge can also serve other objectives. Several examples of application are presented with the main objective not only to show how the tests are carried out but also to introduce which are the main issues to take into account to obtain accurate and reliable results that can be used in the assessment of the actual capacity of the bridge. In the case of static tests, conclusions regarding the measurement stabilization time are presented. For dynamic tests in railway bridges, the extrapolation to lower and higher speeds is also discussed.

In Chapter 12 of Volume 12, Bačinskas, Jakubovskis, and Kilikevičius show how diagnostic load tests can be applied to pedestrian bridges. This chapter contains basic information related to static and dynamic testing of pedestrian bridges. A brief overview of test objectives and test classification is presented. The chapter also covers the stages of preparation for testing, aspects of test program creating, and methods of static and dynamic loading of footbridges. A sequence of test organization and execution is also presented. The second part of the chapter discusses the aspects of processing and evaluation of static and dynamic test results as well as aspects of comparing them with obtained by the theoretical modeling of the bridge. The chapter presents requirements for pedestrian bridges specified in the design codes and recommendations of various countries. These need to be taken into account in assessing the results of the tested bridges. The chapter ends with a discussion on the presentation of the test results and aspects of assessment of the bridge condition according to the test data. The presented material may be useful to researchers and experts involved in the design, construction, and maintenance of pedestrian bridges.

Chapter 1 of Volume 13, by Lantsoght, deals with the methodology for proof load testing. All aspects of proof load testing that are shared with other load testing methods have been discussed in Part II of Volume 12. In this chapter, the particularities of proof load testing are discussed based on the current state of the art. These elements include the determination of the target proof load (which is still a topic of research), the procedures followed during a proof load test (loading method, instrumentation, and stop criteria), and the post-processing of proof load test data, including the assessment of a bridge after a proof load test.

In Chapter 2 of Volume 13, Jauregui, Weldon, and Aguilar show that load rating of concrete bridges with no design plans is currently an issue in the United States and other countries. Missing or incomplete design documentation creates uncertainties in establishing the safe load limits of bridges to carry legal vehicles. Guidance for evaluating planless concrete bridges, in particular prestressed structures, is limited in the AASHTO Manual for Bridge Evaluation and very few state departments of transportation have rating procedures for such bridges. The authors developed a multi-step load rating procedure for planless prestressed bridges that includes (1) estimating the material properties from past specifications and amount of prestressing steel using Magnel diagrams; (2) verifying the steel estimate by rebar scanning; (3) field testing at diagnostic and/or proof load levels based on strain measurements; and (4) rating the bridges using the proof test results. Three prestressed concrete bridges (including a double T-beam, box beam, and I-girder bridge) are evaluated using nondestructive load testing and material evaluation techniques to illustrate the procedures. Rating factors are determined for AASHTO and New Mexico legal loads using the proof test results for the serviceability limit state (SLS) (i.e. concrete cracking). Using load rating software, rating factors are also computed for the strength limit state (i.e. shear or flexural capacity) based on the measured bridge dimensions and estimated material properties. Load ratings for serviceability and strength are finally compared to establish the bridge capacities.

Lantsoght, Hordijk, Koekkoek, and Van der Veen describe a case study of a proof load test in Chapter 3 of Volume 13. The viaduct Zijlweg was proof load tested at a position that is critical for bending moment and at a position that is critical for shear. The viaduct Zijlweg has cracking caused by an alkali-silica reaction, and the effect of material degradation on the capacity is uncertain. Therefore, the assessment of this viaduct was carried out with a proof load test. This chapter details the preparation, execution, and evaluation of viaduct Zijlweg.

In Chapter 4 of Volume 13, Schacht, Bolle, and Marx show how the principles of load testing of bridges also can be used for building constructions by sharing the background and main recommendations from the guideline for load testing of existing concrete structures of the Deutscher Ausschuss für Stahlbeton (DAfStb). Thanks to this guideline, published in 2000, the experimental proof of the load-bearing capacity became a widely accepted method in Germany. Since then, over 2000 proof load tests have been carried out and a lot of experience exists in the usage of the guideline. Nevertheless, there are still reports about load tests carried out with mass loads or using mechanical measurement techniques. To prevent misuse of the guideline in the future, to take into account the great experiences existing in the evaluation of the measuring results and the bearing condition, the DAfStb decided to revise the existing guideline. In recent years there has also been increased research activity focusing on the development of evaluation criteria for possible brittle failures in shear and to extend the safety concept for the evaluation of elements that have not been directly tested. These results will also be considered in the new guideline. The authors give an overview of the rules of the existing guideline and discuss these in the background of the gained experience over the past two decades. The new criteria for evaluating shear capacity are explained, and thoughts about the new safety concept for load testing are introduced.

In Chapter 5 of Volume 13, Alipour, Shariati, Schumacher, Harris, and Riley show that in order to capture the deformation response of a bridge member during load testing, the structure has to be properly instrumented. Image- and video-based measurement techniques have significant advantages over traditional physical sensors in that they are applied remotely without physical contact with the structure, they require no cabling, and they allow for measuring displacement where no ground reference is available. The authors introduce two techniques used to analyze digital images: digital image correlation (DIC) and Eulerian virtual visual sensors (VVS). The former is used to measure the static displacement response and the latter to capture the natural vibration frequencies of structural members. Each technique is introduced separately, providing a description of the fundamental theory, presenting the equipment needed to perform the measurements, and discussing the strengths and limitations. Case studies provide examples of real-world applications during load testing to give the reader a sense of the advances and opportunities of these measurement techniques.

Chapter 6 of Volume 13, by ElBatanouny, Anay, Abdelrahman, and Ziehl, discusses the use of acoustic emission (AE) measurements for load testing. Several methods for analyzing AE data to classify damage in reinforced and prestressed concrete structures during load tests were developed over the past two decades. The majority of methods offer relative assessment of damage – for example, classifying cracked versus uncracked conditions in prestressed members during or following a load test. In addition, significant developments were made in various AE source location techniques, including one-, two-, and three-dimensional source location as well as moment tensor analysis, which allow for accurate location of damage through advanced data filtering techniques. This chapter provides an overview of the acoustic emission technique for detecting and classifying damage during load tests. Recent efforts to apply the method in the field are also presented along with recommended field applications based on the current state of practice.

Chapter 7 of Volume 13, by Casas, Barrias, Rodriguez Gutiérrez, and Villalba, presents the application of fiber optic sensor technology in the monitoring of a load test. First, fiber optic technology is described with main emphasis in the case of distributed optical fiber sensors (DOFS), which have the potential of measuring strain and temperature along the fiber with different length and accuracy ranges. After that, two laboratory tests in reinforced and prestressed concrete specimens show the feasibility of using this technique for the detection, localization, and quantification of bending and shear cracking. Finally, the technique is applied to two real prestressed concrete bridges: in the first case, during the execution of a diagnostic load test; and in the second case, for the continuous monitoring in time and space of a bridge subjected to rehabilitation work. These experiences show the potential of this advanced monitoring technique when deployed in a load test.

In Chapter 8 of Volume 13, Gentile shows that recent advances in radar techniques and systems have favored the development of microwave interferometers suitable to the non-contact measurement of deflections on large structures. The main characteristic of the new radar systems, entirely designed and developed by Italian researchers, is the possibility of simultaneously measuring the deflection of several points on a large structure with high accuracy in either static or dynamic conditions. In this chapter, the main radar techniques adopted in microwave remote sensing are described, and advantages and potential issues of these techniques are addressed and discussed. Subsequently, the application of microwave remote sensing in live-load static and ambient vibration tests performed on full-scale bridges is presented in order to demonstrate the reliability and accuracy of the measurement technique. Furthermore, the simplicity of use of the radar technology is exemplified in practical cases, where the access with conventional techniques is difficult or even hazardous, such as the stay cables on cable-stayed bridges.

In Chapter 9 of Volume 13, Frangopol, Yang, Lantsoght, and Steenbergen review concepts related to the uncertainties associated with structures and discuss how the results of load tests can be used to reduce these uncertainties. When an existing bridge is subjected to a load test, it is known that the capacity of the cross section is at least equal to the largest load effect that was successfully resisted. As such, the probability density function of the capacity can be truncated after the load test, and the reliability index can be recalculated. These concepts can be applied to determine the required target load for a proof load test to demonstrate that a structure achieves a certain reliability index. Whereas the available methods focus on member strength and the evaluation of isolated members, a more appropriate approach for structures would be to consider the complete structure in this reliability-based approach. For this purpose, concepts of systems reliability are introduced. It is also interesting to place load testing decisions within the entire life cycle of a structure. A cost-optimization analysis can be used to determine the optimum time in the life cycle of the structure to carry out a load test.

In Chapter 10 of Volume 13, Val and Stewart demonstrate the determination of the remaining service life of reinforced concrete bridge structures in a corrosive environment after load testing. Reinforced concrete (RC) bridge structures deteriorate with time, and the corrosion of reinforcing steel is one of the main causes of bridge deterioration. Load testing alone is unable to provide information about the extent of deterioration and the remaining service life of a structure. In this chapter, a framework for the reliability-based assessment of the remaining service life of RC bridge structures in corrosive environments will be described. Existing models are considered for corrosion initiation, corrosion-induced cracking, and effects of corrosion on stiffness and strength of RC members. Special attention is paid to the potential effects of a changing climate on corrosion initiation and propagation in these structures. Examples illustrating the framework application are provided.

Chapter 11 of Volume 13, by Elfgren, Täljsten, and Blanksvärd, provides the perspective of bridge owners on load testing by discussing the experience in Sweden. Load testing of new and existing bridges was performed regularly in Sweden up to the 1960s. It was then abandoned due to high costs as opposed to the small amount of extra information obtained. Most bridges behaved well in the serviceability limit range, and no knowledge of the ultimate limit stage could be obtained without destroying the bridge. At the same time, methods for calculating the capacity were developed and new numerical methods were introduced. Detailed rules were given on how these methods should be used. Some decommissioned bridges were tested to their maximum capacity to study their failure mechanisms and to calibrate the numerical methods. In this chapter, some examples are given on how allowable loads have increased over the years and how tests are being performed. Nowadays, load testing may be on its way back, especially to test existing rural prestressed concrete bridges, where no design calculations have been retained.

In Chapter 12 of Volume 13, De Boer gives an overview of load testing within the framework of bridge management in the Netherlands. In the Netherlands, field tests on existing bridges have been carried out over the past 15 years. The tests consist of SLS load level tests (before the onset of nonlinear behavior, also known as proof load tests) on existing bridges, ultimate limit state tests (collapse tests) on existing bridges, and laboratory tests on beams sawn from the existing bridges. The goal of these experiments is to have a better assessment of the existing bridges. The assessment is combined with material sampling (concrete cores, reinforcement steel samples), for a better quantification of the material parameters, and with nonlinear finite element models. In the future, a framework will be developed in which the concrete compressive strength, determined from a large number of similar bridges from which core samples are taken, is first used as input for an assessment with a nonlinear finite element model according to the Dutch non-linear finite element analysis guidelines. Then it can be determined if sampling of the actual viaduct is needed, following recommendations on how to handle concrete cores and how to scan reinforcement, and if a load test should be used to evaluate the bridge under consideration. For this purpose, guidelines for proof load testing of concrete bridges are under development, but further research on the topic of load testing for shear is necessary for the development of such guidelines.

Finally, in Chapter 13 of Volume 13, Lantsoght brings all information from the previous chapters in Volumes 12 and 13 together and briefly restates the topics that have been covered. The topics to which more research energy has been devoted are discussed, as are the remaining open questions. Based on the current knowledge and state of the art, as well as the practical experiences presented by the authors in this book, general practical recommendations are presented.

These 25 chapters across two volumes provide a well-rounded overview of the state of the art internationally and the open research questions related to load testing of bridges, and how these insights can be applied to other structures such as buildings. The target audience of this book is researchers, students, practicing engineers, and bridge owners. With the inclusion of practical examples as well as open research questions, this book could be useful for academics as well as practitioners.

As book editor, I would like to thank all authors for their contributions to this work. Developing one (or more) book chapters is a large commitment in terms of time and effort, and I am grateful for all the authors who collaborated on this work. I would also like to thank the reviewers of the book proposal for their ideas on how to improve the contents of these books. Finally, I would like to acknowledge the work of the team at Taylor & Francis who made the publication of this work possible. In particular, I would like to thank Mr. Alistair Bright for his support throughout the development of this work and his kind guidance along the way.

Eva O. L. Lantsoght
Quito, Ecuador, December 2018

About the Editor

Eva O. L. Lantsoght is a full professor at Universidad San Francisco de Quito in Quito, Ecuador; a part-time researcher at Delft University of Technology, Delft, the Netherlands; and a structural engineer at ADSTREN, Quito, Ecuador. She obtained a Master's Degree in Civil Engineering from the Vrije Universiteit Brussels, Brussels, Belgium, in 2008; a Master's Degree in Structural Engineering from the Georgia Institute of Technology, Atlanta, USA, with scholarships from the Belgian American Educational Foundation (BAEF) and Fulbright in 2009; and a Ph.D. in Structural Engineering from Delft University of Technology, Delft, the Netherlands in 2013.

The field of research of Dr. Lantsoght is the design and analysis of concrete structures and the analysis of existing bridges. Her work has focused on the following topics: second-order effects in reinforced concrete columns, shear in one-way slabs under concentrated loads close to supports, fatigue in high-strength concrete, assessment of reinforced concrete slab bridges, plastic design models for concrete structures, proof load testing of concrete bridges, shear and torsion in structural concrete, and measurement methods and monitoring techniques. She has published more than 50 indexed journal and conference papers on the aforementioned topics.

Dr. Lantsoght is an active member of the technical committees of the Transportation Research Board in Concrete Bridges (AFF-30) and Testing and Evaluation of Transportation Structures (AFF-40). She serves on the technical committees of the American Concrete Institute and Deutscher Ausschuß für Stahlbeton Shear Databases (ACI-DAfStb-445-D), the joint ACI-ASCE (American Society of Civil Engineers) committee on Design of Reinforced Concrete Slabs (ACI-ASCE 421), and the ACI Committee on Evaluation of Concrete Bridges and Concrete Bridge Elements (ACI 342), and she is an associate member of the committees on Shear and Torsion (ACI-ASCE 445), and on Strength Evaluation of Existing Concrete Structures (ACI 437). She is a guest associate editor for *Frontiers in Built Environment – Bridge Engineering* for the research topic "Diagnostic and Proof Loading Tests on Bridges" and was the editor of the ACI Special Publication "Evaluation of Concrete Bridge Behaviour through Load Testing – International Perspective". She has served on the scientific committee of numerous conferences and has organized a number of special sessions and mini-symposia on topics related to load testing and monitoring of bridges. She is a reviewer for many scientific journals, an academic editor for *PLOS One*, and the editor in chief of ACI *Avances en Ciencias e Ingenierías*.

DOI: https://doi.org/10.1201/9780429265426

As a professor, Dr. Lantsoght teaches undergraduate and professional courses in the field of structural engineering and enjoys working with undergraduate and graduate students on their thesis and/or research projects. She is dedicated to sharing knowledge about existing structures to professionals in Ecuador and beyond in order to improve the safety and sustainability of the built environment.

Dr. Lantsoght's interests include also science communication, science outreach, higher education policy, and the improvement of doctoral education. She is the editor and main author of "PhD Talk," a blog for Ph.D. students, and the author of the free e-book *Top PhD Advice from Start to Defense and Beyond* and the textbook *The A–Z of the PhD Trajectory – A Practical Guide for a Successful Journey* in the Springer Texts in Education series.

Author Data

Bačinskas, Darius
Department of Reinforced Concrete Structures and Geotechnics
Vilnius Gediminas Technical University
Sauletekio al. 11
LT-10223 Vilnius, Lithuania
Tel: +370 5 237 0591
Email: darius.bacinskas@vgtu.lt

Bolle, Guido
Hochschule Wismar
Civil Engineering Department
23952 Wismar, Germany
Tel: +49 3841 753 72 90
Email: guido.bolle@hs-wismar.de

Bonifaz, Jonathan
ADSTREN Cia Ltda.
Plaza del Rancho
Av. Eugenio Espejo 2410
Bloque 1, Oficina 203
Quito, Ecuador
Tel: +593 23 957 606
Email: jbonifaz@adstren.com

Casas, Joan R.
Technical University of Catalunya, UPC-BarcelonaTECH
Jordi Girona 1-3. Campus Nord. Modul C1
08034 Barcelona, Spain
Tel: +34-934016513
Email: joan.ramon.casas@upc.edu

Diaz Arancibia, Mauricio
University at Buffalo, the State University of New York
206 Ketter Hall
Buffalo, NY 14260, USA
Tel: +1 716 533 0693
Email: mdiazara@buffalo.edu

DOI: https://doi.org/10.1201/9780429265426

ElBatanouny, Mohamed K.
Wiss, Janney, Elstner Associates, Inc.
330 Pfingsten Road
Northbrook, IL 60062
USA
Tel: (1) 847.753.6395
Email: melbatanouny@wje.com

Harris, Devin K.
Associate Professor – Department of Civil and Environmental Engineering
Director – Center for Transportation Studies
Faculty Director of Clark Scholars Program
University of Virginia
351 McCormick Rd Charlottesville
VA 22904
USA
Tel: (434)924-6373
Email: dharris@virginia.edu

Hernandez, Eli S.
Missouri University of Science and Technology
Department of Civil, Architectural & Environmental Engineering
1304 North Pine Street, Office 205
Rolla, MO 65409
USA
Tel: (832)589-3078
Email: eli.hernandez@mst.edu

Jakubovskis, Ronaldas
Department of Reinforced Concrete Structures and Geotechnics
Vilnius Gediminas Technical University
Sauletekio al. 11
LT-10223 Vilnius, Lithuania
Tel: +370 5 274 5225
Email: ronaldas.jakubovskis@vgtu.lt

Kilikevičius, Arturas
Institute of Mechanical Science
Vilnius Gediminas Technical University,
J. Basanavičiaus g. 28
LT-03224 Vilnius, Lithuania
Tel: +370 5 237 0594
Email: arturas.kilikevicius@vgtu.lt

Lantsoght, Eva O. L.
Politécnico
Universidad San Francisco de Quito
Diego de Robles y Pampite
Sector Cumbaya
EC 170157 Quito, Ecuador
Tel: (+593) 2 297-1700 ext. 1186
Email: elantsoght@usfq.edu.ec

Concrete Structures
Civil Engineering and Geosciences
Delft University of Technology
Stevinweg 1
2628 CN Delft, the Netherlands
Tel: (+31)152787449
Email: E.O.L.Lantsoght@tudelft.nl

ADSTREN Cia Ltda.
Plaza del Rancho
Av. Eugenio Espejo 2410
Bloque 1, Oficina 203
Quito, Ecuador
Tel: +593 23 957 606
Email: elantsoght@adstren.com

Myers, John J.
Missouri University of Science and Technology
305 McNutt Hall; 1400 N. Bishop Ave.
Rolla, Missouri, USA 65409
Tel: 1-573-341-7182
Email: jmyers@mst.edu

Okumus, Pinar
University at Buffalo, the State University of New York
222 Ketter Hall, Buffalo
NY 14260, USA
Tel: +1 716 645 4356
Email: pinaroku@buffalo.edu

Olaszek, Piotr
Road and Bridge Research Institute
ul. Instytutowa 1
03-302 Warszawa, Poland
Tel: +48 22 390 02 85
Email: polaszek@ibdim.edu.pl

Sanchez, Telmo Andres
ADSTREN Cia Ltda.
Plaza del Rancho
Av. Eugenio Espejo 2410
Bloque 1, Oficina 203
Quito, Ecuador
Tel: +593 23 957 606
Email: tasanchez@adstren.com

Schacht, Gregor
Marx Krontal GmbH
Uhlemeyerstraße 9+11
30175 Hannover
Germany
Tel: +49 511 51515423
Email: gregor.schacht@marxkrontal.com

Schmidt, Jacob W.
Technical University of Denmark
Brovej, Bygning 118
2800 Kgs. Lyngby, Denmark
Tel: +45 4525 1773
Email: jws@byg.dtu.dk

Contributors List

Bačinskas, Darius, *Department of Reinforced Concrete Structures and Geotechnics, Vilnius Gediminas Technical University, Vilnius, Lithuania*

Bolle, Guido, *Hochschule Wismar, Wismar, Germany*

Bonifaz, Jonathan, *ADSTREN Cia Ltda., Quito, Ecuador*

Casas, Joan R., *Department of Civil and Environmental Engineering, Technical University of Catalunya, UPC-BarcelonaTECH, Barcelona, Spain*

Diaz Arancibia, Mauricio, *University at Buffalo, State University of New York, Buffalo, NY, USA*

ElBatanouny, Mohamed K., *Wiss, Janney, Elstner Associates, Inc, Northbrook, IL, USA*

Harris, Devin K., *Department of Civil and Environmental Engineering, University of Virginia, Charlottesville, VA, USA*

Hernandez, Eli S., *Missouri University of Science and Technology, Rolla, MO, USA*

Jakubovskis, Ronaldas, *Department of Reinforced Concrete Structures and Geotechnics at Vilnius Gediminas Technical University, Vilnius, Lithuania*

Kilikevičius, Arturas, *Institute of Mechanical Science at Vilnius Gediminas Technical University, Vilnius, Lithuania*

Lantsoght, Eva O. L., *Politecnico, Universidad San Francisco de Quito, Quito, Ecuador; Concrete Structures, Delft University of Technology, Delft, the Netherlands; ADSTREN Cia Ltda., Quito, Ecuador*

Myers, John J., *Missouri University of Science and Technology, Rolla, MO, USA*

Okumus, Pinar, *University at Buffalo, State University of New York, Buffalo, NY, USA*

DOI: https://doi.org/10.1201/9780429265426

Olaszek, Piotr, *Road and Bridge Research Institute, Warsaw, Poland*

Sanchez, Telmo Andres, *ADSTREN Cia Ltda., Quito, Ecuador*

Schacht, Gregor, *Marx Krontal, Hannover, Germany*

Schmidt, Jacob W., *Technical University of Denmark, Kongens Lyngby, Denmark*

List of Tables

2.1	List of bridge collapses during load tests (incomplete) (Stamm, 1952)	16
2.2	Classification of experimental investigations on existing structures	25
3.1	Requirements for crack width w for newly developing cracks and increase in crack width Δw for existing cracks	33
3.2	Determination of K_{b1}	42
3.3	Determination of K_{b2}	42
3.4	Determination of K_{b3}	42
3.5	Determination of parameters per bridge type (Frýba and Pirner, 2001)	55
3.6	Limitations to crack widths that can occur in a load test (Frýba and Pirner, 2001)	55
3.7	Limitations to deviation between measured and calculated frequencies (Frýba and Pirner, 2001)	57
3.8	Recommendations for the execution of dynamic load tests (Ministerio de Fomento-Direccion General de Carreteras, 1999)	63
3.9	Limitations to deviation between measured and calculated deformations (Hungarian Chamber of Engineers, 2013)	67
5.1	Examples of goals that can be achieved with different types of load tests, for new and existing bridges	99
9.1	Truck weights	186
9.2	Midspan vertical deflections	191
9.3	Load distribution factors estimated from deflection measurements	192
9.4	Girders' longitudinal strain	192
9.5	Load distribution factors estimated from strain measurements	193
9.6	Dynamic load allowance	195
9.7	Bridge A7957's design parameters	196
9.8	Summary of AASHTO LRFD girder distribution factors	196
9.9	Bridge A7957 load rating data	198
9.10	Bridge A7957's analytical and experimental evaluation results	198
10.1	Purposes of load cases 1–4	210
11.1	Results of the analysis of the speed of vertical displacements stabilization in phase c of the load test of concrete bridges	229
11.2	The results of quasi-static values measured at 10 km/h (6.21 mph) and estimated with the use of the different filtering methods	243

DOI: https://doi.org/10.1201/9780429265426

| 11.3 | Cutoff frequencies used as filtration parameters compared to first frequency of free vibration and static frequency | 244 |
| 12.1 | Approximate values of logarithmic decrement of structural damping (data taken from EN 1991–1-4) | 285 |

List of Figures

1.1	Overall structure of these two books, Vol. 12 and Vol. 13	6
2.1	Design of bridge over the Neva, planned by Kulibin (Wikimedia)	10
2.2	Loading test at the opening ceremony of the bridge "Blaues Wunder" in Dresden	13
2.3	Example of a typical proof load test with simple (manual) deflection measurement underneath the bridge (von Emperger, 1908)	13
2.4	Picture of the suspension bridge over the Saale near Nienburg, Germany	14
2.5	Picture of the bridge collapse in Münchenstein, Switzerland	17
2.6	Proof load testing using enormous masses and loads (Mörsch, 1908)	17
2.7	Load testing of bridge in Alexisbad, Germany (Denkhahn, 1904)	18
2.8	BELFA-loading vehicle for proof load testing of bridges	19
2.9	Schuylkill River "Permanent Bridge" at Philadelphia, PA. Built by Palmer in 1804–1806	20
2.10	Hudson River at Waterford, NY. Built by Burr in 1804–1806	21
2.11	Mohawk Bridge at Schenectady, NY. Built by Burr in 1808	21
2.12	"Colossus Bridge" at Philadelphia, PA. Built by Wernwag in 1812	22
2.13	Photograph of a load test, New Mexico (Anay et al., 2016) with permission from ASCE	24
3.1	Safety philosophy of the German guideline (Deutscher Ausschuss für Stahlbeton, 2000), showing the two possible scenarios	32
3.2	Cyclic loading protocol from ACI 437.1R-07	45
3.3	Load-deflection curve for two cycles at the same load level, used to determine the repeatability index I_R and the permanency index I_p, from ACI 437.2M-13 (ACI Committee 437, 2013)	47
3.4	Load-deflection curve for six cycles, used to determine the deviation from linearity index I_{DL}, from ACI 437.2M-13 (ACI Committee 437, 2013)	47
3.5	Loading protocol for monotonic load test procedure from ACI 437.2M-13 (ACI Committee 437, 2013)	51
3.6	Illustration of procedure for evaluating the stabilization of measurements after application of target load (Ministerio de Fomento-Direccion General de Carreteras, 1999)	60
3.7	Illustration of procedure for evaluating the stabilization of measurements after unloading (Ministerio de Fomento-Direccion General de Carreteras, 1999)	60

DOI: https://doi.org/10.1201/9780429265426

XXXVIII List of Figures

3.8 Cycle of loading and unloading (Ministerio de Fomento-Direccion General de Carreteras, 1999) — 61

3.9 Determination of residual measurements (Ministerio de Fomento-Direccion General de Carreteras, 1999) — 61

4.1 Flowchart to determine if load testing or other alternative can be used — 75

4.2 Examples of (a) concrete cores and (b) reinforcement samples from an existing concrete bridge — 77

4.3 Truncation of probability density function of resistance after proof load test, based on Nowak and Tharmabala (1988) — 81

4.4 Application of load for collapse test and high magnitude loading of (a) Ruytenschildt Bridge in the Netherlands and (b) Rosmosevej Bridge in Denmark — 82

4.5 Levels of approximation — 83

4.6 Monitoring with LVDTs, lasers, strain gages, acoustic emission sensors, and digital image correlation during a proof (a) and high magnitude (b) load test of a reinforced concrete slab bridge. Only the bottom face can be instrumented for these bridges — 85

4.7 Instrumentation used during a test on a prestressed concrete bridge (a); monitoring preparation for OT (overturned T-section) bridge (b) — 85

4.8 Example of signaling for the traveling public during a collapse test. Translation: do not access – weakened structure caused by testing (a). Closure of a road leading under the bridge enabling a safe working area (b) — 89

5.1 Example of frozen bearing — 100

5.2 Example of map of damages — 102

5.3 Proof load test on viaduct where one lane of traffic (right side of photograph) remains open for traffic (Fennis et al., 2014; Koekkoek et al., 2015) — 103

5.4 Distribution of wheel print to center of cross section, applied to a concrete bridge, showing the tire contact area from NEN-EN 1991–2:2003 — 106

5.5 Example of distribution of shear stresses in transverse direction in reinforced concrete slab tested in the laboratory with seven bearings for the support. Due to small imperfections, the slab did not rest on all bearings at the beginning of the test, resulting in an asymmetric stress profile — 108

5.6 Punching of two wheels: (a) perimeter with three sides; (b) perimeter with four sides. d_{avg} is the average of the effective depths to the x- and y-direction flexural reinforcement, and b_{edge} is the edge distance — 110

5.7 Punching of the entire tandem, showing perimeter with three sides. d_{avg} is the average of the effective depths to the x- and y-direction flexural reinforcement, and b_{edge} is the edge distance — 111

5.8 Example of time schedule for planning, showing one day — 112

5.9 Measurement engineers following measurements in real time during a proof load test — 114

5.10 Load application methods: (a) dead weights on bridge and (b) application of loading rig with hydraulic jacks — 115

5.11	Load application with dump trucks	115
5.12	Example of application of dead weight with water	116
5.13	Example of application of dead weight with cement bags	117
5.14	Example of application of load with load testing vehicle	118
5.15	Use of counterweights, steel spreader beam, and a single hydraulic jack	119
5.16	Overview of traffic situation during a load test. From left to right: one lane open for traffic, one lane used for load testing, temporary bike bridge	120
5.17	Example of sensor plan, showing position of loading tandem, position and range of lasers and LVDTs to determine deflection profiles, acoustic emissions sensors and additional LVDTs for monitoring crack width of cracks that are selected on site	121
5.18	Application of acoustic emissions sensors	123
5.19	Example sensor plan, indicating LVDT1 as a reference strain measurement to compensate for the effects of temperature and humidity on the strain measurements	124
6.1	Example of special loading vehicle	131
6.2	Load testing of a new bridge prior to opening with trucks: Villorita Bridge, Quito, Ecuador	132
6.3	Comparison between results from linear finite element model and strains measured during field test	133
6.4	Example of the loaded area and measured deformations	134
6.5	Real-time data visualization as part of a data acquisition and visualization system	135
6.6	Use of weatherproof boxes for data acquisition system	136
6.7	Means of communication during load test	137
6.8	Safety shoes, life jacket, hard hat, and protective clothing for load test	139
6.9	Signposting for detour during load test	139
7.1	Comparison between four load cells on which the same load should be measured	143
7.2	Example of elastomeric bridge bearings in laboratory conditions	144
7.3	Strain measured at position not influenced by the applied load: strain development over time due to changes in temperature and humidity	145
7.4	Longitudinal deflection profiles for different load levels	146
8.1	Cross section of Los Pajaros Bridge, showing the two separate structures	157
8.2	Load case 1 for Los Pajaros Bridge, Quito, Ecuador	157
8.3	Load case 2 for Los Pajaros Bridge, Quito, Ecuador	158
8.4	Load case 3 for Los Pajaros Bridge, Quito, Ecuador	158
8.5	Load case 4 for Los Pajaros Bridge, Quito, Ecuador	158
8.6	Load case 5 for Los Pajaros Bridge, Quito, Ecuador	159
8.7	Load case 6 for Los Pajaros Bridge, Quito, Ecuador	159
8.8	Load case 7 for Los Pajaros Bridge, Quito, Ecuador	159
8.9	Configuration of trucks used for load test on Los Pajaros Bridge, Quito, Ecuador: (a) elevation; (b) plan view	160

8.10	Predicted deflections for load case 4 on girder 6 of Los Pajaros Bridge	161
8.11	Diagnostic load test with loading vehicles on the Los Pajaros Bridge, Quito, Ecuador	162
8.12	Results of measured deflections after unloading of Los Pajaros Bridge, girder 6	164
8.13	Review of different load cases for girder 6 of Los Pajaros Bridge	165
8.14	Symmetry of load cases left and right for Villorita Bridge in Quito, Ecuador	165
8.15	Example of difference between analytically determined maximum deflection and measured maximum deflection for flexible structure	169
8.16	Comparison between analytically determined deflections and predicted deflections for girder 6 of Los Pajaros Bridge for the case that causes maximum sagging deflection	170
8.17	Comparison between analytically determined deflections and predicted deflections for girder 6 of Los Pajaros Bridge for the case that causes maximum hogging deflection	170
8.18	Total experimental moment, modified from Barker (2001): (a) composite section; (b) measured bending moments M_T; (c) elastic moments M_E	179
9.1	Bridge A7957 elevation	182
9.2	Bridge A7957 plan view	182
9.3	Bridge A7957 cross section	183
9.4	VWSG installation details: (a) cluster locations layout; (b) midspan section sensors; (c) near-end section sensors (Hernandez and Myers, 2016a)	184
9.5	Bridge A7957 ATS Prism locations	185
9.6	Non-contact remote data acquisition systems: (a) automated total station; (b) remote sensing vibrometer (RSV-150)	185
9.7	H20 dump truck average dimensions	186
9.8	Static test configurations (stops)	188
9.9	Distance from safety barrier to trucks' exterior axle: (a) stops 1–6; (b) stops 7–9; (c) stops 10–13	190
9.10	Dynamic and quasi-static vertical deflection	194
10.1	(a) Photograph of the bridge under construction, and (b) cross section of the bridge looking east	202
10.2	(a) Extruded view of preliminary model, and (b) positive flexure and negative vertical shear strain envelopes under two trucks arranged in series	203
10.3	Electrical resistance surface strain gages on prestressed girders after coating to measure (a) bending and (b) shear strains	206
10.4	Vibrating wire gages installed (a) in the deck before deck pour and (b) in a no-stress cylinder enclosure	206
10.5	(a) Displacement transducers on bearing and (b) optical displacement target on girders	207
10.6	Locations of shear and bending strain gages and their data acquisition systems	208
10.7	Vibrating wire gage locations and directions and the data acquisition system location	208

10.8	Displacement transducer and data acquisition system locations and optical displacement target locations	209
10.9	Trucks used for load testing	210
10.10	Locations and number of trucks used in each load case (LC): LC1–LC4	210
10.11	Shear strain influence lines for girders 13, 14, and 15 for load cases 1–4	212
10.12	Bending strain influence lines for girders 13, 14, and 15 for load cases 1–4	213
10.13	Maximum deck strains in absolute value from all load cases at all gages and at deck extreme fibers for all gage locations	214
11.1	An example of displacements observed in the time function recorded during the diagnostic static load test with marked phases of the diagnostic test	224
11.2	Lateral view of tested bridges, from above. (1) Steel bridge with a composite-reinforced concrete bridge deck; (2) reinforced concrete slab bridge; (3) reinforced concrete frame bridge; (4) cable-stayed bridge – steel girders with reinforced concrete deck	225
11.3	Lateral view of tested bridges, from above. (5) Steel arch bridge with a reinforced concrete bridge deck; (6) prestressed concrete box girder bridge; (7) steel girders with reinforced concrete deck; and (8) prestressed concrete girders with reinforced concrete deck	227
11.4	Analysis of the vertical displacements stabilization during static load test in case of the steel bridge with a composite-reinforced concrete bridge deck (bridge 1). Upper figure: time history of the displacements. Lower figure: time history of the displacement increments in phases c and e of the load tests, determined at intervals $\Delta t = 1, 5, 10, 15$ minutes	230
11.5	Analysis of the vertical displacements stabilization during static load test in case of the reinforced concrete slab bridge (bridge 2). Upper figure: time history of the displacements. Lower figure: time history of the displacement increments in phases c and e of the load tests, determined at intervals $\Delta t = 1, 5, 10, 15$ minutes	231
11.6	Analysis of the vertical displacements stabilization during static load test in case of the prestressed concrete girders with reinforced concrete deck (bridge 8). Upper figure: time history of the displacements. Lower figure: time history of the displacement increments in phases c and e of the load tests, determined at intervals $\Delta t = 1, 5, 10, 15$ minutes	232
11.7	Bridge 2: example of using extrapolation to forecast excess displacements. From top to bottom: time history of the displacements with interrupted registration together with an interpolated curve on the basis of a fragment from the last 30 minutes and extrapolated curve; actual time history of the displacements together with an extrapolated curve from the previous figure; summary of results obtained on the basis of an analysis of the actual and extrapolated time history	233
11.8	View of tested bridges, from above. (1) Steel truss bridge with reinforced concrete deck and curved railway track; (2) steel plate girder with orthotropic deck; (3) steel arch bridge with reinforced concrete bridge deck	237

XLII List of Figures

11.9   Bridge 1: measured time histories of the vertical displacements for train passages at the speed $v = 10$ km/h (6.21 mph) (upper) and $v = 200$ km/h (124.27 mph) (lower)   239

11.10   Bridge 2: measured time histories of the vertical displacements for train passages at the speed $v = 10$ km/h (6.21 mph) (upper) and $v = 150$ km/h (93.21 mph) (lower)   240

11.11   Bridge 3: measured time histories of the vertical displacements for train passages at the speed $v = 10$ km/h (6.21 mph) (upper) and $v = 200$ km/h (124.27 mph) (lower)   241

11.12   Bridge 1: an analysis of the quasi-static value estimation using the low-pass FIR filters (cutoff frequency $f_{c1} = 1.35$ Hz, $f_{c2} = 0.80$ Hz, and $f_{c3} = 0.45$ Hz) for the time histories of the displacements registered during the train ride at 200 km/h (124.27 mph)   241

11.13   Bridge 2: an analysis of the quasi-static value estimation using the low-pass FIR filters (cutoff frequency $f_{c1} = 3.20$ Hz, $f_{c2} = 1.60$ Hz, and $f_{c3} = 0.98$ Hz) for the time histories of the displacements registered during the train ride at 150 km/h (93.21 mph)   242

11.14   Bridge 3: an analysis of the quasi-static value estimation using the low-pass FIR filters (cutoff frequency $f_{c1} = 0.83$ Hz, $f_{c2} = f_{c3} = 1.20$ Hz) for the time histories of the displacements registered during the train ride at 200 km/h (124.27 mph)   242

11.15   Bridge 2: an analysis of the accuracy of the extreme displacement amplitude extrapolation for the train ride speed of 150 km/h (93.21 mph) on the basis of the results of the measurements taken for the train ride speeds up to 120 km/h (74.56 mph) (points of two rides of the speed 150 km/h [93.21 mph], practically overlapped)   244

11.16   Bridge 1: analysis of the accuracy of the extreme displacement amplitude extrapolation for the train ride speed of 200 km/h (124.27 mph) on the basis of the results for the speeds up to 160 km/h (99.42 mph)   245

11.17   Bridge 3: analysis of the accuracy of the extreme displacement amplitude extrapolation for the train ride speed of 200 km/h (124.27 mph) on the basis of the results obtained for speeds up to 160 km/h (99.42 mph)   245

12.1   Different types of footbridges   250

12.2   Testing of bridge decks after construction (a) and before installation (b)   253

12.3   Application of destructive technique for the identification of the compressive strength of concrete   256

12.4   Application of nondestructive methods for the identification of mechanical properties of concrete (a) and steel (b)   256

12.5   Crack pattern of the abutment (a) and inaccurate installation welded joint of truss elements (b)   257

12.6   Different types of loading in static field tests of footbridges   260

12.7   Static loading of footbridge using loaded vehicles   261

12.8   Different types of free-vibration excitation of pedestrian bridges   262

12.9   Excitation of forced vibrations of the bridge by walking or running of pedestrian groups   263

12.10   Electromechanical shakers for forced vibration tests   264

List of Figures XLIII

12.11	Measuring of bridge deck displacements using LVDT	266
12.12	Equipment for displacement data processing	266
12.13	Measuring of bridge displacement using leveling instrument (a) and total station (b)	267
12.14	Different gages for strain measurement in bridge elements	267
12.15	Dynamic equipment for measuring vibrations	268
12.16	Locations of measurement points during static (a) and dynamic (b) tests	269
12.17	Example of loading and unloading stages during the static test of bridge deck	270
12.18	LVDT readings during the static test under different loading and unloading stages	273
12.19	Experimental deformed shapes of the footbridge (Bačinskas et al., 2014): measured deflections of the bridge deck in longitudinal (a) and transverse directions (b)	273
12.20	Basis of footbridge dynamics: (a) and (b) typical vertical force function for normal walking (adopted from Živanović et al. [2005]); (c) presentation of harmonic functions in frequency domain; and (d) approximation of vertical force by harmonic functions (Fourier transform)	275
12.21	Signal processing from time to frequency domain and extraction of FRF of steel truss pedestrian bridge	276
12.22	Modal characteristics obtained from FRF	277
12.23	Schematic representation of algorithmic decrement	278
12.24	FE model of the tested footbridge (a), theoretical deformed scheme due to static loading (b), and comparison of theoretical and experimental results (c)	281
12.25	Comparison of experimental (a) and theoretical (b) self-vibration modes and frequencies	282

Part I

Background to Bridge Load Testing

Chapter 1

Introduction

Eva O. L. Lantsoght

Abstract

This chapter introduces the general topic of load testing of structures. In this chapter, the reader can find a short background to the topic of load testing of structures, the scope of this book, the aim of this book, and a short discussion of the structure and outline of this book.

1.1 Background

Load testing of bridges can be considered a very specialized topic as well as a very broad topic. On the one hand, it is a very specialized topic since it is one type of testing that is part of the bridge engineering profession. On the other hand, the amount of information and different views and practices in this book demonstrate that load testing is a broad topic and that load testing can be used to serve a large number of purposes.

Both new and existing bridges can be subjected to a load test. Two types of load tests can be distinguished: diagnostic load tests and proof load tests. This book separates these two topics in separate parts with best practices and examples of application. Diagnostic load tests are used to obtain direct information from a bridge in the field, to verify behavior, or to update analytical models. Proof load tests are used to demonstrate directly that a bridge fulfills the code requirements.

For new bridges, diagnostic load tests are the most common type of load tests. In some countries, diagnostic load tests are required prior to opening of some or all new bridges to demonstrate that the bridge behaves in the way it was designed and to confirm the assumptions made in the analytical models used for the design. In the past, proof load tests on new bridges were used prior to opening to demonstrate to the traveling public that the bridge is "safe" for use. Safety in this context was defined as the bridge being able to carry a large number of heavily loaded trucks.

For existing bridges, diagnostic load tests can be used to update the analytical models developed for the assessment of the bridge. A diagnostic load test can be used to determine the transverse distribution of the actual structure, the actual stiffness of the structure, the amount of restraint at the supports, the effect of unintended composite action in non-composite sections, and so forth. This information can then be used to have a better understanding of the bridge behavior. Proof load tests on existing structures can be used to demonstrate that a given bridge fulfills the code requirements by applying a load that is representative of the factored load combination. Proof load tests can be interesting when

DOI: https://doi.org/10.1201/9780429265426

there are large uncertainties on the structure and its behavior; these may be caused by the effect of material degradation and deterioration on the capacity, the case of planless bridges, or uncertainties on the structural behavior at larger load levels, for example. Since they involve larger loads, proof load tests are more expensive and require a more extensive preparation than diagnostic load tests.

With an increasing need to assess existing structures, methods of field testing including load testing, monitoring of structures, and nondestructive testing have increased in importance over the past decades. The elements of the method of load testing that are discussed in this book are the general procedures regarding deciding on load testing, preparing load tests, executing load tests, and evaluating the results of a load test after the test. These elements are described in general terms, as well as separately for diagnostic and proof load tests, for which different considerations are important. This book contains a number of case studies, showing for specific bridges how these considerations are applied in practice and providing guidance and advice for practicing engineers who are preparing load tests. Besides the tried and tested methods for load testing, this book also discusses novel measurement techniques that can improve the practice of load testing in the future. Additionally, this book shows how load testing can move from a singular test to a method to evaluate the safety of a given bridge by linking the principles of load testing to concepts of structural reliability.

The field of load testing is moving from deterministic approaches based on rules of thumb to practices in which the proven safety after a load test can be quantified. Many researchers have worked on making this step over the past years, often in close collaboration with practicing engineers, bridge owners, and road authorities. With an increasing need to consider structures within their life cycle and as part of a network, there is an incentive to place load testing within the framework of a decision-making process at the network level and to carry out load tests at the optimal time during the life cycle of a structure. These principles are not present yet in the existing codes and guidelines for load testing, but they are topics of research that are discussed in this book.

1.2 Scope of application

As the title indicates, this book discusses load testing of bridges. However, the same procedures can be applied to other structures. For this purpose, a chapter on load testing of buildings has been included. In the introductory chapters on the history of load testing and the current codes and guidelines, both bridges and buildings are discussed. The general principles for load testing of bridges and buildings are the same, but closing a building for a load test has a lower impact than closing a bridge, which results in driver delays and secondary costs. As such, in the past, loading protocols that last more than a day were common for buildings. Another element of particular interest for buildings is that a building may contain a large number of floor spans with similar dimensions. An open question here is how many spans should be tested to have a statistically relevant number of tests for the evaluation of all floor spans in the building. This topic is addressed in Chapter 4 of Volume 13 about load testing of buildings.

This book encompasses all bridge types, and both new and existing bridges. The presented case studies show cases of new bridges that were load tested upon opening because they are built with a novel material, and/or to verify the structural behavior of the new bridge. Other case studies show how load testing can be used for the assessment of existing bridges. The presented case studies of proof load tests all deal with existing structures with large uncertainties, where a proof load test is used to directly demonstrate adequate performance.

The scope includes all building materials. Most of the presented case studies are based on reinforced concrete road bridges, but some applications of steel bridges and prestressed concrete bridges are included as well. The case studies do not include timber, masonry, or plastic composite bridges, but the same principles can be applied to these building materials. In the chapters that deal with general considerations and methodology of load testing, topics of interest related to these building materials are highlighted, and references to relevant articles and reports are included for the reader.

The majority of the information and case studies in this book apply to road bridges, as such bridges make up the majority of the bridges in the transportation network. The chapter with examples of applications of diagnostic load tests from Europe also discusses dynamic load testing of railroad bridges. The same principles are valid for road and railroad bridges, yet the tolerances are different. During the preparation stage of a load test on a railroad bridge, the governing tolerances should be reviewed, and where necessary, the stop and acceptance criteria should be updated to reflect these tolerances. Most codes and guidelines for load testing focus on road bridges. However, the Spanish and French codes also give recommendations for load testing of pedestrian bridges. The applied load is then either dead loads or human pedestrians. One chapter deals with the topic of load testing of pedestrian bridges.

1.3   Aim of this book

Over the past decades, there have been a large number of developments related to improved methods for the assessment of existing structures. This evolution reflects the fact that our built environment is aging and requires suitable methods to evaluate these structures, which are different from the methods used for designing new structures. Topics that have been of interest in this regard are methods of field testing including load testing, monitoring of structures, and nondestructive testing techniques. This increased importance is reflected by the number of mini-symposia, special sessions, paper collections, special publications, and so forth that have been organized and developed over the past years.

This book aims at bringing together the available information and knowledge on the topic of load testing of bridges, and (to a smaller extent) other structures. The gathered knowledge includes practical experience from load tests as well as research insights. It reflects the current state of the art at an international level and aims to show the way forward in research related to load testing of bridges. The reader will notice throughout this book that local practices differ. Overall recommendations on which code or guidelines is the "best" for load testing are not included in this book, since a variety of goals for load tests can be identified. Prior to the load test, the test engineers should identify the goals for the load test, decide if a load test is the appropriate instrument to meet these goals, and then outline how these goals can be met and which practical considerations in terms of measurements and load application should be taken into account to meet these goals. The reader will understand through the case studies that each load test is different and thus requires different particular considerations during the preparation, execution, and post-processing stages.

1.4 Outline of this book

This book is divided in eight parts. Figure 1.1 gives the overall structure of these two books, which begins with the historical background to load testing and how past practices led to the current codes and guidelines, to the current practice of load testing, to topics of research that currently are explored in pilot load tests and that will be implemented in the codes and guidelines of the future.

Part I of Volume 12 focuses on the historical background of load testing and how past practices resulted in the current codes and guidelines. This part covers the historical perspectives and currently governing codes and guidelines. The background is discussed from an international perspective, outlining the history of load testing in North America and Europe, and summarizing the current codes and guidelines for load testing in Germany, the United States, the United Kingdom, Ireland, Poland, Hungary, Spain, the Czech Republic and Slovakia, Italy, Switzerland, and France.

Part II of Volume 12 is the first part that deals with the current practice of load testing. It gives an overview of the general practical aspects of load testing, which are valid for the different types of load tests (diagnostic and proof load tests), and covers different bridge types in terms of structural systems and in terms of construction material. These practical aspects cover the entire stage of a load testing project: from preparation, to execution, to post-processing and reporting about the load test. The first topic in this part contains general considerations for each load testing project: which type of load test is suitable for the project, and can a load test give answers to the open questions regarding the

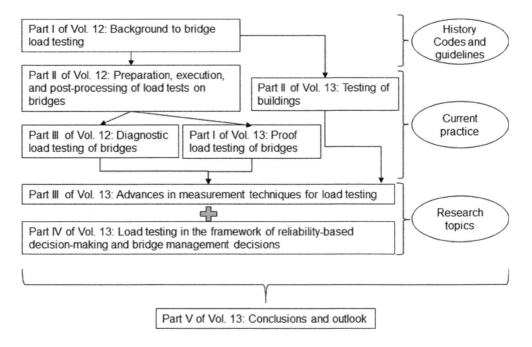

Figure 1.1 Overall structure of these two books, Vol. 12 and Vol. 13

structure? If the decision is made to load test a bridge, the next step is to prepare this load test. Therefore, the next topic discusses the elements of the preparation of a load test: inspection, preliminary calculations, and planning of the project. Then, general aspects of the execution of the load test are discussed with regard to loading equipment, measurement equipment, and practical aspects of communication and safety. The last topic of this part deals with post-processing of load testing data, reporting of load tests, and decision-making after a load test.

Part III of Volume 12 continues the topic of current practice of load testing and focuses on the particularities of diagnostic load testing of bridges. It discusses the aspects of the preparation, execution, and post-processing of a diagnostic load test that are different from other load tests, as well as the general methodology for diagnostic load tests. Chapters describing detailed examples of diagnostic load tests from North America and Europe are included. One chapter discusses diagnostic load testing of pedestrian bridges.

Part I of Volume 13, which is shown adjacent to the third part in Figure 1.1, focuses on proof load testing of bridges. It discusses the specific aspects of proof load testing during the preparation, execution, and post-processing of a proof load test. Important topics in this part are the interpretation of the measurements in real time during the experiment and thresholds to the structural response, and the determination of the target proof load. Since high loads are applied in proof load tests, the risk of collapse or permanent damage to the structure exists. For this reason, careful instrumentation and monitoring of the bridge during the proof load test is of the utmost importance. Criteria based on the structural response that can be used to evaluate when further loading may not be permitted are discussed. Chapters describing examples of proof load test from North America and Europe are included.

Part II of Volume 13 fits in the current practice of load testing and describes how the practice of load testing can also be applied to buildings. Whereas the main focus of this book is on bridges, the same principles can be used for load testing of buildings. Buildings can require a load test prior to opening when there are doubts about the adequate performance of the building, or during the service life of the building when there are doubts regarding the capacity as a result of material degradation or deterioration, or when the use and associated live load of the building are to be changed. This part also discusses current research topics related to the practice of load testing that will be included in future issues of the German guideline for load testing.

Part III of Volume 13 is the first part that focuses on research topics and the future of load testing. This part discusses novel measurement techniques used for load testing that are not yet part of the standard practice for load testing. Methods using non-contact sensors are discussed, such as image- and video-based measurement techniques. With acoustic emission measurements, signals of microcracking and distress can be observed before cracks are visible. Fiber optics are explored and applied as a new measurement technique. The topic of measurements through radar interferometry is also discussed. These chapters contain the background of these measurement techniques and recommendations for practice, and identify topics for further research and improvement for these measurement techniques.

Part IV of Volume 13 also deals with research topics and the future of load testing. This part discusses load testing in the framework of reliability-based decision-making and in the framework of a bridge management program. The first topic of this part is the research-related theme of applying concepts of structural reliability to the practice of load testing in order to quantify "safety" of a structure after a load test. In particular, the first topic includes updating the reliability index of a bridge after load testing and discusses the required proof load magnitude for this purpose, as well as systems reliability considerations. The second topic deals with the effects of degradation, in particular corrosion, and how results of load tests can be used to update the reliability index which diminishes over time when degradation is considered. The final topic of this part deals with the future of load testing, when load testing will be an integral part of a bridge management program instead of an isolated object-specific practice. In this chapter, the bridge owner discusses how load testing fits in the overall framework of a bridge management program.

Part V of Volume 13 brings all information from the previous parts together and summarizes the current state of the art on load testing. An overview of the open questions for research is given along with practical recommendations for load testing.

Chapter 2

History of Load Testing of Bridges

Mohamed K. ElBatanouny, Gregor Schacht and Guido Bolle

Abstract

This chapter provides an overview of the historical development and current practices of bridge load tests in Europe and the United States. The use of load tests to prove the capacity of structures is as old as mankind and plays an important role in the historical development of reinforced concrete design and construction. Historically, load tests provided proof that a structure could carry a certain load and, therefore, they were likely used to convince the people of the bearing capability of bridges. With the development of static calculations and of acceptable design rules, load tests became unnecessary for new structures. However, currently, strength evaluation of existing structures is becoming more and more important and the advantages of experimental assessments are used. The chapter describes the development of the technology of load testing and shows the European and American way of practice.

2.1 Introduction

The origin of civil engineering lies in the building trade, in which experiences gained by trial and error were of particular importance. This way of proving that something works as intended is not only the etymological but also the historical origin of proof loading. The first load tests performed by humans, for example to determine the tensile resistance of a liana or the load-carrying capacity of a fallen tree, were carried out many millions of years ago. Load tests are as old as humanity itself. They served to confirm that structures were capable of withstanding the relevant loads and to instill trust in the load-bearing capacity of structures of critical importance. Supporting this trust in the load-bearing capacity is the main goal of all proof load tests. It is achieved by carrying out a visual demonstration of the structural safety of the construction. New construction methods or novel structural elements had (and still have) to earn this trust by passing proof load tests. In the Middle Ages, master builders passed on their lifelong experiences and those of their teachers to their students. It was only around 1650 that documenting experiments and tests and investigating their theoretical background began. This resulted in an improved understanding of the associated mechanical fundamentals, which is illustrated in the following example.

In the 18th century, Ivan Petrovich Kulibin developed the design of an arch bridge over the Neva River with a span of 300 m (Fig. 2.1) – an inconceivable span length at the time (Bolle et al., 2010a). His carefully prepared drawings and static calculations, which he submitted to the supreme council of the Academy of Sciences of Saint Petersburg, were met with astonishment and suspicion. Kulibin determined the optimal shape of the arch from the deflection

DOI: https://doi.org/10.1201/9780429265426

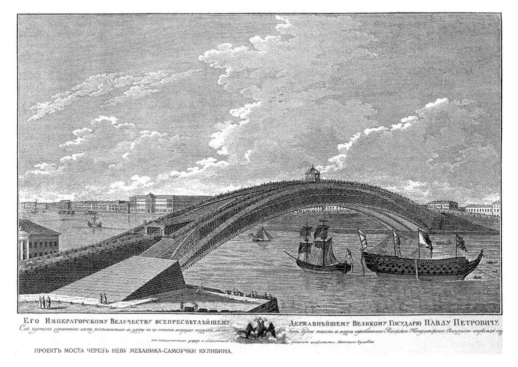

Figure 2.1 Design of bridge over the Neva, planned by Kulibin (Wikimedia)

curve of an open polygon subjected to point loads at its vertices. The fundamental principle of this approach was used in Europe in the 19th century and was subsequently called the "thrust line method" (Bolle et al., 2010b). The committee deliberated for a long time on Kulibin's design. It was only due to the support of French committee member Daniel Bernoulli and the mathematician Leonhard Euler that it was decided to build a model to the scale of 1:10 in order to test the structural capacity of the designed structure. The model of the timber bridge was erected on the shores of the Neva River, and on 27 December 1776, the day of the test, the members of the academy's committee congregated around Bernoulli and Euler to witness the event. Using similar conditions, Kulibin had determined that the bridge model had to safely carry 10 times its own weight in order to pass the test. Initially, a single ox-drawn cart was driven across the bridge. When the bridge appeared to withstand this load without any sign of distress, the number of carts was increased until finally 12 carts had been moved onto the bridge and were stopped at its apex (Bernstein, 1957).

Even so, the committee was not fully convinced. Therefore, all available materials, such as spare ferrules and a block of bricks, were also placed on the bridge. When the bridge safely withstood this additional load as well, Euler approached Kulibin, who was standing on the bridge, and offered his hand in a gesture of deep appreciation. Kulibin was celebrated by the spectators and praised by Bernoulli for his "remarkable construction." In the test protocol it was noted that the model had been designed and built correctly, and hence the construction of a 300 m bridge was possible. Unfortunately, due to lack of government funds the bridge was never built, despite the successful execution of the proof load test and the recognition of Ivan Kulibin's achievement (Bernstein, 1957).

This example demonstrates the potential and importance of a successfully conducted test for confirming the structural capacity of a construction. Proof load tests provide not only an ultimate and deterministic confirmation of the structural capacity but also form the basis for systematic evaluations of experimental observations and for theoretical modeling. For the engineers, tests serve multiple purposes: they allow investigation of the unknown structural behavior of novel structure types or unusually dimensioned structural elements for which no information exists, and they are used to verify analytical models. These two traditional objectives of experimental investigations are still as important today as they were in the beginning, particularly because we aim to use any structural reserve of existing constructions. This is achieved by carrying out proof load tests or system identifications with accompanying fitting of the numerical models (Bernstein, 1957).

The type and importance of tests in the area of civil engineering have changed significantly over human history. The focus used to be on very large and exceptional structures which were important for the public's sense of security: bridges.

Initially, load tests served only to confirm and check whether the structure was able to carry the required load. As the understanding of the structural behavior grew and more measuring technologies were developed, a quantitative rather than purely qualitative assessment of the structural response was increasingly sought. However, initially no empirical data or previous knowledge was available with respect to what constituted allowable or critical structural states. Simple, undifferentiated criteria often offered a false sense of security, which gave rise to increased criticism of this type of testing. Even so, tests were required to verify analytical approaches and were therefore very useful: this was the only way to gather empirical data. As the accuracy of analytical approaches increased, experimental investigations ceased to be a necessity and even became obsolete in some cases. Because of the shift in priorities towards the structural assessment of existing constructions, the importance of experimental investigations is on the rise again now, as structural reserves that can only be discovered on the "live patient" need to be activated; it is often impossible to solve this type of problem using theoretical considerations only. The continuously improving analytical approaches and technologies, however, also allow a combined verification of the structural fitness to be carried out in the form of a hybrid verification (using experimental investigations to calibrate and verify an analytical model) (Bernstein, 1957).

2.2   Bridge load testing in Europe

The Romans were the first European civilization to erect a large number of bridges as part of the infrastructure required for their economic and military success. Aided by the knowledge garnered from previously executed projects, the basic arch shape was continuously refined and the span length was increased. Structures were built according to the experience of the executing engineer. In order to demonstrate the quality of his work, a Roman master builder was required to stand underneath his bridge as the support structure was removed (Kortschinski, 1955).

At the end of the 18th century, masonry arch bridges and timber bridges were the most commonly built bridge types, and their designs were based exclusively on empirical knowledge. Bridge construction in Europe changed only when the option of producing steel of predefined quality in larger quantities became available. While steel bridges built in the early 19th century are still reminiscent of timber and arch bridge construction, new structural systems and distinct construction types were developed in rapid succession. These structures allowed for larger span lengths while minimizing material usage. However, no empirical knowledge or analytical models were available, which is why these structures had to undergo load tests for confirmation of the structural capacity.

Load tests as part of the commissioning of a new bridge have a long tradition. On the one hand, this is due to the capability of load tests to confirm the required load-bearing capacity in a way that even laymen can understand. On the other hand, it was also used to compensate for the deficits in the existing (still not very sophisticated) calculation methods and to dispel all remaining doubts with respect to the structural capacity of the finished structure. Also, material defects could often be detected only during a load test.

In many cases, proof load testing was carried out as part of the inaugural celebrations for the bridge, and the public was invited to take part in the tests in order to be convinced of the safety of the structure (Fig. 2.2).

To produce the sometimes very heavy loads required for the load tests, ballast masses were placed directly on the structures. Depending on the location and function of the bridge and the availability of the required materials, different types of ballast were used. Depending on the individual circumstances of every bridge, water tanks, sandbags, steamrollers, or heavy trucks were used to produce the loads.

For example, for the test of the bridge over the Thur River at Oberbüren, Switzerland, in 1886, water from the river was pumped with fire hoses into containers positioned on the bridge (Bersinger, 1886). For railway bridges such as the bridge at Tübingen, Germany, the load placed on the bridge generally consisted of locomotives and heavily loaded freight wagons. In 1895, the road bridge at Walsburg, Germany, was uniformly loaded with sandbags. Additionally, two carts, each weighing 5.3 tons (5.8 short tons), were pulled across the bridge by a total of eight horses (Paul, 1895). Steamrollers were also frequently used in load tests. Von Emperger (von Emperger, 1908) reports the proof load testing of a bridge in Rotterdam in 1901 during which trucks were suspended by chains from the structure and loaded from the road with sand.

Only the deflections were generally measured, as they were considered the most important effect in the structure due to the load. If they were sufficiently small or close to the calculated deflections, proof of sufficient load-bearing capacity was considered to have been provided. The measurements were often carried out by very simple means (Bolle et al., 2010b). In some cases, for example, an auxiliary wooden construction containing a linear scale was erected underneath the bridge. A weight with an attached indicator was suspended from the bridge by a wire (see Fig. 2.3).

Early on, it was understood that the measured elastic deformation was not a reliable parameter for signaling the imminent failure of the bridge, as some modes of failure are not preceded by measurable deflections. However, no other reliable criteria were available. Low deflections measured during proof load testing also provided no guarantee for safe operation of the structure in the long term. The steel truss bridge over the Morava River at Ljubicevo, Kosovo, exhibited construction and material deficiencies even before its completion. Proof load testing was carried out on the most carefully executed opening of the bridge (according to the opinion of all the experts) on 21–22 September 1892. River sand was used to apply load to the bridge. On the second day of the test, the bridge failed suddenly due to the buckling of the under-dimensioned top chords. The deflections

Figure 2.2 Loading test at the opening ceremony of the bridge "Blaues Wunder" in Dresden
Source: © SLUB Dresden.

Figure 2.3 Example of a typical proof load test with simple (manual) deflection measurement underneath the bridge (von Emperger, 1908)

at the time of failure were within the normal range and did not give any indication of a critical state having been reached (Tetmajer, 1893).

The road bridge near Salez (in the canton of St. Gallen, Switzerland) collapsed during a load test in 1884, at a measured elastic deformation of only 10 mm (3/8 in.) (Zimmermann, 1884). Failure was caused by buckling of the top chords of the steel trusses due to a missing bracing and poor construction of the junction plates. The bridge collapsed abruptly, even though the theoretical permitted maximum deflection had been determined to be 17.5 mm (2/3 in.).

Other sudden failures occurred because of imprudence or gross negligence during testing. The Rhône bridge near Peney (Switzerland), having a span of 100 m (328 ft), passed a load test in which sandbags were used as ballast. After the test, however, heavy rain started and the sandbags absorbed so much water that the anchorages failed and the bridge collapsed. In 1873, the bridge over the Broye near Payerne also passed a load test, but it collapsed when workers carelessly began throwing the water tanks used for the test off the bridge and damaged a main girder, because they wanted to get to the bridge opening party.

It was also argued that a successfully conducted load test did not guarantee the structural safety in the long term. This was definitely true for the bridge over the Saale River near Nienburg, Germany (see Fig. 2.4). The bridge, which was 118 m (387 ft) long and 7.6 m (25 ft) wide, was completed in 1825 and represents an early version of a cable-stayed bridge. Because this was a completely new bridge type, the execution of a load test was particularly important. A carriage drawn by 10 horses and carrying 1202 bricks drove over the bridge, and the measured deflections were so small that the construction was immediately deemed to be safe. Some months later the duke visited the bridge. A torchlight parade was organized in his honor and the public was dancing on the bridge and celebrating. When the duke's convoy was about to cross the bridge, the crowd congregated on one side of the bridge, which overloaded the bridge chains. The bridge collapsed and more than 55 people died. The previously conducted load test had instilled such a sense of security that the maximum allowable load of about 9.5 tons (10.5 short tons) was exceeded (at the time of the collapse the bridge was carrying about 13 tons = 14.3 short tons) (Nebel, 2018).

A similar tragedy occurred in Salford (Great Britain), just as 74 British soldiers were crossing the Broughton Suspension Bridge. This was one of the first suspension bridges

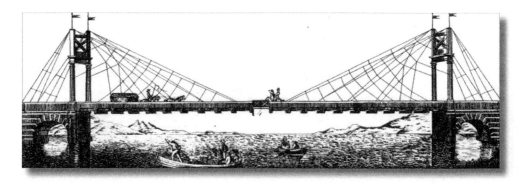

Figure 2.4 Picture of the suspension bridge over the Saale near Nienburg, Germany
Source: Drawings from Christian G. H. Bandhauer.

in Europe, and it became famous for collapsing on 12 April 1831 due to resonance vibrations caused by marching troops. On that day, two groups of soldiers had already crossed the bridge when a further group of soldiers walking in four rows passed over the bridge on their return to the barracks. Initially, the group must not have been marching in lockstep. However, the vibrations of the bridge encouraged some soldiers to whistle a marching song and the soldiers probably adjusted their steps to the song and the vibrations, which further amplified the vibrations. When the first soldiers had already reached the other side of the bridge, there was a loud bang and one pylon collapsed onto the bridge, causing parts of the deck to fall into the river approximately five meters (16 ft) below. About 40 soldiers were swept against the chain fragments or into the river, which was very shallow at the time. There were no fatalities, but about 20 soldiers were injured, six of them very badly. Even though this bridge collapse did not prevent other suspension bridges being built, the effect of resonance vibrations had been brought to the attention of the public. The British military subsequently issued a general directive prohibiting troops from marching in lockstep across bridges.

The lack of rules for the execution and evaluation of load tests resulted in numerous collapses. They are listed in Table 2.1.

Because of some spectacular bridge collapses, such as the collapse of the Münchensteiner Bridge (Switzerland) in 1891, shown in Figure 2.5, the Swiss railway department demanded regular inspections of all railway bridges (Waldner, 1891). During these inspections, simple load tests had to be carried out in which the bridge was loaded according to the load assumptions of the previously performed structural analysis. Thus load tests became standard procedure, with the consequence that the results of the tests were often not even analyzed. The bridge was considered to have sufficient load-bearing capacity if it did not collapse during the load test and if some further easily confirmed criteria were fulfilled. These limit criteria had been derived from long experience and personal assessments and affected mainly the serviceability of the construction (i.e. deflections and vibrations).

Initial efforts to create rules for load tests were undertaken in the Bavarian civil service after 1878. According to these rules, bridges were to be loaded with two cars with 5-ton (5.5 short tons) axle loads driving side by side, as well as 3 persons per square meter (1 person per 3.6 ft^2) marching in lockstep (Bavaria, 1878). The deflection of the main girders at midspan and at the supports were measured. The permanent midspan deflection after unloading, measured in an identical test carried out three months later, was to be smaller than 1/4000 of the span length. Similar rules were issued by the Swiss railway administration in 1879 (Swiss Railway, 1879). Load testing became a formal means for testing the structural safety of bridges only after 1892 (as a consequence of the incident at Münchenstein; Fig. 2.5). This was achieved by establishing detailed descriptions of the requirements for the deflections measured during the load tests. The rules demanded the execution of static tests of the bridge subjected to the amount of load assumed in the structural analysis, as well as dynamic bridge tests. Further criteria stipulated a maximum allowable ratio of permanent to elastic deflection of 20% or 15%.

Other countries in Europe and the Western world followed the Swiss railway and defined similar regulations for the load testing of bridges at the end of the 19th century.

The limited practical relevance of some of the criteria quickly led to a general discussion about the value of the now obligatory proof load tests for iron bridges. Despite all the critical voices, Ritter stated: "Load tests must always be carried out, if only to reassure the

16 Load Testing of Bridges

Table 2.1 List of bridge collapses during load tests (incomplete) (Stamm, 1952)

Year	Type of bridge	Location	Length	Failure	Personal injuries
1825	"Chain-stayed" bridge over the Saale River	Nienburg, Germany	118 m	Rupture of the chains due to a crowd congregating on one side of the bridge	55 fatalities; 36 injured
1852	Cable suspension bridge over the Rhône River	Peney, Switzerland	100 m	Rain-saturated sandbags caused the inadequate anchorage of the cables to fail	none
1873	Iron truss bridge over the Broye River	Payerne, Switzerland	30 m	Some parts of the bridge were damaged as the load was removed	unknown
1883	Semiparabolic truss bridge over the Töss River	Between Rykon and Zell, Switzerland	21 m	A typical trough bridge issue: buckling of the top chord members due to insufficient lateral stiffness	1 fatality 5 injured
1883	Suspension bridge over the Charente River	Tonnay-Charente, France	623 m	Collapse during a trial load test prior to structural rehabilitation	none
1884	Iron bridge	Salez, Switzerland	35 m	Insufficient buckling strength of the compression diagonal	2 injured
1892	Semiparabolic truss bridge over the Morava River	Ljubicevo, Kosovo	185 m	Partial collapse under partial loading: buckling of the compression chord due to deficient connections in the compression members consisting of two iron profiles	none
1931	Suspension bridge over the Igle River	Bordeaux, France	76 m	Total collapse under the load of nine trucks carrying sand. One hanger was damaged due to the collision of a truck; progressive failure of other hangers ensued	15 fatalities 40 injured
1986	Steel arch bridge	Latvia	n.s.	Collapse as the 14th truck carrying gravel crossed the bridge	10 fatalities

Conversion: 1 m = 3.3 ft.

public" ("*Belastungsproben sind immer durchzuführen, schon um die Laien zu beruhigen*"; quotation translated by the authors) (Ritter, 1892).

Thus bridge load tests remained the standard method for confirming the structural capacity of concrete and reinforced concrete bridges during the late 19th and early 20th centuries (see Fig. 2.6). And still no objective assessment criteria were available. The engineers' trust in the structural capacity of their bridges was therefore demonstrated by the engineer standing underneath the bridge during the load test (see Fig. 2.7).

When concrete and reinforced concrete construction became more common in the early 20th century, the need for rules to evaluate load tests of concrete bridges increased. From 1904 onwards, nearly all European countries developed rules regarding the load to be applied to the bridge and the deformations to be assessed. Despite the criticism about using only deflection measurements to assess a structure, this criterion was also adopted

Figure 2.5 Picture of the bridge collapse in Münchenstein, Switzerland
Source: Bulacher and Kling.

Figure 2.6 Proof load testing using enormous masses and loads (Mörsch, 1908)

Figure 2.7 Load testing of bridge in Alexisbad, Germany (Denkhahn, 1904)

for concrete bridges. Because initially no empirical data were available, the allowable limits were defined rather vaguely; for example, it was stipulated that no "noteworthy permanent changes in shape" ("*nennenswerte bleibende Formänderungen*"; quotation translated by the authors) (Bolle et al., 2010b) were allowed to occur. By 1920 the assessment of bridges by evaluation of the ratio of the permanent deflection after unloading to the measured maximum deflection of 1/4 (in Germany, Austria, Norway, and Poland) or 1/3 (in Hungary, Czechoslovakia, and Russia) had become the norm in all the relevant European standards. In some standards an additional condition was given with the ratio of the maximum measured to the maximum calculated deflection, tacitly assuming that the deflection of reinforced concrete members can be calculated precisely. Permitted was an exceedance of the calculated deflection of 10% (Italy), 20% (Sweden, Poland), or 25% (Czechoslovakia, Norway) (von Emperger, 1928).

Graf and Bach already showed in 1914 (Bach and Graf, 1914) that the ratio of the deflections highly depends on the loading history of a construction. They also showed that beams with and without appropriate shear reinforcement have similar deflections under the permitted loads. They concluded, therefore, that "the passing of a loading test because of not exceeding a deflection limit, which would have been defined appropriate, in general doesn't ensure a total robustness of a construction" (p. 18). Despite all criticism, the ratio of residual to maximum deflection remained the only criterion for the assessment of a loading test for many years. To compensate the influence of a previous loading, some foreign standards allowed repeating a failed loading test once (provided that the construction did not collapse). Then the same (Great Britain, USA) or a slightly stricter criterion (France, Austria) was used to assess the deflection ratio.

In the 1930s and 1940s, EMPA Zurich published several reports on systematic proof load tests of bridge structures. More than 50 load tests document the status of bridge load testing at the time. The extent of the executed measurements (e.g. strains and temperatures), which far exceeded the normally executed deflection measurements, was remarkable. One objective of the static and dynamic investigations was to collect more empirical information of certain structural systems and bridge types and to use them for further developments (Ros, 1939).

In the 1950s, intense development of calculation methods began, primarily due to the increased use of computer technologies. Professionals and laymen became confident that it was possible to calculate the load-bearing capacity of bridges with sufficient accuracy. As a result of this development and the discussion about the usefulness of load tests, no rules for load tests have been included in design standards since DIN 1045:1972 was issued in Germany (Schacht et al., 2017). Other countries didn't remove their rules and kept them up to today.

Load tests of bridges have since then only been carried out in exceptional cases and mainly to verify the assumptions made in calculation. During the past few decades, when it became necessary for engineers to assess and evaluate an increasing number of existing bridge structures, it became clear that rules for load tests for the assessment of existing bridges needed to be included in the standards. However, it was also clear that load tests had to be updated in line with the state of the art technically as well as theoretically.

The necessary work to include rules for load tests in the design standards took place almost simultaneously in East and West Germany in the 1980s and ended after reunification of Germany in a guideline for load testing published in 2000 (DAfStb, 2001). The technical and theoretical basics of this development are described in Part II of Volume 13 (Schacht et al., 2018).

Besides the development of a guideline for load tests in 2000, research projects for the development of mobile bridge load-testing equipment were carried out. BELFA (for road bridges) and BELFA DB (for railway bridges) mobile loading trucks were developed. These trucks are equipped with a complete servo-hydraulic system and were developed for load testing of bridges having short and medium spans (see Fig. 2.8).

The BELFA allows load testing of bridges up to loads at the level of the ultimate limit state, because of its self-securing loading techniques – that is, that the pressure of the jacks, and hence the load on the structure, automatically decreases with increasing deformations. A disadvantage is that for larger structures, the anchoring of loads is usually difficult and

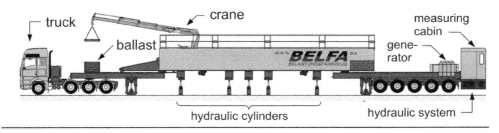

Figure 2.8 BELFA-loading vehicle for proof load testing of bridges

very time- and resource-intensive. The BELFA has a hydraulic loading system but does not need additional anchorage constructions. Therefore, only a short track possession time is needed for the test. The BELFA is limited to medium-span bridges up to a span of 20 m (66 ft) and maximum load of 1580 kN (355.3 kips). For larger bridges, a proof load test with hydraulic loading systems is extremely expensive and complicated. Therefore, mainly loading tests at the serviceability load level are carried out in most European countries today (diagnostic load tests).

2.3 Bridge load testing in North America

Early bridge construction in the United States mainly consisted of timber and masonry bridges. Most of the timber bridges built before the 18th century were short-span bridges. Long-span bridges constructed during this period were mainly trestle bridges which consist of beams supported on closely spaced piers. One of the first significant timber truss bridges built in the United Sates is the "geometry-work" bridge built by John Bliss in the 1760s in Norwich, Connecticut (Rutz, 2004). Another timber bridge to provide long, clear spans was completed in 1785 by Colonel Enoch Hale. The bridge was 365 feet long (111 m) and was constructed over the Connecticut River at Bellows Falls, Vermont (Ritter, 1990).

Many of the timber truss bridges constructed in the 19th century were designed using the trial-and-fail method by local bridge builders (Ritter, 1990). This method implicitly uses load testing as means to confirm the design: if the bridge can sustain applied loads, then it is safe. At this time, with the expansion of the railroad, the increasing demand for bridges led to academic progress by the formation of civil engineering departments in a number of schools, with the US Military Academy at West Point and Rensselaer School forming civil engineering programs in 1823 and 1824, respectively (Rutz, 2004).

The use of trusses in combination with arches to construct bridges was mainly introduced by three notable bridge builders: Timothy Palmer, Louis Wernwag, and Theodore Burr. All three builders built bridges at the end of the 18th century and beginning of the 19th century, including:

- Piscataqua Bridge ("The Great Arch") at Portsmouth, NH. Built by Palmer in 1794.
- Schuylkill River "Permanent Bridge" at Philadelphia, PA. Built by Palmer in 1804–1806 (Fig. 2.9).
- Hudson River at Waterford, NY. Built by Burr in 1804–1806 (Fig. 2.10).

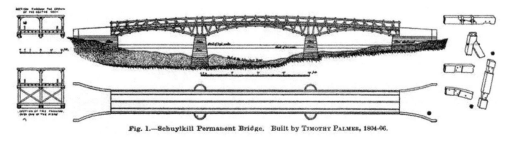

Figure 2.9 Schuylkill River "Permanent Bridge" at Philadelphia, PA. Built by Palmer in 1804–1806
Source: www.catskillarchive.com/rrextra/brcoo0.html.

History of Load Testing of Bridges   21

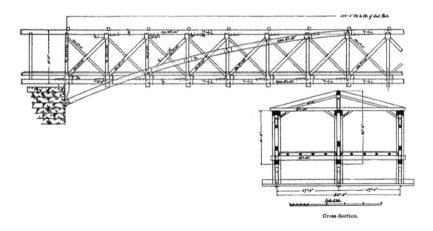

Figure 2.10 Hudson River at Waterford, NY. Built by Burr in 1804–1806
Source: www.catskillarchive.com/rrextra/brcoo0.html.

Figure 2.11 Mohawk Bridge at Schenectady, NY. Built by Burr in 1808
Source: www.catskillarchive.com/rrextra/brcoo0.html.

- Mohawk Bridge at Schenectady, NY. Built by Burr in 1808 (Fig. 2.11).
- Susquehanna River at Harrisburg, PA. Built by Burr in 1812–1816.
- Schuylkill River "Colossus Bridge" at Philadelphia, PA. Built by Wernwag in 1812 (Fig. 2.12).
- Delaware River at New Hope, PA. Built by Wernwag in 1814.

Following this period, the bridge industry was influenced by several inventors. Ithiel Town developed a timber lattice truss that was patented in 1820. Stephen Long invented the Long truss and patented bridges in 1830, 1836, and 1839. It is noted that Long's trusses were proportioned using Navier's analogy of the parallel chord truss as a beam, which may be considered as the first attempt to design trusses based on analytical methods (Gasparini and Simmons, 1997). William Howe proposed a new design in 1840 that improved upon Long's trusses through the use of wrought iron, which is considered the first truss to use iron in conjunction with timber trusses. It is noted that Howe's patent included a complete stress analysis of the design (Ritter, 1990). New designs were also introduced in the following years by, among others, Nathaniel Rider, Stephen Moulton, Thomas and Caleb Pratt, and Squire Whipple, who is also known as "the father of iron bridges" (Griggs, 2002).

During this time, many bridge companies were constructing bridges across the United States. Each of these companies typically had a set of guidelines or standards for bridge

22 Load Testing of Bridges

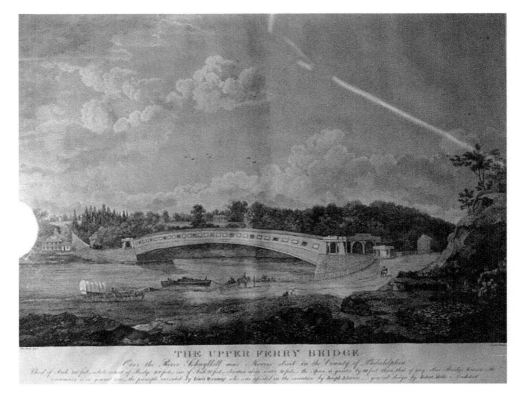

Figure 2.12 "Colossus Bridge" at Philadelphia, PA. Built by Wernwag in 1812
Source: https://en.wikipedia.org/wiki/Colossus_Bridge.

construction. In 1889, Theodore Copper published a book on the history of bridges in the United States titled *American Railroad Bridges*. In his discussion, Cooper indicated that most bridge companies in the United States were capable of load testing samples of the materials or full-sized bridge members. This led to better understanding of the material properties, which along with advances in knowledge of bridge analysis helped in designing better bridges. Cooper also indicated that bridge collapses have occurred in the United States under train service, giving examples of some bridge collapses.

In 1890, Theodore Copper published what can be considered the first national standard for bridge construction, titled "General Specifications for Iron and Steel Railroad and Viaducts." This standard was followed by revised and updated versions in 1896 and 1901. The specifications called for a final test where

> the Engineer may make a thorough test by passing over each structure the specified loads, or their equivalent, at a speed not exceeding 45 miles an hour [72 km/h], and bringing them to a stop at any point by means of the air or other brakes, or by resting the maximum load upon the structure for twelve hours.
>
> (Copper 1890, p. 23, paragraph 156, section Final Test)

A qualitative acceptance criterion was established requiring the bridge to return to its original position without showing permanent damage.

The American Association of State Highway Officials (AASHO), which was later renamed the American Association of State Highway and Transportation Officials (AASHTO) published its first specification in 1931, followed by versions in 1935, 1941, 1944, 1949, 1953, 1957, 1961, 1965, and 1969. Many other versions followed until the most recent version, the *AASHTO LRFD Bridge Design Specifications*, 8th Edition, published in 2018 (AASHTO, 2018). The standards define design loads, materials, and material tests, among others. Although the standards did not call for load testing of full-scale structures, the AASHO Road Test program which started in 1951 included bridge load testing. Referred to as the Ottawa test, this test provided fundamental data using realistic bridges and vehicles to evaluate the strength, dynamic response, and fatigue characteristics of the bridge. The testing included 16 bridges between 1958 and 1960, and its results helped with the development and calibration of AASHO design provisions (Fisher et al., 2006).

The Silver Bridge collapse of 1967 prompted public concern and brought inspection and condition evaluation of highway bridges to the attention of designers and highway officials. This collapse was the catalyst behind the creation of the National Bridge Inspection Standards (NBIS). Following this tragedy, procedures to conduct load rating were presented in the AASHTO *Manual for Maintenance Inspection of Bridges* (AASHTO, 1970). Currently, the AASHTO *Manual for Bridge Evaluation* (2010) includes provisions to load rate structures using theoretical methods and/or load tests. The theoretical methods are consistent with three AASHTO design philosophies, as follows (Hernandez Ramos, 2018):

- Allowable stress rating (ASR): This load rating method is consistent with the allowable stress design method adopted by AASHTO until the early 1970s. This was included in the 1970 version of the AASHTO *Manual for Maintenance Inspection of Bridges* (AASHTO, 1970).
- Load factor rating (LFR): This load rating method is based on the load factor design philosophy, which was adopted by AASHTO following the ultimate strength and plastic design philosophies for reinforced concrete and steel, respectively. This was included in the 1994 version of the AASHTO *Manual for Condition Evaluation of Bridges* (AASHTO, 1970).
- Load and resistance factor rating (LRFR): This load rating method is based on the load and resistance factor design (LRFD) presented in 1998. This was included in the 2003 version of the *Manual for Condition Evaluation and Load and Resistance Factor Rating (LRFR) of Highway Bridges* (MCE) (AASHTO, 2003).

The current state of the practice for experimental load rating includes two nondestructive load testing techniques, which can be implemented depending on the nature of the issue for a particular bridge: diagnostic load tests and proof load tests. Diagnostic load tests use service loads and are usually conducted for the case of multiple span and high profile bridges. This technique includes a rigorous investigation to assess the condition and the safe load capacity of the bridge where the bridge is loaded and instrumented with sensors, usually strain gages, and then a finite element model (FEM) is developed and calibrated based on the test results. The results of this test can be used to adjust the calculated ratings. On the other hand, proof load tests can be used with multiple- or single-span bridge girders to verify a maximum safe load for the tested bridge (Fig. 2.13). These tests use

Figure 2.13 Photograph of a load test, New Mexico (Anay et al., 2016) with permission from ASCE

higher loads than the diagnostic load tests. The AASHTO Manual for Bridge Evaluation (AASHTO, 2010) allows for calculating a target live-load factor, which is used in conjunction with the rating vehicle weight to determine the test load. This method accounts for the uncertainties associated with the live load.

General consensus indicates that typically the bridge strength is higher than the estimated capacity calculated from analytical load ratings. This is attributed to many factors including actual live-load distribution, actual section dimensions, and unintended composite action, among other factors (Stallings and Yoo, 1993; TRB, 1998; Barker, 1999, 2001; Cai and Shahawy, 2003; Hernandez Ramos, 2018). Therefore, experimental evaluation of the strength through nondestructive load tests offers a solution for bridge engineers to increase the reliability of load ratings while ensuring the safety of the public.

2.4 The potential of load testing for the evaluation of existing structures

As described in the previous sections, load tests have historically been used as ultimate proof of structural capability or proof of safety for new structures. This is no longer necessary since a lot of experience is obtained in the calculation, design, and construction of new structures.

Load tests can be used to validate structural calculations and designs or to calibrate numerical models, to proof a sufficient structural safety, or to determine the bearing strength by failure tests. Table 2.2 provides a summary of the various types of load tests (Schacht et al., 2017). Even though a lot of experimental load tests have been carried out until today, many open questions remain to be answered.

History of Load Testing of Bridges 25

Table 2.2 Classification of experimental investigations on existing structures

Type of load test		Purpose of load test	Comments
Diagnostic load testing (characteristic service loads)	Short term F_k	Determination of deformations and strains as well as the structural behavior under service loads; calibration of numerical models; verification of the assumptions of structural analyses	No temperature influence; numerical models can be improved; difficult extrapolation of the load-bearing capacity; dynamic effects
	Monitoring F_k	Determination of time-dependent deformations and strains; verification of the assumptions of structural analyses; assessment of damage development, durability	Long-term deformation behavior (creep, temperature effects); stable measuring equipment; big data
Proof load test (factored design loads)	F_{target}	Proof of sufficient structural safety under a defined test load without the occurrence of critical deformations	Standard test situation for evaluation of the load-bearing reserve of existing structures
	F_{lim}	Determination of the maximum load under which no critical deformations and strains are detected	Increased measurement efforts; danger of irreversible damage to the structure
Failure load test	F_{ult}	Proof of the structural capacity	Very helpful for the evaluation of structural safety

2.5   Summary and conclusions

Load tests of bridges are an old tradition and have historically been used to demonstrate the load-bearing safety of bridges. Today the value of load testing is its application as an effective tool for in-situ evaluation of the structural behavior and assessment of existing structures. The development of new measuring techniques and the possibilities to use the information about the structural behavior for numerical investigations have a high potential to further improve the design, construction, and maintenance of bridges.

References

American Association of State Highway and Transportation Officials. (AASHTO) (1970) *Manual for Maintenance Inspection of Bridges*. American Association of State Highway and Transportation Officials, Washington, DC.

American Association of State Highway and Transportation Officials. (AASHTO) (1994) *Manual for Condition Evaluation of Bridges*. American Association of State Highway and Transportation Officials, Washington, DC.

American Association of State Highway and Transportation Officials. (AASHTO) (2003) *Manual for Condition Evaluation and Load and Resistance Factor Rating (LRFR) of Highway Bridges*. American Association of State Highway and Transportation Officials, Washington, DC.

26 Load Testing of Bridges

American Association of State Highway and Transportation Officials. (AASHTO) (2010) *The Manual for Bridge Evaluation (2nd Edition) with 2011, 2013, 2014 and 2015 Interim Revisions.* American Association of State Highway and Transportation Officials, Washington, DC.

American Association of State Highway and Transportation Officials. (AASHTO) (2018) *AASHTO LRFD Bridge Design Specifications*, 8th edition. American Association of State Highway and Transportation Officials, Washington, DC.

Anay, R., Cortez, T. M., Jáuregui, D. V., ElBatanouny, M. K. & Ziehl, P. (2016) On-site acoustic-emission monitoring for assessment of a prestressed concrete double-tee-beam bridge without plans. *Journal of Performance of Constructed Facilities*, 30(4).

Bach, C. & Graf, O. (1914) *Gesamte und bleibende Einsenkungen von Eisenbetonbalken. Verhältnis der bleibenden zu den gesamten Einsenkungen.* Deutscher Ausschuss für Eisenbeton, Berlin. p. 27.

Barker, M. G. (1999) Steel girder bridge field test procedures. *Construction and Building Materials*, 13(4), 229–239.

Barker, M. G. (2001) Quantifying field-test behavior for rating steel girder bridges. *Journal of Bridge Engineering*, 6(4), 254–261. https://doi.org/10.1061/(ASCE)1084-0702(2001)6:4(254)

Bavaria (1878) Vorschriften über Entwurf, Ausführung und Prüfung von Straßenbrücken mit eisernem Überbau. Königliches Staatsministerium des Innern des Königreichs Bayern.

Bernstein (1957) *Очерки по истории строительной механики.* Moscow. pp. 99–108.

Bersinger (1886) Die neue eiserne Straßenbrücke über die Thur bei Oberbüren, Canton St. Gallen. *Schweizerische Bauzeitung*, 8(25), 147–150.

Bolle, G., Schacht, G. & Marx, S. (2010a) Geschichtliche Entwicklung und aktuelle Praxis der Probebelastung – Teil 1: Geschichtliche Entwicklung im 19. und Anfang des 20. *Jahrhunderts. Bautechnik*, 87(11), 700–707. https://doi.org/10.1002/bate.201010047

Bolle, G., Schacht, G. & Marx, S. (2010b) Geschichtliche Entwicklung und aktuelle Praxis der Probebelastung – Teil 2: Entwicklung von Normen und heutige Anwendung. *Bautechnik*, 87 (12), 784–789. https://doi.org/10.1002/bate.201010052

Cai, C. S. & Shahawy, M. (2003) Understanding capacity rating of bridges from load tests. *Practice Periodical on Structural Design and Construction*, 209–216. https://doi.org/10.1061/(ASCE) 1084-0680(2003)8:4(209)

Cooper, T. (1889) American railroad bridges. *Transactions of the American Society of Civil Engineers*, 21, 1–58, 566–607.

Cooper, T. (1890) *General Specifications for Iron and Steel Railroad Bridges and Viaducts.* Engineering News Pub., New York.

DAfStb (2001) *Richtlinie für Belastungsversuche.* Deutscher Ausschuss für Stahlbeton, Berlin, Germany.

Denkhahn (1904) *Brücken und Decken.* System Möller, Braunschweig.

Fisher, J., Hall, D., McCabe, R., Price, K., Seim, C. & Woods, S. (2006) Steel Bridges in the United States Past, Present, and Future, 50 Years of Interstate Structures, Steel Bridge Committee. pp. 27–48.

Gasparini, D. & Simmons, D. (1997) American truss bridge connections in the 19th century. I: 1829–1850. *Journal of Performance of Constructed Facilities*, 11(3), 119–129.

Griggs, F. E., Jr. (2002) Squire Whipple: Father of iron bridges. *Journal of Bridge Engineering*, 7(3), 146–155.

Hernandez Ramos, E. S. (2018) *Service Response and Evaluation of Prestressed Concrete Bridges through Load Testing.* Doctoral Dissertation, Missouri University of Science and Technology, Rolla, Missouri. p. 190.

Kortschinski, J. L. (1955) Belastungsprüfungen von Bauwerken und Konstruktionen. Fachbuchverlag Leipzig.

Mörsch, E. (1908) *Der Eisenbetonbau – seine Theorie und Anwendung*, 3rd edition. K. Wittwer, Stuttgart, Germany.

Nebel, B. (2018) Available from: www.bernd-nebel.de [accessed 4 April 2018].

Paul, W. (1895) Straßenbrücke bei Walsburg a. d. Saale nach Monier-Bauweise und ihre Belastungsprobe. Centralblatt der Bauverwaltung. pp. 32–33.

Ritter, M. A. (1990) *Timber Bridges: Design, Construction, Inspection, and Maintenance.* United States Department of Agriculture, Washington, DC. 944 p.

Ritter, W. (1892) Ueber den Werth der Belastungsproben eiserner Brücken. *Schweizerische Bauzeitung*, 20(3), 14–17.

Ros, M. (1939) Versuche und Erfahrungen an ausgeführten Eisenbeton-Bauwerken in der Schweiz. EMPA.

Rutz, F. R. (2004) *Lateral Load Paths in Historic Truss Bridges.* Doctoral Dissertation, University of Colorado, Denver.

Schacht, G., Wedel, F. & Marx, S. (2017) *Load Testing of Bridges in Germany: ACI SP 323 Evaluation of Concrete Bridge Behavior through Load Testing: International Perspectives.* American Concrete Institute, Farmington Hills, MI, USA.

Schacht, G., Bolle, G. & Marx, S. (2018) *Load Testing of Concrete Building Constructions: Load Testing of Bridges* – Vol II. Series: Structures and Infrastructures. CRC Press, Boca Raton, FL.

Stallings, J. M. & Yoo, C. H. (1993) Tests and ratings of shortspan steel bridges. *Journal of Structural Engineering*, 2150–2168. https://doi.org/10.1061/(ASCE)0733-9445(1993)119:7(2150)

Stamm (1952) *Brückeneinstürze und ihre Lehren. ETH Zürich.* Verlag Leeman, Zürich.

Swiss Railway (1879) Entwurf für die Verordnung über die technische Einheit im schweizerischen Eisenbahnwesen. Swiss Railway Department.

Tetmajer, L. (1893) Ueber die Ursachen des Einsturzes der Morawa-Brücke bei Ljubitschewo. *Schweizerische Bauzeitung*, (21/22), 55–58.

Transportation Research Board. (TRB) (1998) *"Manual for Bridge Rating through Load Testing" Research Results Digest No. 234.* Transportation Research Board, Washington, DC.

von Emperger, F. (1908) *Handbuch für Eisenbeton, Several Books.* Ernst & Sohn, Berlin.

von Emperger, F. (1928) *Handbuch für Eisenbetonbau, Band 9 – Die in – und ausländischen Eisenbetonbestimmungen.* Ernst & Sohn, Berlin.

Waldner, A. (1891) Das Eisenbahnunglück bei Mönchenstein. *Schweizerische Bauzeitung*, 17(25, 26), No. 1, 2, 3.

Zimmermann, H. (1884) Einsturz einer Straßenbrücke bei Salez in der Schweiz. *Centralblatt der Bauverwaltung*, 548–549.

Chapter 3

Current Codes and Guidelines

Eva O. L. Lantsoght

Abstract

This chapter reviews the existing codes and guidelines for load testing of structures. A summary of the main requirements of each existing code is provided, with a focus on the determination of the required load and measurements. The requirements for load testing of bridges and buildings are revised, for new and existing structures. An international perspective is given, revising the practice from Germany, the United Kingdom, Ireland, the United States, France, Switzerland, the Czech Republic, Slovakia, Spain, Italy, Switzerland, Poland, and Hungary. The chapter concludes with a short overview of the current developments and with a discussion of the different available codes and guidelines.

3.1 Introduction

Load testing practices originated as part of the bridge engineering profession to demonstrate to the traveling public that a given bridge is safe for use. Over time, load testing practices for the assessment of existing bridges were also developed. This chapter gives an overview of the load testing practices and regulations internationally. Not all countries have guidelines for diagnostic and proof load testing of bridges and buildings. In some countries, the guidelines are only applicable to structures of a certain material; for example, the German guidelines were originally developed for proof load testing of concrete buildings, and the ACI guidelines are only applicable for proof load testing of concrete buildings. These guidelines have been included in this section since few bridge codes deal in detail with proof load testing.

In a proof load test, a high load is used that is representative of the factored live load. The determination of the magnitude of this load is different across the available guidelines for proof load testing. Therefore, in the description of the codes that allow for proof load testing, the determination of the target proof load is discussed. Since high loads are used in proof load tests, it is important to identify when the test should be terminated, even though the target proof load has not been reached yet. The criteria that are used to identify if a test should be terminated are called "stop criteria." In the ACI codes and guidelines, the term "acceptance criteria," for criteria that should be fulfilled and checked after the test, is used. These criteria are defined based on the measurements on the structure. Stop criteria generally are defined based on strain, deflection, stiffness, crack width, and other responses that can be monitored in real time during the proof load test. If a stop criterion is exceeded, further loading is not permitted. The structure is then approved for the highest load level that was achieved before exceeding a stop criterion.

DOI: https://doi.org/10.1201/9780429265426

30 Load Testing of Bridges

Since the goal of diagnostic load testing is to gather information about the structural response of the tested bridge, the magnitude of the applied load is less important in a diagnostic load test. The load should be large enough so that the structural response can be measured by the applied sensors and be representative of service load levels. The measured structural response is then used to update the analytical model that was used to develop predictions of the structural responses. The measurements should be followed during the load test, but since the applied loads are much smaller than in a proof load test, the risk for causing irreversible damage to the structure is smaller. Therefore, codes typically do not define stop criteria for diagnostic load tests.

Even though the German and ACI guidelines have been developed for buildings, their contents are discussed in this chapter since the proposed stop and acceptance criteria in these guidelines have been important for the further development of stop criteria for concrete bridges. Whereas the focus of this chapter and book is on load testing of bridges regardless of their building material, for concrete structures the definition of stop criteria may be more convoluted than for steel structures, where strain measurements can indicate directly how far away from the yield strain the occurring strain is, provided that the strain caused by permanent loads can be closely estimated. For fracture- and fatigue-critical steel structures, the current codes do not permit testing. Future research should focus on developing safe guidelines for testing of these types of steel structures. For shear-critical concrete bridges, recent research has focused on developing stop criteria that allow for the safe execution of proof load tests on these types of structures.

3.2 German guidelines

3.2.1 *General*

In Germany, a guideline for load testing (*Deutscher Ausschuss für Stahlbeton*, 2000) of plain and reinforced concrete structures is available. The guideline cannot be used when a brittle failure mode can occur in the proof load test, and thus does not allow testing of shear-critical structures. The original scope of the guideline was proof load testing of buildings.

Prior to the load test, the structural system, geometry, and material properties need to be known. The data should either be taken from the available documentation or should be determined by measurements and/or testing of the actual structure. A visual inspection with additional destructive and nondestructive testing is required prior to the test. The structural capacity needs to be determined prior to the load test, using the following partial factors:

- Permanent loads: $\gamma_G = 1.15$
- Concrete: $\gamma_C = 1.4$
- Reinforcement steel: $\gamma_S = 1.1$

According to the German guideline, load testing is permitted in case of insufficient knowledge on the calculation models, the composite action and load distribution among structural members, the effect of material damage, and the effect of repair actions. During the preparations of the load test, the following elements need to be determined: the measurements expected during the test, the effect of changes to the state or system (effect of uncracked versus cracked section, effect of changes in temperature), the expected stresses and strains for the applied load, and the effect of the load test on the substructure.

The position of the load in a proof load test has to be such that it is representative of the most unfavorable loading position. The German guideline prescribes a cyclic loading protocol for proof load tests. In a proof load test, the load has to be applied in at least three steps, and after each step unloading is required at least once.

After the load test, the results need to be analyzed and compared with the calculations made before the test. Similar structures to the tested structure can also be assessed based on the load test when their equivalence can be shown through all essential details and properties.

3.2.2 Safety philosophy and target proof load

The guideline specifies the following loads:

- The limiting load level F_{lim}: the load at which a stop criterion is reached, indicating that further loading would cause permanent damage to the structure.
- The total target load F_{target}: the planned maximum load prior to the load test.
- The applied target load $ext.F_{target}$: the externally applied target proof load, without the present permanent loads.

The German guideline prescribes two scenarios that can occur in a load test:

1 The target load $ext.F_{target}$ is applied to the structure, and none of the stop criteria is exceeded. The load test has shown successfully that the structure can carry the applied load.
2 Prior to reaching the target load $ext.F_{target}$, a stop criterion is exceeded at a load level $ext.F_{lim}$. Further loading is thus not permitted. The results of the load test may be used to conclude that the structure can carry loads at a load level up to F_{lim}.

This safety philosophy is illustrated in Figure 3.1. The concept is illustrated based on the load-displacement diagram, here more generally shown as the relation between action and effect in the experiment. The value of F_{lim} is shown as the onset of nonlinear behavior. The permanent loads G_1 are shown on the diagram, as well as the total capacity $effR_u$ which corresponds to the maximum action. The value of F_{lim} can be subdivided into the effect of the permanent loads G_1 and the externally applied load $ext.F_{lim}$. The target load should correspond to the considered load combination of the factored permanent and live loads. The applied load $ext.F_{target}$ is then the target load F_{target} minus the occurring permanent load G_1: $G_{dj} + Q_d$. As shown in Figure 3.1, two scenarios are possible: $F_{target} \leq F_{lim}$, which means that the structure has been shown to fulfill the requirements; and $F_{target} > F_{lim}$, which means that the load test has to be terminated at a lower load level than the target load F_{target}.

In the German guideline, the applied target proof load $ext.F_{target}$ is determined as follows:

$$ext.F_{target} = \sum_{j>1} \gamma_{G,j} G_{k,j} + \gamma_{Q,1} Q_{k,1} + \sum_{i>1} \gamma_{Q,i} \psi_{Q,i} Q_{k,i} \tag{3.1}$$

with

$$0.35 G_{k,1} \leq ext.F_{target} \leq ext.F_{lim} \tag{3.2}$$

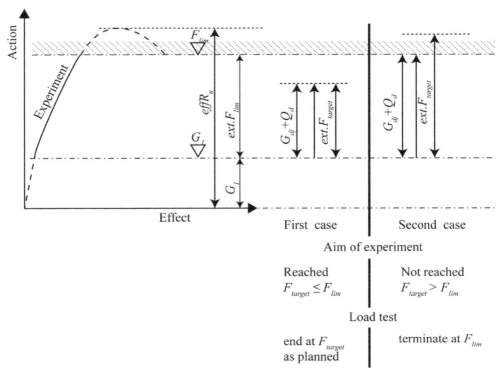

Figure 3.1 Safety philosophy of the German guideline (Deutscher Ausschuss für Stahlbeton, 2000), showing the two possible scenarios

Source: Reprinted with permission from Lantsoght et al. (2017).

The defined loads are $G_{k,1}$, the characteristic value of the permanent loads occurring in the load test; $G_{k,j}$, the characteristic value of the permanent loads occurring after the load test; and $Q_{k,1}$ and $Q_{k,i}$, the characteristic values of the live load for the governing load 1 and other loads i. The defined load factors are $\gamma_{G,j}$, the partial factor for the permanent loads G; $\gamma_{Q,1}$ and $\gamma_{Q,i}$, the load factor for the governing live load 1 and other live loads i; and $\psi_{Q,i}$, the combination factor for the live loads Q. The partial factors and combination factor should be taken from the governing German codes. The values of $Q_{k,1}$, $Q_{k,i}$, and $G_{k,j}$ are no longer valid when changes are made to the studied structure.

3.2.3 Stop criteria

The German guideline prescribes stop criteria to be used for proof load tests for structures that are expected to be flexure-critical. The limiting load F_{lim} from the safety philosophy illustrated in Figure 3.1 is reached when a stop criterion is exceeded. In total, five stop criteria are defined. The first stop criterion describes a limiting strain in the concrete:

$$\varepsilon_c < \varepsilon_{c,lim} - \varepsilon_{c0} \tag{3.3}$$

In Equation (3.3), ε_c is the strain measured in the concrete during proof loading. This strain has to be smaller than the limiting strain of $\varepsilon_{c,lim} - \varepsilon_{c0}$. The value of $\varepsilon_{c,lim}$ equals 0.6‰ in general, and can be increased to 0.8‰ when the concrete compressive strength is larger than 25 MPa (3.6 ksi). The strain ε_{c0} is the analytically determined short-term strain in the concrete caused by the permanent loads that act on the structure before the application of the proof load.

The second stop criterion describes a limiting strain in the reinforcement steel. The value is determined as:

$$\varepsilon_{s2} < 0.7 \frac{f_{ym}}{E_s} - \varepsilon_{s02} \tag{3.4}$$

When the stress-strain relationship of the steel bars is known, Equation (3.4) can be replaced by:

$$\varepsilon_{s2} < 0.9 \frac{f_{0.01m}}{E_s} - \varepsilon_{s02} \tag{3.5}$$

In Equations (3.4) and (3.5), ε_{s2} refers to the measured strain in the steel, and ε_{s02} is the analytically determined strain (assuming that the concrete cross section is cracked) in the reinforcement caused by the permanent loads that act on the structure before the application of the proof load. The value of f_{ym} is the average yield strength of the tension steel. If the stress-strain relationship of the bars is known, the more precise $f_{0.01m}$ can be used, with $f_{0.01m}$ the average yield strength based on a strain of 0.01% in the steel. E_s is the Young's modulus of the reinforcement steel.

The third stop criterion limits the width w of new cracks that can occur during the load test, as well as the increase in crack width Δw of existing crack widths during the load test. Limitations are given both for the maximum value of the crack width and the residual crack width after removal of the load. This stop criterion is given in Table 3.1.

The fourth stop criterion is related to the measured deflection. This stop criterion is either exceeded when a clear increase of the nonlinear part of the deformation is observed, or when more than 10% permanent deformation is found after removing the load.

A final stop criterion is only applicable to beams with shear reinforcement. For this case, the strains in the shear span are further limited: for the stop criterion from Equation (3.3), the maximum strain is taken as 60% of the maximum strain given in Equation (3.3), and for the stop criterion from Equation (3.4), the maximum strain is taken as 50% of the maximum strain given in Equation (3.4).

Additional conditions that require the termination of a load test in the German guideline are (Deutscher Ausschuss für Stahlbeton, 2000):

- When the measurements (e.g. the load-deflection curve or acoustic emission measurements) indicate critical changes in the structure, which are expected to cause damage when the load is further increased;

Table 3.1 Requirements for crack width w for newly developing cracks and increase in crack width Δw for existing cracks

	During load testing	After load testing
Existing cracks	$\Delta w \le 0.3$ mm = 0.12 in	$\le 0.2\ \Delta w$
New cracks	$w \le 0.5$ mm = 0.20 in	$\le 0.3\ w$

34 Load Testing of Bridges

- When the stability of the structure cannot be further guaranteed;
- When critical displacements occur at the supports.

To evaluate the stop criteria, the instrumentation of the structure needs to be able to measure the input necessary for verifying these stop criteria. Furthermore, the effect of the environmental conditions (temperature, humidity, and wind) needs to be known.

3.3 British guidelines

3.3.1 *General*

In the United Kingdom, a guideline for load testing is available (*The Institution of Civil Engineers: National Steering Committee for the Load Testing of Bridges*, 1998). This guideline was developed as a consequence of the implementation of the 40-tonne (88 kip) vehicle in 1988, for which a large number of existing bridges did not have sufficient capacity. The guideline is suitable for showing sufficient load-carrying capacity of apparently understrength bridges and for checking the performance of newly constructed bridges. Load tests according to this guideline can be used to aid assessment by calculation.

The nomenclature used in the British guidelines slightly differs from what is used throughout this book. The British guidelines describe the following types of load tests:

- Supplementary load tests: load tests to supplement the analytical methods of assessment based on calculation and the use of codes of practice (called "diagnostic load tests" in this book).
- Proof loading: load tests to validate the design method and design assumptions for newly constructed bridges, with loading levels to the serviceability limit state (SLS) (called "diagnostic load tests" in this book).
- Proving load testing: load tests to provide a safe load-carrying capacity without further theoretical analysis, where the load is increased to some predetermined maximum or until the structure shows signs of deterioration or distress (called "proof load tests" in this book).
- Dynamic load testing: load tests using ambient or forced vibrations to measure the stiffness (called "dynamic load tests" in this book).

The British guideline is limited to supplementary load testing (diagnostic load testing) as an integral part of the overall assessment procedure for existing bridges. The results of such a load test can be used to improve the existing finite element model based on the field measurements. The updated model is then used for the assessment.

The described procedures can be used for structures and structural elements. The guideline does not recommend testing when a brittle failure mode can occur, such as failure in shear or bearing at the support. Additionally, a structure in poor condition (with excessive deterioration or significant deflections) should not be load tested. It must be verified prior to deciding if a load test is suitable that the instrumentation can be applied where required and that the results of the load test can be used to improve the analysis. While the British guideline encompasses all types of bridges, it seems to be based mostly on work on masonry arch bridges. The reinforced concrete bridges that were tested (some to failure) to calibrate the proposed method were the Dornie Bridge (a beam-

and-slab-type bridge) (Ricketts and Low, 1993) and a series of filler beam bridges (Low and Ricketts, 1993).

If possible, load testing should be carried out at night when there is less traffic and bridge deck temperatures are most stable. At a time of the year when the ambient temperatures are high, the surfacing stiffness will be low. This consideration should be kept in mind.

3.3.2 Preparation and application of loading

An assessment is required prior to the field test in the British guideline. For this assessment, the actual dimensions of the bridge elements are to be used. The actual dimensions should be based on site measurements with due allowances for corrosion and other forms of deterioration.

One difficulty in load testing that is highlighted in the British guideline is that it is often difficult to separate the influence of the different effects that contribute to the overall response. However, supplementary load testing (diagnostic load testing) does provide a way of improving the accuracy of the analytical model used in structural analysis so that it more closely models the behavior of the real structure. The following effects on the structural behavior can be assessed through load testing:

- Transverse load distribution;
- Composite action: for example, when no shear connectors are provided between the girders and the deck;
- Restraints at the supports: continuous surfacing over buried joints or friction in the bearings can provide some rotational and translational restraint;
- Pin-joint fixity: in truss members, there is always a certain amount of restraint in the connections;
- Transverse compression, or arching action: this effect can occur in concrete deck slabs of beam-and-slab-type bridges.

The applied loading during the load test should reflect the current vehicle construction and use regulations, including allowances for overloading and the dynamic behavior of the vehicles. The magnitude of the load used during the load test is determined based on the following criterion: the applied load should produce measurable elastic deformations, such as deflections or surface strains, without causing permanent damage. Possible methods of load application that are suggested in the guidelines are:

- Dead weights or kentledge blocks as distributed or concentrated loads, applied directly on the deck or on a frame spanning the structure, to apply concentrated loads to the deck by jacking against the dead weight;
- Flexible water bags that provide dead weight for testing;
- Jacking systems reacting against ground or rock anchors;
- An HB single-axle trailer which holds kentledge units symmetrically about the axle and can provide single-axle loading of maximum 45 tonnes (99 kips);
- Loaded vehicles, indicated as the most commonly used method of bridge testing, usually 30 or 32 tonne (66 or 73 kips) four-axle rigid aggregate trucks, filled to the desired load and weighed at a weighbridge or with portable weigh pads;
- Railway loading for railroad bridges with locomotives as static or moving loads.

3.3.3 *Evaluation of the load test*

For the interpretation of the test results, the obtained measurements (of strains and/or displacements) are used to determine an improved value of the safe load-carrying capacity of the structure. A first step in the analysis deals with the comparison of measured and calculated results in terms of:

- Linearity: changes in the loading should correspond to prorated changes in deflections and strains. All measured responses should return to zero when the load is removed.
- The effect of the dead load on the structure and the proportion of the load-carrying capacity required to carry these loads needs to be determined.
- The tested stiffness needs to be compared to the stiffness assumed in the calculations and finite element model.
- Local effects should be considered. For example, if concentrated loads are applied on the deck, and strain measurements are taken around the load, it can be evaluated if membrane action occurs.

In terms of estimating the structural capacity and ultimate limit state (ULS), the guideline clearly states: "It is this relatively complex and unique nature of most bridge structures that makes it impossible to derive a safe load capacity directly from load tests in the elastic range with any degree of confidence. The most effective role of supplementary load testing is in providing a better understanding of the global and local behavior of a particular structure and hence in improving the analytical model so that it more closely mirrors that of the real structure." Extra strength can be attributed to the structure for the following cases:

- If relevant data from collapse tests is available, the structural actions could be arrived at from these collapse data, and they can be reliably used at the ULS.
- If the structural actions are identified by analysis and can be shown to be reliable.
- If the structural actions are identified by analysis, and the unreliable actions can be retrofitted.
- If the structural actions are identified by analysis, and the unreliable actions can be monitored.
- If the structural actions are identified by analysis, and the unreliable actions can be verified periodically with load tests.

3.4 Irish guidelines

3.4.1 *General*

In Ireland, a manual for load testing is available (NRA, 2014). This manual is suitable for older metal bridges and older concrete bridges that may be found to be substandard when assessed using the calculation methods from the currently governing assessment standards. Load tests can be used to increase the assessed capacity so that sources of reserve capacity can be taken into account. Only diagnostic load tests are permitted by the Irish guidelines. Such load tests should be considered an accompaniment to the assessment calculations. When upon assessment a bridge is found to be insufficient for shear, load testing is not recommended, or the load levels should be limited. Proof load tests, which are a self-supporting alternative to a theoretical assessment, are not permitted.

Load testing should only be used for bridges that have insufficient capacity according to an assessment and where there is a realistic possibility of improving the assessed capacity to a level of significant benefit.

3.4.2 Recommendations for applied loading

The applied loading in the field test can be one or more vehicles, axle or patch loads, or combinations thereof. A single procedure that should be followed for the prescribed diagnostic load tests is not provided and each case has to be treated individually. The loading can be used to generate the bending moments produced by the assessment live loading, for example, and should be applied in increments. The structural responses should be kept within the elastic range of the bridge flexural behavior.

The loading needs to reflect the traffic of the day in a safe and conservative way. Increases in the assessment live load to take into account include the following:

- Axle impact effects;
- Overloading of vehicles;
- Bunching of vehicles for bridges with loaded lengths less than 50 m (164 ft).

The load levels in the tests should not exceed those caused by the loads carried by the bridge on a day-to-day basis.

3.4.3 Evaluation of the load test

After the field test, the measurements of strains and deflections can be used to assess the hidden reserve capacity in the bridge, and each possible source of hidden strength should be examined, possibly based on knowledge gained in earlier collapse tests on similar bridges. This additional source of strength can then be implemented into the assessment calculations. Extrapolation from load tests at fairly low load levels to ULS conditions is not recommended unless earlier collapse tests have been carried out on bridges with similar materials and details.

3.5 Guidelines in the United States

3.5.1 *Bridges:* Manual for Bridge Rating through Load Testing

3.5.1.1 *General*

The *Manual for Bridge Rating through Load Testing* (NCHRP, 1998), used in North America, links the concepts of load testing and bridge rating. The manual is valid for all types of bridges except long-span bridges, shear-critical concrete beam bridges, bridges with an extremely low capacity analytically, bridges with frozen joints that could cause a sudden release of energy, steel bridges with fracture-critical members, and bridges on poor soil and foundation conditions. The concepts from this manual are also repeated in the *Manual for Bridge Evaluation* (MBE) (AASHTO, 2011).

38 Load Testing of Bridges

The manual describes diagnostic load tests as tests in which the load is placed at designated locations and the effects of this load on individual members of the bridge are measured by the instrumentation attached to these members. The measurements are then compared to computed effects. Proof load tests are described as tests in which the bridge is loaded up to its elastic limit, when the test is stopped and the maximum load and position are recorded. In some cases, a target proof load is established prior to the test, and the load test is stopped when this goal is reached. Note that all other codes and guidelines require the determination of the target proof load prior to a proof load test.

The goal of the manual is to establish realistic safe service live-load capacities for bridges. For load rating, the MBE (AASHTO, 2011) currently prescribes the following expression to determine the rating factor RF for LRFR (load and resistance factor rating):

$$RF = \frac{C - (\gamma_{DC})(DC) - (\gamma_{DW})(DW) \pm (\gamma_P)(P)}{(\gamma_{LL})(LL + IM)} \tag{3.6}$$

The capacity C for the strength limit states is determined as:

$$C = \varphi_c \varphi_s \varphi R_n \quad \text{with} \quad \varphi_c \varphi_s \geq 0.85 \tag{3.7}$$

In Equation (3.7), R_n is the nominal member resistance as inspected. For the SLSs, the capacity C is determined as:

$$C = f_R \tag{3.8}$$

with f_R the allowable stress specified in the LRFD Code (AASHTO, 2015). In Equation (3.6), DC is the dead load effect due to structural components and attachments, DW is the dead load effect due to wearing surface and utilities, P is the effect from the permanent loads other than dead loads, LL is the live-load effect, and IM is the dynamic load allowance. The following load and resistance factors are defined in Equations (3.6), (3.7), and (3.8): γ_{DC} is the LRFD load factor for structural components and attachments; γ_{DW} is the LRFD load factor for wearing surfaces and utilities; γ_P is the LRFD load factor for permanent loads other than dead load, which equals 1.0; γ_{LL} is the evaluation live-load factor; ϕ_c is the condition factor; ϕ_s is the system factor; and ϕ is the LRFD resistance factor.

In a diagnostic load test, the live-load effect from Equation (3.6) is measured directly during the test in one or more critical bridge members. These values are then compared with the values computed by an analytical model. The difference between the theoretical and measured load effects is used to update the analytical model and then to determine the load rating for a bridge member. The difference is caused by uncertainties about bridge behavior (material properties, boundary conditions, effectiveness of repairs, unintended composite action, and effect of damage and deterioration) or as part of routine parametric determinations (e.g. load distribution, impact factors). The contribution from nonstructural components (non-composite deck slabs or parapets) may cease at higher load levels, so that the applied load should be "sufficiently high."

In a proof load test, on the other hand, the capacity of the bridge to carry live load, which is the numerator of Equation (3.6), is measured. According to the *Manual for Bridge Rating through Load Testing*, a test should be terminated when:

1   A predetermined maximum load (the target proof load) has been reached; or
2   The bridge exhibits the onset of nonlinear behavior or other visible signs of distress.

Moreover, the *Manual for Bridge Rating through Load Testing* mentions other types of nondestructive load tests, such as:

* Load identification (based on WIM data);
* Tests for unusual forces (effects of stream flow, ice, wind pressure, seismic action and thermal response, which are not part of the usual rating procedures);
* Dead load effects (by using residual stress gages for steel members, or by jacking, which is a dangerous procedure);
* Dynamic effects to determine the bridge frequency and damping (by using moving loads, portable sinusoidal shakers, sudden release of applied deflections, sudden stopping of vehicles by braking, or impulse devices);
* Impact (influence by the surface roughness of the deck and bumps on the bridge approach);
* Fatigue for steel highway bridges.

3.5.1.2 *Preparation of load tests*

The first step in preparing a load test is identifying if a load test is a suitable procedure to determine the unknown information necessary for the bridge load rating. The types of bridges that have been identified in the *Manual for Bridge Rating through Load Testing* as interesting for load testing are:

* Slab bridges, for which a load test can be used to determine the transverse load distribution;
* Multi-stringer bridges, to determine the distribution of the load to the stringers, the effect of composite action, and the effect of restraints at the stringer and girder supports;
* Two-girder bridges, where the structure is made partially continuous because the deck slab is made continuous over the transverse floor beams;
* Truss bridges, to quantify the restraint in the joints;
* Masonry arch bridges, which typically are rather old;
* Rigid frame bridges, where proof load tests can establish a safe service load;
* Timber bridges, in which the material properties are time dependent.

When a load test seems to be useful for the bridge under consideration, the next step required by the *Manual for Bridge Rating through Load Testing* is an inspection of the bridge. This inspection should determine the condition of the bridge, determine occurring damage, and help assess the actual dead loads by, for example, measuring the thickness of the asphalt layer. These results are then used for a preliminary rating, according to Equation (3.6). In this phase, it is determined if load testing is a feasible alternative for establishing the load rating. If the rating factor is larger than one, a load test is unnecessary. If the rating factor is much smaller than one, it is unlikely that a load test will

40 Load Testing of Bridges

be able to bring the rating factor up to a value larger than or equal to one. Therefore, the most suitable candidate bridges for load testing are those with a rating factor smaller than one, but close to one. The initial calculations are also necessary for defining the sensor plan.

3.5.1.3 Execution of load tests

The elements of the execution of load tests that are discussed in the *Manual for Bridge Rating through Load Testing* are the target load, the loading system, and the possibilities for sensors.

The required load depends on the type of load test. For a diagnostic load test, the load levels are around service levels and usually involve one or two loaded dump trucks. The test load should stress all critical elements that need to be evaluated in the load test. The required load during a proof load test is higher, and is closely related to the rating factor. It will be discussed in Section 5.1.5.

The loading system can be stationary loading with heavy blocks or jacks, or moveable loading with test vehicles that simulate the legal vehicles and that move at crawl speed or at normal operating speeds. The loading system should fulfill the following requirements:

- It should be representative of the rating vehicles;
- The load should be adjustable in magnitude;
- Loads should be maneuverable;
- Loads should allow for repeatability to check the linearity of the bridge response and return of response to zero.

The manual also states that for multiple-lane bridges, a minimum of two lanes should be loaded concurrently.

The *Manual for Bridge Rating through Load Testing* does not rigidly prescribe a certain loading protocol. Critical test load cases should be repeated at least two times or until correlation between each repetition is obtained. This correlation between the results for repeated load positions generally indicates elastic behavior and provides assurance that the test instrumentation is performing correctly. For proof load tests, the first-stage loading should not exceed 25% of the target load, and the second stage should not exceed 50% of the target load. Smaller increments of loading may be warranted, particularly when the applied proof load approaches the target load.

During a load test, displacements, strains, rotations, and crack widths must be measured. These measurements are taken at the start of the load test and at the end of each load increment. The measurements should then be compared to the predicted response from the preliminary calculations to detect unusual behavior. The load-deformation response and deflection recovery should be monitored very closely to determine the onset of nonlinear behavior, which is used as a stop criterion for proof load tests in the *Manual for Bridge Rating through Load Testing*. In a standard proof load test according to the manual, only the applied loads and resulting displacements are monitored. If more extensive instrumentation is applied, the test is considered a mixed test, including elements from diagnostic load testing. The manual also contains an extensive description of possibilities for load test instrumentation, albeit outdated.

3.5.1.4 Determination of the rating factor after a diagnostic load test

For a diagnostic load test, the rating factor after the test can be updated based on the observations made during the load test. One has to be careful when extrapolating the results of the diagnostic load test to higher load levels. The *Manual for Bridge Rating through Load Testing* proposed the following equation to update the rating factor prior to the test (RF_C) from Equation (3.6) to the rating factor based on the test results (RF_T):

$$RF_T = RF_c \times K \tag{3.9}$$

In Equation (3.9), K is an adjustment factor resulting from the comparison of measured test behavior with the analytical model, given as:

$$K = 1 + K_a \times K_b \tag{3.10}$$

The factor K_a accounts for both the benefit derived from the load test, if any, and considerations of the section factor resisting the applied test load. K_a captures the test benefit without the effect of unintended composite action, which cannot be extrapolated to higher load levels. The value of K_a can be determined as follows:

$$K_a = \frac{\varepsilon_c}{\varepsilon_T} - 1 \tag{3.11}$$

In Equation (3.11), ε_T is the maximum member strain measured during the load test, whereas ε_c is the corresponding theoretical strain due to the test vehicle and its position on the bridge which produced ε_T. The value of ε_c can be determined analytically as:

$$\varepsilon_c = \frac{L_T}{(SF)E} \tag{3.12}$$

In Equation (3.12), L_T is the calculated theoretical load effect in member corresponding to the measured strain ε_T, SF is the member appropriate section actor (area, section modulus, etc.), and E is the member modulus of elasticity.

The second factor in Equation (3.10) is the factor K_b. This factor K_b takes into account differences between the actual behavior of the bridge and the revised analytical model, specifically with regard to lateral and longitudinal load distribution and the participation of other members, and consists of several elements:

$$K_b = K_{b1} \times K_{b2} \times K_{b3} \tag{3.13}$$

The factor K_{b1} in Equation (3.13) takes into account the analysis performed by the load test team and their understanding and explanations of the possible enhancements to the load capacity observed during the test. $K_{b1} = 0$ reflects the inability of the test team to explain the test behavior or validate the test results, whereas $K_{b1} = 1$ means that the test measurements can be directly extrapolated to performance at higher loads corresponding to the rating levels. Intermediate values can be derived based on the test vehicle effect T and the gross rating load effect W (see Table 3.2).

The second factor, K_{b2}, takes into account the ability of the inspection team to find problems in a timely manner to prevent changes in the bridge condition that could invalidate the test result. The value of K_{b2} depends on the type and frequency of inspection, and can be found in Table 3.3.

42 Load Testing of Bridges

Table 3.2 Determination of K_{b1}

Can member behavior be extrapolated to 1.33W?		Magnitude of test load			K_{b1}
Yes	No	T/W < 0.4	$0.4 \leq T/W \leq 0.7$	T/W > 0.7	
x		x			0
x			x		0.8
x				x	1
	x	x			0
	x		x		0
	x			x	0.5

Table 3.3 Determination of K_{b2}

Inspection		K_{b2}
Type	Frequency	
Routine	between 1 and 2 years	0.8
Routine	less than 1 year	0.9
In-depth	between 1 and 2 years	0.9
In-depth	less than 1 year	1.0

Table 3.4 Determination of K_{b3}

Fatigue controls?		Redundancy?		K_{b3}
No	Yes	No	Yes	
	x	x		0.7
	x		x	0.8
x		x		0.9
x			x	1.0

The last factor, K_{b3}, accounts for the presence of critical structural features which cannot be determined in a diagnostic test and which could contribute to sudden fatigue, fracture, or instability failure of the bridge. The value of K_{b3} is determined as given in Table 3.4.

3.5.1.5 Determination of the rating factor after a proof load test

For a proof load test, the maximum applied load is a lower bound on the live-load capacity of the bridge. The target proof load should cover uncertainties, in particular the possibility of bridge overloads during normal operations, as well as the impact allowance. The goal of the target proof load is to result in a rating factor of one after the proof load test. The rating factor is one if the test safely reaches the legal rating plus the impact allowance magnified by the target live-load factor X_{pA} during the proof load test, which is determined as:

$$X_{pA} = X_p \left(1 + \frac{\Sigma\%}{100} \right) \tag{3.14}$$

In Equation (3.14), X_p is the factor prior to adjustments, which equals 1.4. This factor is adjusted as follows (represented by $\Sigma\%$ in Equation (3.14)):

- X_p needs to be increased by 15% if one lane load controls the response.
- For spans with fracture-critical details, X_p shall be increased by 10%.
- If routine inspections are performed less than every two years, X_p should be increased by 10%.
- If the structure is ratable (i.e. has no hidden details), X_p can be reduced by 5%.
- Additional factors including traffic intensity and bridge condition may also be incorporated in the selection of the live-load factor X_p.

The resulting value for X_{pA} lies between 1.3 and 2.2. Once X_{pA} is known, the target proof load L_T can be calculated as:

$$L_T = X_{pA} L_R (1 + I) \tag{3.15}$$

L_R is the comparable live load due to the rating vehicle for the lanes loaded, and I is the AASHTO specifications impact allowance.

The maximum load applied during the proof load test is L_p. This value can be the target proof load L_T when the test is successful, or a lower load level if nonlinearity occurred prior to reaching L_T. The capacity at the operating level is then determined based on L_p as follows:

$$OP = \frac{k_O L_p \text{ and } X_{pA}}{X_{pA}} \tag{3.16}$$

The value of k_O equals 1.0 if the target proof load was achieved during the test and 0.88 if the test had to be terminated because nonlinear behavior was observed. The rating factor at the operating level is then calculated as:

$$RF_O = \frac{OP}{L_R(1 + I)} \tag{3.17}$$

From the previous analyses, it can be seen that for $X_p = 1.4$ the rating factor at the operating level equals 1.0. This analysis at the operating rating level corresponds to a reliability index of 2.3. The lower beta value is justified as it reflects past rating practice at the operating levels. The factor $X_p = 1.4$ is derived to match this reliability index, based on a first-order reliability approach.

3.5.2 *Buildings*

3.5.2.1 *ACI 437.1: "Load tests of concrete structures: methods, magnitude, protocols, and acceptance criteria"*

For load tests on buildings in the United States, ACI provides a number of documents that will be discussed in this chapter since these documents have influenced the discussion on the determination of the target proof load and the stop criteria for proof load testing of bridges. For new and existing structures, ACI 437.1R-07 (ACI Committee 437, 2007) describes the procedures, the target proof load, the load testing protocol, and the acceptance criteria. Similar provisions can be found in ACI 318-14 (ACI Committee 318, 2014) for new buildings. For existing structures, ACI 437.2M-13 (ACI Committee 437, 2013) describes the procedures, the target proof load, the load testing protocol, and the acceptance

44 Load Testing of Bridges

criteria. Similar provisions can be found in ACI 562-16 (ACI Committee 562, 2016) for existing buildings. ACI 562-16 fully refers to ACI 437.2M-13 for load testing.

This section summarizes the provisions for load tests on structures that are given in ACI 437.1R-07 (ACI Committee 437, 2007). This report provides the background to a new proposal of ACI Committee 437 with regard to the target proof load, so that the proof load testing procedures can be aligned to changes made to the ACI building code in its 2002 version. The recommendations from the report can be applied to normal strength concrete structures and buildings, but not to bridges. The goal of the proof load tests described by ACI 437.1R-07 (ACI Committee 437, 2007) is to "show that a structure can resist the working design loads in a serviceable fashion where deflections and cracking are within limits considered acceptable by ACI 318."

In ACI 437.1R-07 (ACI Committee 437, 2007), the target proof load, called "test load magnitude," is redefined. Parameter studies showed that defining the test load as a constant percentage of the required design strength results in the fact that the relationship between the proof load applied to the structure and the service live load is not apparent and is not a reasonably constant ratio. Moreover, defining the test load as a combination of a factored dead and live load makes the relationship between the target proof load and the service live loads variable. For this reason, it is recommended to separate the dead load into the components that do not vary (the dead load due to self-weight D_w) and those that vary (the dead load due to weight of construction and other building materials, i.e. the superimposed dead load D_s). For the dead load due to self-weight D_w, a load factor of 1.0 is proposed, whereas for the superimposed dead load D_s a factor greater than 1.0 is proposed, since these loads may change over time depending on the owner's use of the facility, and construction and maintenance means and methods. If only portions of the suspect areas of a structure can be tested, a higher test load is recommended to improve the level of confidence that significant flaws or weaknesses in the design, construction, or current condition of the structure are made evident by the load test. The resulting recommendation for the test load magnitude TLM for the case when all suspect portions of a structure are to be load tested or when the members to be tested are determinate and the suspect flaw or weakness is controlled by flexural tension is that the test load magnitude TLM shall not be less than:

$$TLM = 1.2(D_w + D_s) \tag{3.18}$$

or

$$TLM = 1.0D_w + 1.1D_s + 1.4L + 0.4(L_r \text{ or } S \text{ or } R) \tag{3.19}$$

or

$$TLM = 1.0D_w + 1.1D_s + 1.4(L_r \text{ or } S \text{ or } R) + 0.9L \tag{3.20}$$

In Equations (3.19) and (3.20), L_r is the live load on the roof, S is the snow load, and R is the rain load.

When only part of suspect portions of a structure is to be load tested and members to be tested are indeterminate, the test load magnitude TLM, including the dead load already in place, shall not be less than:

$$TLM = 1.3(D_w + D_s) \tag{3.21}$$

or

$$TLM = 1.0D_w + 1.1D_s + 1.6L + 0.5(L_r \text{ or } S \text{ or } R) \tag{3.22}$$

or

$$TLM = 1.0D_w + 1.1D_s + 1.6(L_r \text{ or } S \text{ or } R) + 1.0L \tag{3.23}$$

The load can be applied as a uniformly distributed load with dead weights, or with a series of concentrated loads to simulate the effects (bending moment and shear forces) of a uniformly distributed load by using patch or strip equivalent loads. Two loading procedures are allowed: monotonic loading and cyclic loading (see Fig. 3.2). For monotonic loading, at least four load increments are used. No sketch of the monotonic loading protocol is given in ACI 437.1R-07. Cyclic loading has the advantage that parameters such as the linearity of the structural deflection response, repeatability of the load-deflection response, and permanency of deflections can be measured. The load cycles at low load levels also help the engineer to better understand end fixity and load transfer characteristics of the tested member by comparing the actual deflection responses with the calculated deflection responses. The duration of loading at the maximum load can be 24 hours, which is the traditional ACI approach, or a shorter duration (approximately 2 minutes). Planning a cyclic load test involves more engineering effort and interpretation and requires more insight in the structural behavior, including effects of load sharing and end fixity. The advantage of a cyclic loading test is that it takes less time and that it allows the engineer a real-time assessment of the performance of the structure. At least six loading and unloading cycles should be used. Each load cycle consists of five load steps (see Fig. 3.2). A minimum load P_{min} of at least 10% of the total test load should be maintained between the cycles to keep the test devices engaged.

The acceptance criteria for the 24-hour monotonic load test are based on a set of visual parameters (no spalling or crushing of compressed concrete is evident), and one of the two

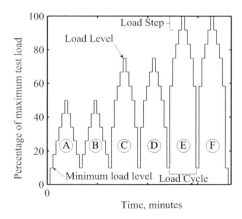

Figure 3.2 Cyclic loading protocol from ACI 437.1R-07
Source: Reprinted with permission from ACI.

46 Load Testing of Bridges

following expressions with regard to deflections must be satisfied:

$$\Delta_{max} \leq \frac{l_t^2}{20000h} \tag{3.24}$$

$$\Delta_{r,max} \leq \frac{\Delta_{max}}{4} \tag{3.25}$$

In Equations (3.24) and (3.25), Δ_{max} is the maximum measured deflection, $\Delta_{r,max}$ is the maximum residual deflection, l_t is the span length, and h is the member depth. Equation (3.24) is unrelated to modern material strengths, deflection limits, or degree of fixity that may be present in the structural member being tested. Furthermore, ACI 437.1R-07 mentions that new acceptance criteria for deflection should be developed, which include the maximum deflection under the full test load compared with the calculated theoretical maximum deflection at that load level, recovery of deflection upon full removal of the load, and linearity of deflection response during loading and unloading. Moreover, it is recommended in ACI 437.1R-07 to replace Equation (3.24) with:

$$\Delta_{max} \leq \frac{l_t}{180} \tag{3.26}$$

It is also proposed that the residual deflection should be less than 25% of the corresponding absolute maximum deflection immediately upon unloading or 24 hours afterwards. No check on the deflection recovery is required if the absolute deflection is less than 1.3 mm (0.05 in.) or if the deflection as a percentage of span length is less than $l_t/2000$.

For a cyclic load test, three acceptance criteria are defined. The first acceptance criterion is the repeatability index I_R, which measures the similarity of behavior of the member or structure during two twin load cycles at the same load level. It is a measure of the recoverable elastic deflection and load-deflection response in general. The repeatability index is calculated with the definitions from Figure 3.3 as follows:

$$I_R = \frac{\Delta_{max}^B - \Delta_r^B}{\Delta_{max}^A - \Delta_r^A} \times 100\% \tag{3.27}$$

A value of I_R in the range of 95%–105% is considered satisfactory.

The second acceptance criterion is the permanency index I_P, which is the relative value of the residual deflection compared with the corresponding maximum deflection. The value of I_P should be less than 10%. Higher values indicate that load application has damaged the member/structure and that nonlinear effects are taking place. The value of the permanency index I_P is determined based on the deflections shown in Figure 3.3 as:

$$I_P = \frac{\Delta_r^B}{\Delta_{max}^B} \times 100\% \tag{3.28}$$

The last acceptance criterion for cyclic load tests is the deviation from linearity, which represents the measure of the nonlinear behavior of a member/structure being tested. Linearity is defined as the slopes of two secant lines intersecting the load-deflection envelope as shown in Figure 3.4. The linearity at any point i on the load-deflection envelope is the percent ratio of the slope of the secant line to point i, expressed by $\tan(\alpha_i)$ to the slope of the

Current Codes and Guidelines 47

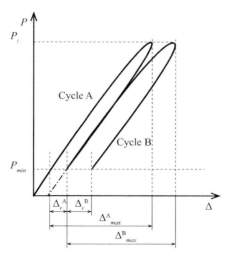

Figure 3.3 Load-deflection curve for two cycles at the same load level, used to determine the repeatability index I_R and the permanency index I_p, from ACI 437.2M-13 (ACI Committee 437, 2013)

Source: Reprinted with permission from ACI.

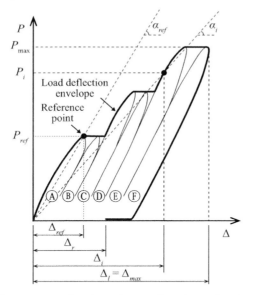

Figure 3.4 Load-deflection curve for six cycles, used to determine the deviation from linearity index I_{DL}, from ACI 437.2M-13 (ACI Committee 437, 2013)

Source: Reprinted with permission from ACI.

reference secant line, expressed by $\tan(\alpha_{ref})$, with the angles as shown in Figure 3.4. The linearity is thus expressed as:

$$Linearity_i = \frac{\tan(\alpha_i)}{\tan(\alpha_{ref})} \times 100\% \qquad (3.29)$$

48 Load Testing of Bridges

The deviation from linearity index I_{DL} is then given as:

$$I_{DL} = 100\% - Linearity_i \tag{3.30}$$

The value of I_{DL} should be less than 25% and should be monitored continuously during the cyclic load test. When a member/structure is initially uncracked and becomes cracked during the load test, the change in flexural stiffness as a result of a drastic change in moment of inertia can result in a very high deviation from linearity that is not necessarily related to degradation in strength. It is suggested that I_{DL} should only be computed for the member/structure under cracked conditions.

An additional advantage of the cyclic loading protocol is that it can be used for load rating. Loading can be continued until one of the acceptance criteria fails. That load level can then be used to calculate the safe live-load level. A qualitative acceptance criterion is also formulated: the structure should show no signs of impending failure, such as concrete crushing in the compressive zone or concrete cracking exceeding a preset limit. The maximum deflection at the second load cycle that reaches the target proof load should be less than the deflection calculated according to the ACI 318 building code, to make sure the engineer has made a prediction given the available information and that such a prediction be used to interpret the experimental results.

At the service load level, crack spacing and width should be analyzed. The variable nature of cracking and the challenges in accurately measuring and predicting crack width make the corresponding limits difficult to implement. The engineer should determine a limiting crack width. The maximum measured deflection should not exceed the permissible values from ACI 318.

3.5.2.2 New buildings: ACI 318-14

The ACI 318-14 Building Code (ACI Committee 318, 2014) discusses strength evaluation by load testing in section 27.4, where the focus is on the evaluation of new buildings through load tests. These provisions are based on the recommendations from ACI 437.1R-07 (ACI Committee 437, 2007). The total test load T_t, including dead load already in place, shall be at least the greatest of:

$$T_t = 1.3D \tag{3.31}$$

$$T_t = 1.15D + 1.5L + 0.4(L_r \text{ or } S \text{ or } R) \tag{3.32}$$

$$T_t = 1.15D + 1.5(L_r \text{ or } S \text{ or } R) + 0.9L \tag{3.33}$$

with D the dead load, L the live load, L_r the live load on the roof, S the snow load, and R the rain load. The live load L can be reduced in accordance with the general building code.

ACI 318-14 (ACI Committee 318, 2014) describes a monotonic loading protocol. The test load T_t is applied in at least four approximately equal increments. If a uniformly distributed load is used, arching must be avoided, since this effect results in a reduction of the load near the midspan. T_t must remain on the structure for at least 24 hours, unless signs of distress are observed. Response measurements such as deflection, strain, slip, and crack width shall be made at locations where maximum response is expected, and these measurements must be recorded after each load increment and after T_t has been applied to the structure for at least 24 hours. The last set of measurements is taken 24 hours after T_t is removed.

The acceptance criteria in ACI 318-14 are the following:

1   The portion of the structure tested shall show no spalling or crushing of concrete, or other evidence of failure, including distress (cracking, spalling, or deflection) of such magnitude and extent that the observed result is obviously excessive and incompatible with the safety requirements of the structure. Local spalling or flaking of the compressed concrete in flexural members related to casting imperfections need not indicate overall structural distress.
2   Members tested shall not exhibit cracks indicating imminent shear failure, that is, when crack lengths increase to approach a horizontal projected length equal to the depth of the member and widen to the extent that aggregate interlock cannot occur, and as transverse stirrups, if present, begin to yield or display loss of anchorage so as to threaten their integrity.
3   If no transverse reinforcement is available, structural cracks inclined to the longitudinal axis and having a horizontal projection greater than the depth of the member shall be evaluated, as these may lead to brittle failure.
4   In regions of anchorage and lap splices of reinforcement, short inclined cracks or horizontal cracks along the line of reinforcement shall be evaluated. These cracks can indicate high stresses associated with the transfer of forces between the reinforcement and the concrete, and may be indicators of impending brittle failure of the member.
5   The measured deflections Δ_l (maximum deflection after 24 hours of load application) and Δ_r (residual deflection 24 hours after removal of the test load) shall satisfy:

$$\Delta_1 \leq \frac{l_t^2}{20000h} \tag{3.34}$$

$$\Delta_r \leq \frac{\Delta_1}{4} \tag{3.35}$$

In Equation (3.34) l_t is the span of the member, taken as the shorter span for two-way slab systems, and h is the overall thickness of the member.

Retesting 72 hours after removal of the loads is allowed if the deflection criteria are not satisfied in the first test. The acceptance criterion for deflection in the retest becomes a function of Δ_2, the maximum deflection during the retest after 24 hours of load application:

$$\Delta_r \leq \frac{\Delta_2}{5} \tag{3.36}$$

If the deflection criteria are not satisfied, the structure shall be permitted for use at a lower load rating.

3.5.2.3   Existing buildings: ACI 437.2M-13

ACI 437.2M-13 (ACI Committee 437, 2013) gives the code requirements for load testing of existing concrete structures. This code prescribes procedures and acceptance criteria for load testing of existing concrete structures, and is valid for application where ACI

50   Load Testing of Bridges

562-13 governs. It cannot be applied to bridges. The code is valid for prestressed and reinforced concrete, provided that the concrete compressive strength is not larger than 55 MPa (8.0 ksi).

The test load magnitude TLM, when only part of the portions of a structure are suspected of containing deficiencies or that have been repaired or rehabilitated and whose adequacy is to be verified, and the members are statically indeterminate, is the larger of:

$$TLM = 1.3(D_W + D_S) \tag{3.37}$$

$$TLM = 1.0D_W + 1.1D_S + 1.6L + 0.5(L_r \text{ or } S \text{ or } R) \tag{3.38}$$

$$TLM = 1.0D_W + 1.1D_S + 1.6(L_r \text{ or } S \text{ or } R) + 1.0L \tag{3.39}$$

In Equations (3.37), (3.38), and (3.39), D_w is the self-weight based on a density of 24 kN/m^3 (150 lb/ft^3), D_s is the superimposed dead load, L is the live load due to use and occupancy of the building, S is the snow load, R is the rain load, and L_r is the live load on the roof produced during maintenance by workers, equipment, and materials, or during the life of the structure by moveable objects such as planers and people.

When all suspect portions of a structure are to be load tested, or when the elements to be tested are determinate, and the suspect flaw or weakness is controlled by flexural tension, the following equations are used to determine the test load magnitude TLM:

$$TLM = 1.2(D_W + D_S) \tag{3.40}$$

$$TLM = 1.0D_W + 1.1D_S + 1.4L + 0.4(L_r \text{ or } S \text{ or } R) \tag{3.41}$$

$$TLM = 1.0D_W + 1.1D_S + 1.4(L_r \text{ or } S \text{ or } R) + 0.9L \tag{3.42}$$

For the evaluation of serviceability, the following test load level has to be included in the loading cycles:

$$TLM_{SLS} = 1.0D + 1.0L + 1.0(L_r \text{ or } S \text{ or } R) \tag{3.43}$$

If the ratio of service live loads to service dead loads exceeds 2.0, and if the suspect deficiency is tension-controlled, the load factor applied to the live load L in Equation (3.38) can be reduced to 1.4 and in Equation (3.41) to 1.3. Similarly, the load factor applied to roof live loads, snow load, or rain loads (L_r or S or R) in Equation (3.39) can be reduced to 1.4 and in Equation (3.42) to 1.3.

As in ACI 437.1R-07 (ACI Committee 437, 2007), two loading protocols are described: a monotonic loading protocol and a cyclic loading protocol. The monotonic loading protocol uses at least four approximately equal increments, see Figure 3.5. The applied sustained load should be ±5% of the full applied test load ATL, which is applied for 24 hours. The deflections are measured after applying each load level until stabilization of the deflections, that is, when the difference between successive deflection readings maximum 2 minutes apart does not exceed 10% of the initial deflection. Each load step is held constant at least for 2 minutes. Measurements are taken at the beginning and end of the 24-hour loading time, and the last set of measurements is taken 24 hours after removing the load.

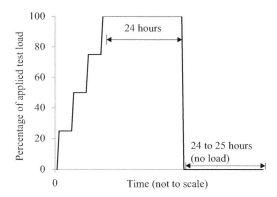

Figure 3.5 Loading protocol for monotonic load test procedure from ACI 437.2M-13 (ACI Committee 437, 2013)
Source: Reprinted with permission from ACI.

The cyclic loading protocol is identical to the protocol described in ACI 437.1R-07 and is shown in Figure 3.2. At least six cycles are used, with cycles A and B at the serviceability load level or 50% of *ATL*, cycles C and D halfway between cycles A and B and 100% *ATL*, and cycles E and F at 100% *ATL*. A ±5% tolerance for the applied load for each load cycle is acceptable. A minimum load level of 10%–15% of *ATL* should be used to keep the test devices engaged. The load is typically applied through hydraulic devices. This method allows to understand end fixity and load transfer characteristics of the tested member, especially for statically indeterminate structures. The same criterion for the stabilization of the deflections governs as for the monotonic loading protocol.

For a load test, the measurements must be taken where maximum responses are expected, and must have a resolution no larger than 1/100 of the expected deflection. A sampling rate of 1/s during the test and 1/min for the 24-hour loading time is recommended. The structure must be visually inspected after each load level.

The acceptance criteria that are used depend on the chosen loading protocol. For both protocols, the first acceptance criterion is that the structure shows no evidence of failure. When deflections exceed the calculated deflections, when cracking is observed, or when distress is observed that could result in a brittle failure type (anchorage failure, bond failure, shear failure, etc.), the test should be terminated.

For monotonic loading, the quantitative acceptance criterion is based on deflections, and uses the maximum deflection Δ_l, the span length l_t, and the residual deflection Δ_r:

$$\Delta_r \leq \frac{\Delta_l}{4} \tag{3.44}$$

$$\Delta_l \leq \frac{l_t}{180} \tag{3.45}$$

The deflections are considered as non-existent if the maximum deflection during the test Δ_l is less than 1.3 mm (0.05 in.) or $l_t/2000$. If the criteria from Equations (3.44) and (3.45)

52 Load Testing of Bridges

are not met, retesting can be permitted a minimum 72 hours after removal of the test load for the first test. For a retest, the deflection Δ_{rrt} at least 24 hours after removal of the load has to fulfill:

$$\Delta_{rrt} \leq \frac{\Delta_{l2}}{10} \tag{3.46}$$

with Δ_{l2} the maximum deflection measured during the second test relative to the position of the structure at the beginning of the second test.

For a cyclic loading protocol, acceptance criteria are defined that require the use of the loading protocol as shown in Figure 3.2. These criteria follow the same basic idea of the acceptance criteria from ACI 437.1R-07 (ACI Committee 437, 2007), but are defined slightly different. The first acceptance criterion is the deviation from linearity index, I_{DL}, which is based on the slope of the reference secant line for the load deflection envelope, $\tan(\alpha_{ref})$ and the secant stiffness of any point i on the increasing loading portion of the load-deflection envelope, $\tan(\alpha_i)$ (see Fig. 3.4). The acceptance criterion for the deviation from linearity index I_{DL} is:

$$I_{DL} = 1 - \frac{\tan(\alpha_i)}{\tan(\alpha_{ref})} \leq 0.25 \tag{3.47}$$

If the load test induces cracking, the test may be restarted. The second stop criterion is based on the permanency ratio I_{pr}. I_{pr} is determined for each pair cycles of a cyclic loading test. The acceptance criterion for the permanency ratio I_{pr} equals:

$$I_{pr} = \frac{I_{p(i+1)}}{I_{pi}} \leq 0.5 \tag{3.48}$$

where I_{pi} and $I_{p(i+1)}$ are the permanency indexes calculated for the ith and $(i+1)$th load cycles, see Figure 3.3 (replace "A" with "i" and "B" with "$i + 1$" in the figure):

$$I_{pi} = \frac{\Delta_r^i}{\Delta_{max}^i} \tag{3.49}$$

$$I_{p(i+1)} = \frac{\Delta_r^{(i+1)}}{\Delta_{max}^{(i+1)}} \tag{3.50}$$

The third stop criterion that is used for a load test with a cyclic loading protocol deals with the residual deflection Δ_r, measured at least 24 hours after removal of the load. This deflection should fulfill Equation (3.44). If the acceptance criteria are exceeded, the structure shall be considered to have failed the load test. If the maximum deflection from Equation (3.45) is fulfilled, retesting is permitted at least 24 hours after complete removal of the test load. Then, the new residual deflection should fulfill Equation (3.46). If a structure does not fulfill the acceptance criteria, the structure can be used at a lower load rating, except if the load test caused significant damage.

3.6 French guidelines

3.6.1 *General*

In France, all new bridges have to be load tested to show that the delivered work complies with the contract between the owner and the building party (Cochet et al., 2004). The French National Annex to the Eurocode includes the requirements of quality control of a structure, including load tests. Since the aim of the load test is to relate the structural response to the calculations and modeling, the load tests described by the French guidelines are diagnostic load tests.

For bridges of less than 10 m (33 ft), simplified load tests can be carried out. For these tests, it is not required to measure load effects. When it is necessary to monitor all responses, the following effects can be measured:

* Deflections
* Settlements (of the supports)
* Horizontal displacements at the supports, abutments, or pylon heads
* Flexural rotations (at the top or bottom of supports)
* Strains (on steel members)
* Curvatures
* Tension forces (in hangers).

For rigid frame bridges, slab bridges, and girder bridges, general guidelines are provided. For slab bridges, two trucks of 26 tons (57 kips) per lane are required. The trucks have to be placed so that the maximum support moments and maximum span moments for each support and span are reached. For bridges with spans more than 40 m (131 ft), special requirements are in place and the preparation has to be carried out on a case-by-case basis.

When a large number of similar bridges are being delivered, at least 10% of these and at least one has to be load tested. For the other bridges, the results of the tested bridges can be extrapolated. For repaired bridges, a load test can be used to compare the behavior of the bridge to its original state.

3.6.2 *Recommendations for load application*

In France, both dead weight (ballast blocks) and trucks are used for load testing. The maximum load that is applied during the load test should correspond to traffic with a return period between one week and one year. In practice, this load is situated between the frequent traffic loads on the structure and 75% of the live-load model from NEN-EN 1991–2:2003 (CEN, 2003). The load should never be smaller than the uniformly distributed load on the carriageway of 2.5 kN/m^2 (0.05 kip/ft^2). For bridges that carry exceptional very heavy convoys (French load classes D and E), a different type of load test may be more suitable.

The critical section in the span and over the support needs to be determined to define the path for the loading vehicle. To limit the necessary number of test vehicles, the maximum load effect caused by the test vehicles as compared to the frequent traffic can be 10% lower. The dynamic forces are covered by using vehicles, and the effect of braking forces can be verified if vehicles of more than 19 ton (42 kips) are used.

54 Load Testing of Bridges

Unlike other guidelines which typically focus on highway bridges, the French guidelines also include pedestrian bridges. Of all bridge types, the sidewalks, if present, also should be load tested. These types of load tests are typically carried out with ballast blocks.

3.6.3 Evaluation of the load test

For the evaluation of the load test, the responses from the analytical models are compared with the measured responses. To make this comparison, an extensive preparation is required. The calculations and hypotheses need to be mentioned, and the sensor plan and driving scheme for the vehicles should be developed based on the critical cross sections. It also has to be determined how the loads and the measurements will be correlated and how the load position will be measured. If the inspection prior to the test has identified defects, these can be instrumented during the load test.

After the load test, the analytical model and the measured responses are compared. The measured load effects cannot be larger than 1.5 times the calculated load effects. An inspection has to be carried out after the load test, and the results of this inspection should be compared to the inspection carried out prior to the load test.

3.7 Czech Republic and Slovakia

3.7.1 General requirements

In the Czech Republic and Slovakia, load tests of bridges are used to check the quality of structures (Frýba and Pirner, 2001). Since the tests compare theoretical assumptions with the actual behavior of the bridge, the types of tests that are carried out are diagnostic load tests (CSN, 1996). The code prescribes static, dynamic, and long-term tests of bridges.

Static load tests are required for new bridges with spans greater than 18 m (59 ft). They can also be carried out if demanded by government authorities or by the designer of the bridge. The applied load is determined together with the designer, and the load distribution, measurement points, and experimental method are defined prior to the test. An inspection prior to the test is required. The vertical deflections are measured where the effects are expected to be largest, as well as the settlements of supports and the squeeze of bearings. Additionally, it is recommended to measure the temperature of the ambient air and the structure, the strains and stresses, the deformations of other parts of the structure, the development of cracks, and the stability of compressed elements of the bridge.

The load can be applied with locomotives, wagons, rail cranes, and so forth for railway bridges, and trucks, track vehicles, building machines, water cisterns, and so forth for highway bridges. The applied load should be between 50% and 100% of the standard load, including the standard dynamic impact factor. The loading protocol requires a minimum loading time of 30 min on concrete bridges and 15 min on steel bridges. The measurements are recorded at least twice before loading, immediately after loading, and at maximum 10-minute intervals during the test. The loading is usually repeated twice.

3.7.2 Acceptance criteria

The Czech and Slovak code describes three acceptance criteria, which are expressed based on the total S_{tot}, the permanent S_r, and the elastic components S_e of all measured values.

The relation between these values is:

$$S_{tot} = S_r + S_e \tag{3.51}$$

The first acceptance criterion for a load test is the condition of elastic deformation, with S_{cal} the calculated value:

$$\beta < \frac{S_e}{S_{cal}} \leq \alpha \tag{3.52}$$

The values of α and β depend on the bridge type and are given in Table 3.5. The second acceptance criterion looks at the condition of permanent deformation:

$$\frac{S_r}{S_{tot}} \leq \alpha_1 \tag{3.53}$$

with α_1 as given in Table 3.5. The third acceptance criterion is related to maximum crack widths for concrete bridges, as given in Table 3.6.

Equation (3.53) can be modified for new bridges. The loading may be repeated if the effects of the first loading comply with:

$$\alpha_1 < \frac{S_r}{S_{tot}} < \alpha_2 \tag{3.54}$$

with α_1 and α_2 as given in Table 3.5. The condition of the permanent deformation now becomes:

$$\frac{S_r}{S_{tot}} < \alpha_3 \tag{3.55}$$

Table 3.5 Determination of parameters per bridge type (Frýba and Pirner, 2001)

Bridge Type	α	α_1	α_2	α_3	β
Prestressed	1.05	0.2	0.5	0.1	0.7
Reinforced	1.10	0.25	0.5	0.125	0.6
Steel	1.05	0.1	0.3	0.05	0.8

Table 3.6 Limitations to crack widths that can occur in a load test (Frýba and Pirner, 2001)

Bridge type	Environmental class	Maximum crack width
Reinforced concrete	1 (dry)	0.4 mm
	2, 3 (humid)	0.3 mm
	4, 5 (aggressive)	0.1 mm
Partially prestressed	1 (dry)	0.2 mm
	2, 3 (humid)	0.1 mm for post-tensioning
		0 mm for prestressing
	4, 5	0 mm
Fully prestressed	any	0 mm

Conversion: 1 mm = 0.04 in.

56 Load Testing of Bridges

with α_3 as given in Table 3.5. If Equation (3.55) is not fulfilled upon retesting, a third test can be executed, which needs to fulfill:

$$\frac{S_r}{S_{tot}} \leq \frac{\alpha_1}{6} \qquad (3.56)$$

If the measured values do not comply with any of the acceptance criteria, a special investigation and/or long-term observation and/or dynamic testing becomes necessary.

3.7.3 Dynamic load tests

Dynamic load tests in the Czech Republic and Slovakia are carried out by using railway or highway vehicles or a group of vehicles moving at a constant speed along the bridge. The speed is first 5 km/h (3 mph) and is increased for the next runs of the vehicles or the bridge. Other options for loading are an exciter with a varying frequency for modal analysis, a rocket motor which gives a controlled impulse, and a standard track irregularity for testing highway bridges. For pedestrian bridges, a group of pedestrians can be used.

The acceptance criteria for dynamic tests are as follows. The product of the standard dynamic impact factor δ and the k_{dyn} must fulfill:

$$\delta k_{dyn} \leq 1 \qquad (3.57)$$

with

$$k_{dyn} = \frac{U_{dyn}}{U} \qquad (3.58)$$

U_{dyn} is the response to the test load and U is the response to the standard load without δ at the measured point. The acceptance criterion for the natural frequencies is based on the calculated deviation:

$$\Delta_{(j)} = \frac{f_{(j)theor} - f_{(j)obs}}{f_{(j)theor}} \times 100 \qquad (3.59)$$

with $f_{(j)theor}$ the calculated natural frequency and $f_{(j)obs}$ the measured frequency. The limits to the deviation are given in Table 3.10. The last acceptance criterion related to the dynamic impact factor is used for railway bridges:

$$(\delta_{obs} - 1)k_{dyn} \leq \delta - 1 \qquad (3.60)$$

with δ_{obs} the measured dynamic impact factor:

$$\delta_{obs} = \frac{S_{max}}{S_m} \qquad (3.61)$$

S_{max} is the maximum dynamic response due to the load at the measured point and S_m is the maximum static response due to the same load at the same point. For the acceptance criterion from Equation (3.60), a minimum of 10 runs of railway vehicles is required, for which 90% of the tests have to fulfill the acceptance criterion.

Current Codes and Guidelines 57

Table 3.7 Limitations to deviation between measured and calculated frequencies (Frýba and Pirner, 2001)

Frequency	$f_{(1)}$	$f_{(2)}$	$f_{(3)}$	$f_{(4)}$	$f_{(5)}$
$\Delta_{(j)}$ (%)	-15 to $+5$	-15 to $+10$	±15	±20	±25

3.8 Spanish guidelines

3.8.1 *General considerations*

The Spanish guidelines (Ministerio de Fomento-Direccion General de Carreteras, 1999) focus on diagnostic load testing of bridges prior to opening to confirm proper functioning. A representative service load should be applied. The guidelines mention that the same methods can also be used for existing structures. The guidelines cover static and dynamic load testing. For railway bridges (Ministerio de Fomento, 2010), diagnostic load testing prior to opening is also required.

The Spanish guidelines cover highway bridges and pedestrian bridges. Static load tests are compulsory prior to opening a new bridge, whereas dynamic load tests can be used for checking vibrations in structures that require such checks. For all bridges with a span length of greater than or equal to 12 m (39 ft) load testing is obligatory, whereas for shorter bridges the decision lies with the project director. Dynamic testing is required for all concrete bridges with a span length larger than or equal to 60 m (197 ft), as well as for bridges with unusual design, bridges built with new materials, and footbridges. The same criteria are valid for steel and composite bridges. Load tests are required prior to opening as well as after widening or rehabilitation of bridges.

For railway bridges (Ministerio de Fomento, 2010), a diagnostic load test is required for every bridge with a span larger than or equal to 10 m (33 ft). The load test should be static and dynamic. When the outcome of the load test is that the observed behavior does not correspond to the analytically predicted behavior, further analyses are required before the bridge can be opened to the traveling public. Moreover, a load test of the bridge prior to installation of the railway track is recommended (Ministerio de Fomento, 2009) to verify the behavior of the structure. Since the railway track is not yet installed, this load test is carried out with trucks.

At the end of the load test, a report needs to be developed, which contains the following information:

- The date of the load test, start and end time of the load test, and personnel present during the load test;
- Reference to the project number of the structure and of the load test;
- Description of the structure and its condition prior to the load test;
- Detailed description of the vehicles used and the different load configurations;
- Description of the measurements in terms of range of sensors, selected instrumentation, and numbering and positions of sensors;
- Information about the development of the test (starting time of each loading configuration, time between loading and unloading, number of load steps, etc.);
- Measurements obtained during the test;

58 Load Testing of Bridges

- Comparison between analytically determined structural responses and measurements, and verification of acceptance criteria;
- Elements observed during the technical inspection before, during, or after the load test;
- Additional information: photographs, weather conditions, reference points of surveying if determined, incidents during load test, etc.

The report lies at the basis of the official minutes of the load test, which should be signed by the director of the construction project, the director of the load test, and a representative of the contractor.

3.8.2 Loading requirements

The load is applied with vehicles. Two standard vehicles can be used: a 26-ton (52 kip) truck with three axles or a 38-ton (76 kip) truck with four axles. The structure type and geometry, as well as the goals of the load test, then define the number and position of the trucks that are used during the test.

The load should be representative of service load levels with a return period of the load of 5 years. The load should be causing around 60% of the ULS load effect based on the load combination prescribed in the code, and should never cause load effects exceeding 70% of the ULS load effect. A deviation of $\pm 5\%$ from the planned load is accepted for the trucks when they are weighed prior to the load test.

The load should be applied in at least two steps, and complete unloading of the structure after each load step is necessary. The total load (i.e. the target load) should remain on the bridge until stabilization of the measurements occurs.

For pedestrian bridges, distributed loads are recommended, but these loads can be replaced by concentrated loads provided that they cause the same load effect.

For simply supported bridges, the load should be applied span per span, and every span should be subjected to a load test. For continuously supported bridges, the loading configuration should consider the overall bridge behavior. Loading can be carried out by placing trucks in contingent spans or by loading alternate spans.

For curved and skewed bridges, the position of the load in the transverse direction becomes important. For such structures, asymmetric loading cases should be added to study the torsional effects.

Where necessary, loading cases to study the transverse distribution should be included in the load test.

Simplified load tests can be carried out for certain cases. For bridges with a number of similar simply supported spans, one in every four spans should be tested with a full load test, and the other spans can be subjected to a simplified load test. For bridges with a number of similar continuous spans, the two end spans should be tested with a full load test, as well as one in every four of the center spans. The remaining spans can be subjected to a simplified load test. Spans are considered similar if no more than 10% of difference in the span length occurs. The full load test should study the maximum flexural response at both supports of the span as well as at midspan. When a number of similar bridges are built during a project, $\leq 50\%$ of these can be subjected to a simplified load tests, whereas the rest should be tested with a full load test in at least two spans. A simplified load test has the following characteristics:

Current Codes and Guidelines 59

- The load is applied in one load step;
- A minimal amount of measurements are taken;
- The number of loading configurations is reduced and can be simplified to a quasi-static test with a truck speed of 5 km/h (3.1 mph).

3.8.3 Stop and acceptance criteria for static load tests

The Spanish guidelines describe recommended stop and acceptance criteria. Moreover, the guidelines mention that different acceptance criteria are allowed if they are defined during the preparation stage of the test.

The required measurements are the deflection at midspan and the support settlements. For box girder bridges, it is recommended to measure the deflections in the transverse direction for the evaluation of torsional effects. For slab-on-girder bridges, it is required to measure the deflections at three positions in the transverse direction: on the tested girder and then on each side of this girder. For steel and composite bridges, strain measurements should be added. Real-time monitoring of the measurements is recommended.

Besides the measurements of the structural responses, two reference sensors outside of the loaded span(s) should be used to measure the influence of temperature and humidity. Surveying of the bridge prior to opening should be done after the load test.

As mentioned in section 3.8.2, the load should be applied until the measurements are stabilized. Stabilization of the measurements should be evaluated when the total load is on the structure and after unloading. Stabilization of the measurements is first evaluated after 10 minutes with the following expression:

$$|f_{10} - f_0| < 0.05|f_0| \tag{3.62}$$

with f_{10} the measurement after 10 minutes and f_0 the instantaneous measured structural response. If the criterion from Equation (3.62) is fulfilled, then the measurements are considered as stabilized and the loading case can be considered as finished, the load can be removed, and the next loading case can be executed. If the criterion from Equation (3.62) is not fulfilled, then the stabilization of the measurements is evaluated after 20 minutes with the following expression:

$$|f_{20} - f_{10}| < 0.2|(f_{20} - f_{10})| \tag{3.63}$$

with f_{20} the measurement after 20 minutes. If the criterion from Equation (3.63) is fulfilled, the measurements are considered as stabilized and the loading case can be considered as finished, the load can be removed, and the next loading case can be executed. If the criterion from Equation (3.63) is not fulfilled, the director of the load test decides if the load should be kept for a longer period of time until stabilization occurs or if the load should be removed and the test terminated. In the latter case, further analyses are required to study why no stabilization of the structural response was obtained, and retesting may be necessary to open the bridge to the traveling public. An example of the evaluation of the stabilization of measurements after application of the load is shown in Figure 3.6 and after unloading in Figure 3.7. A loading cycle is illustrated in Figure 3.8.

For static load tests, the stop criteria are defined based on the observed measurements during a load cycle, see Figure 3.8. The remaining measurement, or permanent

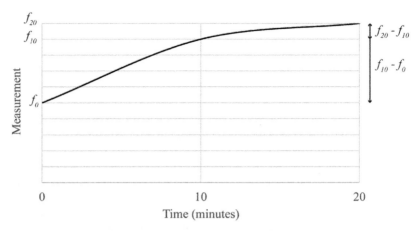

Figure 3.6 Illustration of procedure for evaluating the stabilization of measurements after application of target load (Ministerio de Fomento-Direccion General de Carreteras, 1999)

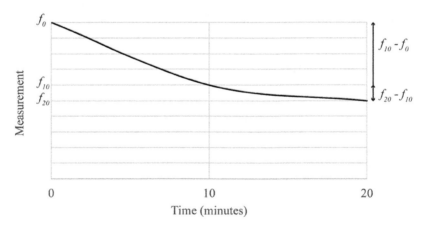

Figure 3.7 Illustration of procedure for evaluating the stabilization of measurements after unloading (Ministerio de Fomento-Direccion General de Carreteras, 1999)

measurement, f_r (abbreviated from $f_{remaining}$ in Fig. 3.8) is used to calculate the remanence α:

$$\alpha = 100 \frac{f_r}{f} \qquad (3.64)$$

with f the total measurement. The following limits are defined:

- $\alpha_{lim} = 20\%$ for reinforced concrete bridges
- $\alpha_{lim} = 15\%$ for prestressed concrete bridges or composite bridges
- $\alpha_{lim} = 10\%$ for steel bridges.

Current Codes and Guidelines 61

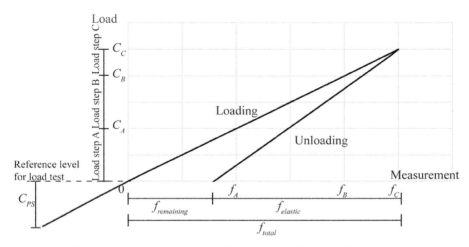

Figure 3.8 Cycle of loading and unloading (Ministerio de Fomento-Direccion General de Carreteras, 1999)

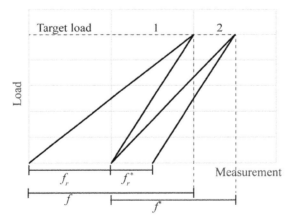

Figure 3.9 Determination of residual measurements (Ministerio de Fomento-Direccion General de Carreteras, 1999)

In Figure 3.9, two load cycles are shown, with the measurement in the first cycle f, the remaining measurement after the first cycle f_r, the measurement in the second cycle f^*, and the remaining measurement after the second cycle f_r^*. The stop criterion for the remanence is defined as follows:

- If $\alpha \leq \alpha_{lim}$, the stop criterion is fulfilled;
- If $\alpha_{lim} < \alpha \leq 2\alpha_{lim}$, a second load cycle has to be applied (see Fig. 3.9);
- If $\alpha > 2\alpha_{lim}$, the load test must be aborted because the stop criterion is exceeded.

62 Load Testing of Bridges

When a second load cycle is required, the stop criterion becomes:

- If $\alpha^* \leq \alpha/3$, the stop criterion is fulfilled;
- If $\alpha^* > \alpha/3$, the stop criterion is exceeded and further load testing is not permitted.

For this stop criterion, α is the remanence from the first load cycle and α^* is the remanence from the second load cycle.

Four acceptance criteria for static load tests are given. The following acceptance criteria are defined in the Spanish guidelines:

- For prestressed and steel bridges, the maximum measured deflection should not be more than 10% larger than the analytically determined deflection. For composite and reinforced concrete bridges, the maximum measured deflection should not be more than 15% larger than the analytically determined deflection. If the measured deflection is less than 60% of the analytically determined deflection, this discrepancy should be checked and explained.
- For simplified load test, the obtained results should not differ by more than 10% of the results obtained in a full load test. This result should be corrected for the difference in span length, if any. If this acceptance criterion is not fulfilled, a complete load test should be carried out.
- The maximum crack width should not exceed the crack width limits from the SLS of cracking prescribed by the governing code.
- No signs of distress and exhaustion of the bearing capacity of the structure should be observed.

3.8.4 Acceptance criteria for dynamic load tests

Dynamic load tests can be carried out to determine the following properties:

- Influence lines
- Accelerograms
- Frequency spectra
- Modes of vibration
- Impact factor (dynamic amplification)
- Damping.

The results of a dynamic load test can be analyzed in the time domain or in the frequency domain. The Spanish guidelines define three speeds that can be used in a dynamic load test, depending on the goals of the test:

- Slow speed: $\leq$ 5 km/h (3 mph) (quasi-static test);
- Medium speed: between 30 km/h (19 mph) and 40 km/h (25 mph);
- Fast speed: > 60 km/h (37 mph), provided that the site conditions allow driving at this speed.

An overview of the different goals of dynamic load tests and recommendations for their execution is given in Table 3.8.

Current Codes and Guidelines 63

Table 3.8 Recommendations for the execution of dynamic load tests (Ministerio de Fomento-Direccion General de Carreteras, 1999)

Goal of test	Loading configuration	Speed of load	Application of obstacle
influence lines	I truck	slow	no
frequency	I truck	medium and fast	optional
damping	I truck	medium and fast	optional
impact factor	I truck	slow, medium, and fast	no
acceleration	I or more trucks	medium and fast	no

For special cases, the loading with trucks can be replaced by the application of a forced vibration or braking action or by dropping a weight. For pedestrian bridges, the applied load during a dynamic test should be a group of pedestrians. These pedestrians should cross the bridge by walking and by running. One person of average weight should be used for each meter (3.3 ft) width of the bridge.

The same instrumentation should be applied during a dynamic load test as during a static load test. Additionally, accelerometers should be used. The sampling rate of the sensors should be fast enough to capture the structural responses.

For dynamic load tests, no acceptance criteria are prescribed by the guidelines, given the diversity of types of structures and factors that may affect the structural response. Nevertheless, a few recommended acceptance criteria are summarized in the Spanish guidelines. These recommended acceptance criteria are subdivided between dynamic characteristics of the structure itself and dynamic characteristics that depend on the applied load.

For the dynamic characteristics of the structure, frequencies and damping are analyzed. For frequencies, the recommended acceptance criterion is that the obtained principal frequency in the experiment should not differ significantly from the calculated value. For damping, the following definition is used:

$$\delta = \frac{1}{n} \ln \left(\frac{A_0}{A_n} \right) \tag{3.65}$$

with n the number of cycles of the interval with respect to the fundamental frequency, A_0 the dynamic response at the beginning of the interval, and A_n the dynamic response at the end of the interval. It is good practice to use at least 5 to 10 cycles. The values of A_0 and A_n should be obtained from the zone of free vibration of the structure, preferably based on deflections, and based on strains or accelerations if no deflection measurements can be obtained. A typical value of the damping, obtained according to Equation (3.65), lies between 0.03 and 0.12. Damping can also be expressed as a percentage:

$$\zeta = \frac{\delta}{2\pi} \tag{3.66}$$

When damping is expressed based on Equation (3.66), the values typically lie between 0.5% and 2%.

For the dynamic characteristics that depend on the applied load (or source of excitation), recommended acceptance criteria for accelerations and impact (dynamic amplification) are

64 Load Testing of Bridges

given in the Spanish guidelines. The limits to accelerations for pedestrian bridges and pedestrian walkways are classified as follows:

- The accelerations lie between imperceptible and easily perceptible when $a \leq 0.025$ g;
- The accelerations are clearly perceptible to disturbing when 0.025 g $< a \leq 0.075$ g;
- The accelerations are disturbing to very disturbing when 0.075 g $< a \leq 0.125$ g.

with a the maximum peak acceleration in gravity direction and g the gravitational constant.

The second recommended acceptance criterion for the dynamic characteristics that depend on the applied load (or source of excitation) is the impact factor Φ, which is defined as:

$$\Phi = \frac{f_{dyn}}{f_{sta}} \tag{3.67}$$

with f_{dyn} the maximum dynamic response and f_{sta} the maximum static or quasi-static response for both responses caused by the same action. The structural response can be based on deflections or based on strains. The impact factor is categorized as follows:

- Low: $\Phi \leq 1.10$
- Medium: $1.10 < \Phi \leq 1.30$
- High: $\Phi > 1.30$

3.9 Other countries

3.9.1 Italy

In Italy (Veneziano et al., 1978, 1984a, 1984b), all road bridges are proof loaded prior to their opening to traffic to verify that the as-built stiffness and resistance of the deck is conform to the design specifications and applicable standards. Usually, heavy trucks are placed on the deck in a longitudinal arrangement to maximize the bending moment at midspan. To check the torsional stiffness of the deck and the transverse redistribution of the load, the trucks are often placed with maximum eccentricity in the transverse direction. During the load test, the applied stresses should be close to the design values. The measured response is the vertical displacement at various points of the deck, measured during and after loading. The measurement points are usually located across the midspan sections and near the supports. The difference between the support settlement and the deflection at midspan is the net deformation, which can be compared to theoretical predictions.

3.9.2 Switzerland

In Switzerland, provisions (SIA 269:2011 [SIA, 2011]) are available for load testing of existing structures (Brühwiler et al., 2012). In addition to this, Switzerland has a long experience in testing new bridges prior to opening (Moses et al., 1994), since a load test is required for every major bridge. Sometimes, these bridges are then also tested after several years in service to check their behavior. The interesting element of the Swiss

Current Codes and Guidelines   65

practice is that it combines elements from diagnostic load testing and proof load testing. The applied load during a load test in Switzerland is 80%–85% of the unfactored live load (SLS load level). Typically, four to eight dump trucks of 250 kN (56 kips) are used. This load is still significantly above the expected lifetime maximum traffic loading. During the load test, the displacements are compared to the analytical predictions for the displacements. Additionally, a dynamic test is carried out, in which moving vehicles and artificially created bumps are used to determine the impacts and frequencies. The deck is checked for cracks, and crack width measurements are carried out. In Swiss practice, the following acceptance criteria are defined:

1   There should be an agreement between the analytical and measured displacements. The measurements should be taken at several points along each girder.
2   The behavior should be linear and the residual displacements should be zero.
3   The measured crack widths should be within acceptable limits.

3.9.3   *Poland*

In Poland, the RIRB requirements (Research Institute of Roads and Bridges, 2008) deal with load testing of bridges. This code requires the measurement of deflections (Halicka et al., 2018). The support deflections should be measured as well to find the net deflection of the bridge. In practice, during load tests the settlements of the support and the compression of the bearings are also typically measured (Filar et al., 2017).

The stop criterion that is prescribed in Poland is that no nonlinear behavior is permitted to occur during the test. The residual deformations need to be verified as well. These are limited to 20% of the maximum deformation for reinforced concrete bridges and 10% for prestressed concrete bridges.

The target load is determined based on a vehicle of abnormal weight Q_{abn}, which is larger than the design live load Q_{design}. This load is applied by using six load levels:

1   The load in the first step is maximum equal to the design load: $L_1 \leq Q_{design}$
2   The second step is an intermediate step with a load of $L_2 \leq Q_{design} + \frac{1}{4}(Q_{abn} - Q_{design})$
3   The third step is an intermediate step with a load of $L_3 \leq Q_{design} + \frac{1}{2}(Q_{abn} - Q_{design})$
4   The fourth step is an intermediate step with a load of $L_4 \leq Q_{design} + \frac{3}{4}(Q_{abn} - Q_{design})$
5   The fifth step reaches the abnormal vehicle weight $L_5 \leq Q_{abn}$
6   The last step goes beyond the abnormal vehicle weight $L_6 \leq Q_{design} + 1\frac{1}{4}(Q_{abn} - Q_{design})$.

3.9.4   *Hungary*

In Hungary, load testing is used for verifying the serviceability of existing buildings (Hungarian Chamber of Engineers, 2013). Two possible requirements can be checked for: if the structure is adequate or if it is acceptable. A structure is considered adequate when all standard requirements are satisfied, and a structure is considered acceptable if:

*   Only minor defects can be observed;
*   No brittle failure is expected;

66 Load Testing of Bridges

- The structure has enough capacity to withstand at least the characteristic level of loads, regardless the fulfillment of deflection (stiffness) and crack width requirements;
- The deterioration of the structure is not faster than in usual cases.

An acceptable structure has a limited time for future operation, a limitation of its function, or should be periodically assessed by an independent expert. If a structure does not meet the requirements for being adequate or acceptable, it is considered dangerous and action should be taken.

The target proof load to verify the adequate condition is calculated as:

$$P_{max,d} = (1 + \beta(a + b\gamma))P_d \tag{3.68}$$

and for the acceptable condition it is calculated as:

$$P_{max,k} = (1 + \beta(a + b\gamma))P_k \tag{3.69}$$

with $\beta = 1$ if a ductile failure mode is expected and $\beta = 1.5$ if a brittle failure mode is expected. The value of a is determined as:

$$a = 0.08\left(1 - \frac{n}{2N^{0.5}}\right) \geq 0 \tag{3.70}$$

with n the tested number of elements and N the total number of structural elements in the complete structure identical to the tested component. The value of b can be determined as:

$$b = 5a + 0.16 \tag{3.71}$$

Equations (3.70) and (3.71) are replaced with $a = b = 0$ if $n = N$. If the loaded components are the weakest parts of the complete structure, $a = 0$ and $b = 0.15$. The value of γ is determined as:

$$\gamma = \frac{G}{G + P} \tag{3.72}$$

with G the permanent loads and P the proof load (P_d or P_k). The value P_k is the characteristic value of the proof load, which corresponds to the characteristic load intensity. The value P_d is the design value of the proof load, which corresponds to the design load intensity.

The stop criteria in the Hungarian guidelines are the following: fracture, rupture, yielding, damage of concrete under compression, buckling, deflections larger than 1/50 between points of inflection, cracks in concrete larger than 1 mm (0.04 in.), cracks in steel, excessive deformations of the cross section, extensive shell-buckling, and masonry cracks larger than 1 mm (0.04 in.). For these cases, the structure will be assumed as failed.

In addition to the stop criteria, acceptance criteria are also formulated. The first acceptance criterion is a limit of the residual deformation to the maximum deformation, depending on the structure type (see Table 3.9). The second acceptance criterion is that the deflection under the characteristic proof load should not exceed the SLS criteria from the Eurocode. For concrete structures, a third acceptance criterion is that the crack width under the characteristic proof load should not exceed the limitations from Eurocode 2 EN 1992–1-1:2005 (CEN, 2005).

Table 3.9 Limitations to deviation between measured and calculated deformations (Hungarian Chamber of Engineers, 2013). The values between brackets are valid for $\gamma < 0.5$

Type of structure	Ratio of residual and total deformation (in %)	
	$P_{max} = P_{max,k}$	$P_{max} = P_{max,d}$
Riveted steel structure	15	20
Welded steel structure	12	15
Steel with bolted connections	20 (25)	25 (30)
Prestressed concrete	20	25
Reinforced concrete	25 (30)	30 (35)
Steel-concrete composite	20	25
Timber structure	30	40

3.10   Current developments

In several countries, currently efforts are geared towards the improvement of the current guidelines for load testing. These efforts include the following:

- Unifying existing guidelines so that these cover all bridge types (reinforced concrete bridges, prestressed concrete bridges, steel bridges, timber bridges, masonry bridges, etc.);
- Unifying existing guidelines so that these cover both diagnostic and proof load testing;
- Including guidelines for existing bridges;
- Including guidelines for bridges that can fail in a brittle manner, such as shear-critical concrete bridges and fracture-critical steel bridges.

To achieve these goals, the following institutions and groups are currently revising codes and guidelines with regard to load testing:

- TRB Committee AFF40 is preparing an e-circular on load testing.
- The German guidelines for load testing are currently being revised, and the aim for the revision is to include testing for shear (Schacht et al., 2016).
- For the 2020 *fib* Model Code, which will include provisions for existing structures, load testing will perhaps be included. These provisions can then possibly lie at the basis for requirements that will be published as a Eurocode.

3.11   Discussion

Load testing practices, as well as the available codes and guidelines, differ across countries and are typically determined by the local demands and the country's construction industry. In the past, most countries focused on developing guidelines for load testing of new bridges to ensure the traveling public that the structure is safe. Nowadays, load testing of the existing infrastructure becomes increasingly interesting as a method for assessment. Some countries, such as the United States, have aimed at covering load testing in a holistic manner by including diagnostic and proof load testing for all bridge types. Other countries have geared their efforts towards a particular problem encountered upon bridge assessment. In Germany, the basis for proof load testing was laid by prescribing procedures for testing concrete buildings. Most other countries have focused on diagnostic load testing of bridges.

3.12 Summary

This chapter gives an overview of the currently available codes and guidelines for load testing of bridges. Where the guidelines for buildings have played an important role for the practice of bridge load testing, the requirements from these guidelines have also been mentioned. The guidelines from Germany, the United Kingdom, Ireland, the United States (bridges and buildings), France, Italy, the Czech Republic, Slovakia, Spain, Switzerland, Poland, and Hungary have been summarized in this chapter.

From the presented discussions, it can be concluded that none of the available guidelines covers all bridge types, new and existing bridges, diagnostic load testing and proof load testing, a clear description of the target load, and stop criteria. For these reasons, several institutions are currently working on revisions of their guidelines and/or developing new guidelines.

References

AASHTO (2011) *The Manual for Bridge Evaluation*. American Association of State Highway and Transportation Officials, Washington, DC.

AASHTO (2015) *AASHTO LRFD Bridge Design Specifications, 7th Edition with 2015 Interim Specifications*. American Association of State Highway and Transportation Officials, Washington, DC.

ACI Committee 318 (2014) *Building Code Requirements for Structural Concrete (ACI 318–14) and Commentary*. American Concrete Institute, Farmington Hills, MI.

ACI Committee 437 (2007) Load Tests of Concrete Structures: Methods, Magnitude, Protocols, and Acceptance Criteria (ACI 437.1R-07). Farmington Hills, MI.

ACI Committee 437 (2013) Code Requirements for Load Testing of Existing Concrete Structures (ACI 437.2M-13) and Commentary. Farmington Hills, MI.

ACI Committee 562 (2016) Code Requirements for Assessment, Repair, and Rehabilitation of Existing Concrete Structures and Commentary ACI 562–16. American Concrete Institute, Farmington Hills, MI.

Brühwiler, E., Vogel, T., Lang, T. & Luechinger, P. (2012) Swiss standards for existing structures. *Structural Engineering International*, 22, 275–280.

CEN (2003) *Eurocode 1: Actions on Structures: Part 2: Traffic Loads on Bridges, NEN-EN 1991–2:2003*. Comité Européen de Normalisation, Brussels, Belgium.

CEN (2005) *Eurocode 2: Design of Concrete Structures – Part 1–1: General Rules and Rules for Buildings, NEN-EN 1992–1-1:2005*. Comité Européen de Normalisation, Brussels, Belgium.

Cochet, D., Corfdir, P., Delfosse, G., Jaffre, Y., Kretz, T., Lacoste, G., Lefaucheur, D., Khac, V. L. & Prat, M. (2004) *Load Tests on Highway Bridges and Pedestrian Bridges*. Sétra – Service d'Etudes techniques des routes et autoroutes, Bagneux-Cedex, France.

CSN (1996) *CSN 73 6209: Load Testing of Bridges (in Czech)*. Czech Republic, Prague.

Deutscher Ausschuss für Stahlbeton (2000) DAfStb-Guideline: Load tests on concrete structures (in German). Deutscher Ausschuss fur Stahlbeton.

Filar, Ł., Kałuża, J. & Wazowski, M. (2017) Bridge load tests in Poland today and tomorrow: The standard and the new ways in measuring and research to ensure transport safety. *Procedia Engineering*, 192, 183–188.

Frýba, L. & Pirner, M. (2001) Load tests and modal analysis of bridges. *Engineering Structures*, 23, 102–109.

Halicka, A., Hordijk, D. A. & Lantsoght, E. O. L. (2018) Rating of concrete road bridges with proof loads. *ACI SP 323 Evaluation of Concrete Bridge Behavior through Load Testing – International Perspectives*, 16.

Hungarian Chamber of Engineers (2013) Guidelines for interventions in Hungary (in Hungarian). Budapest, Hungary.

The Institution of Civil Engineers: National Steering Committee for the Load Testing of Bridges (1998) *Guidelines for the Supplementary Load Testing of Bridges*. The Institution of Civil Engineers, London, UK.

Lantsoght, E. O. L., Van Der Veen, C., Hordijk, D. A. & De Boer, A. (2017) State-of-the-art on load testing of concrete bridges. *Engineering Structures*, 150, 231–241.

Low, A. M. & Ricketts, N. J. (1993) *The Assessment of Filler Beam Bridge Decks Without Transverse Reinforcement*. Transport Research Laboratory, Harmondsworth, UK.

Ministerio de Fomento (2009) *Instrucciones para la Puesta en carga de estructuras (pruebas de carga provisionales)*. Ministerio de Fomento, Madrid, Spain.

Ministerio de Fomento (2010) *Instrucción de acciones a considerar en puentes de ferrocarril (IAPF)*. Ministerio de Fomento, Madrid, Spain

Ministerio de Fomento-Direccion General de Carreteras (1999) *Recomendaciones para la realización de pruebas de carga de recepción en puentes de carretera*. Ministerio de Fomento, Madrid, Spain.

Moses, F., Lebet, J. P. & Bez, R. (1994) Applications of field testing to bridge evaluation. *Journal of Structural Engineering*, 120.

NCHRP (1998) Manual for Bridge Rating through Load Testing. Washington, DC.

NRA (2014) *Load Testing for Bridge Assessment*. National Roads Authority, Dublin, Ireland.

Research Institute of Roads and Bridges (2008) *The Rules for Road Bridges Proof Loadings (in Polish)*. RIRB, Warsaw, Poland.

Ricketts, N. J. & Low, A. M. (1993) *Load Tests on a Reinforced Beam and Slab Bridge at Dornie*. Transport Research Laboratory, Harmondsworth, UK.

Schacht, G., Bolle, G., Curbach, M. & Marx, S. (2016) Experimental evaluation of the shear bearing safety (in German). *Beton- und Stahlbetonbau*, 111, 343–354.

SIA (2011) *Existing structures – Bases for examination and interventions SIA 505 269:2011, (in German)*, SIA Schweizerischer Ingenieur- und Architektenverein. Zurich, Switzerland, p. 269.

Veneziano, D., Meli, R. & Rodriguez, M. (1978) Proof loading for target reliability. *Journal of the Structural Division, ASCE*, 104, 79.

Veneziano, D., Galeota, D. & Giammatteo, M. M. (1984a) Analysis of bridge proof-load data I: Model and statistical procedures. *Structural Safety*, 2, 91–104.

Veneziano, D., Galeota, D. & Giammatteo, M. M. (1984b) Analysis of bridge proof-load data II: Numerical results. *Structural Safety*, 2, 177–198.

Part II

Preparation, Execution, and Post-Processing of Load Tests on Bridges

Chapter 4

General Considerations

Eva O. L. Lantsoght and Jacob W. Schmidt

Abstract

This chapter discusses aspects that should be considered prior to every load test. The first questions that needs to be answered are "Is this bridge suitable for load testing?" and "If so, what are the goals of the load test?" To answer these questions, information must be gathered and preliminary calculations should be carried out. In order to evaluate if a bridge is suitable for load testing, different types of testing are shown, the topic of when to load test a bridge is discussed, and structure type considerations are debated. Finally, some safety precautions that should be fulfilled during a load test are discussed.

4.1 Initial considerations

4.1.1 Introductory remarks

Load testing of bridges can be a desirable method to show:

- That a newly built bridge behaves in the way it was designed;
- To improve the analytical assessment of existing bridges;
- To directly show that a bridge can carry the code-prescribed loads without causing permanent undesirable damage.

This part of the book deals with the topic of preparation, execution, and post-processing of load testing. The particularities of diagnostic load tests are discussed further in Part III. Part I of Vol. 13 focuses on proof load testing.

Prior to preparing load testing, the decision needs to be taken whether or not such testing can answer the open questions regarding the overall performance or specific response of the structure. In addition, sufficient preparation related to these questions is essential for the execution and post-processing of information leading to decision linking to the stop criteria and target load (depending on the type of load test). Several investigations are needed before the final decisions about field testing can be done. This information provides the decision basis and are essential elements that need to be addressed prior to deciding the execution methodology.

4.1.2 Load test types and their goals

It must be decided which type of load test is required to meet the aims of the test. To evaluate the design assumptions prior to opening of a new bridge and to update the analytical

DOI: https://doi.org/10.1201/9780429265426

74 Load Testing of Bridges

model used for the assessment of an existing test with field measurements of unknown properties, diagnostic load testing can be used. To demonstrate directly that a bridge can carry the code-prescribed (factored) loads, a load that is representative for the code-prescribed (factored) loads can be applied to the bridge in a proof load test. If the bridge can carry this target proof load without signs of distress, than it has been shown directly that the code requirements are fulfilled. Only collapse tests (also called failure tests), which are a type of destructive testing, can answer questions with regard to the ultimate capacity of a given structure.

4.1.3   Type of bridge structure or element

In-situ full-scale load testing is a method which can be used for several types of bridge structures. However, the vast number of bridge type combinations containing new bridges, existing bridges, bridge sections, damaged/deteriorated bridges, and so forth (as well as the large number of reasons why one could decide to load test a bridge) often makes projects unique where full-scale testing is used. However, the loading procedures which reflect traffic loading, described in guidelines or codes, are unaffected by the structural type. The way of applying a load configuration with a related target load, including the evaluation of a stop criterion, should be uniquely addressed for each full-scale bridge type or bridge element, however. It is normally seen in such applications that target loading is used for proof loading whereas stop criterions are used to find the highest acceptable loading rate. Diagnostic testing can address other uncertainties on the structural behavior.

Some of the considerations concerning steel bridges, reinforced concrete bridges, prestressed concrete bridges, masonry (arch) bridges, and timber bridges are given in Section 4.4. The limitations and points of attention for diagnostic testing and proof load testing of the individual structure types are discussed separately for the given bridge types. Further considerations per bridge type are given in Part III for diagnostic load tests and in Part I of Vol. 13 for proof load tests.

4.1.4   Structural inspections, background codes, and literature

Prior to deciding to proceed with a load test, a few elements need to be checked, which are also discussed in the British guidelines (The Institution of Civil Engineers: National Steering Committee for the Load Testing of Bridges, 1998). This guide provides a flowchart to assess if a bridge is a good candidate for a load test. During the preparation stage, the engineer should decide if load testing is the most suitable option to answer the open questions regarding the structural performance. In some cases, advanced modeling tools such as probabilistic analyses or nonlinear finite element models can be used to analyze the critical elements in more detail. When ambient traffic loading can be sufficient to create the required structural responses, short- or long-term monitoring can be applied to the structure. For other cases, load testing is the best method to address the concerns related to the structure. The decision-making process prior to deciding on a load test is reflected in Figure 4.1.

In addition to the flowchart in Figure 4.1, it should be noted that a technical inspection of the object and an inspection of its surroundings are essential as background knowledge. If there are site-specific restrictions that inhibit the application of the required sensors or access to the structure, it must be evaluated if the load test can still be used to answer the open questions regarding the performance of the structure. When that is not the case, the

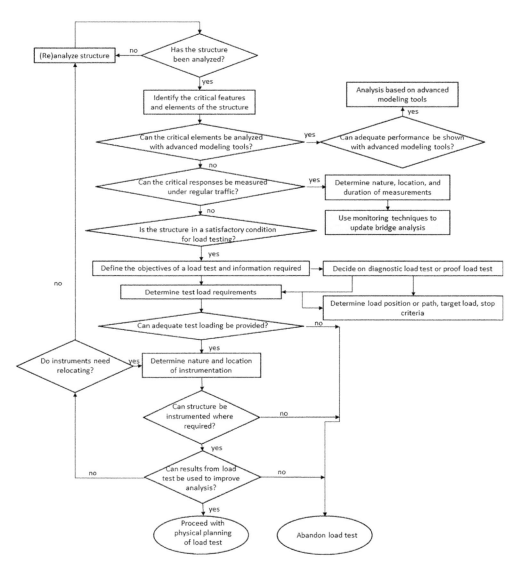

Figure 4.1 Flowchart to determine if load testing or other alternative can be used

Source: Extended from the flowchart by The Institution of Civil Engineers: National Steering Committee for the Load Testing of Bridges (1998).

load test should be abandoned. When the risks for the structure, the personnel on site, and the traveling public are too large, the load test should be abandoned or re-evaluated.

For proof load tests, the Unity Check in an assessment of the critical failure mode, should be only slightly larger than 1. If the differences between the Unity Check (or Rating Factor) in an analytical assessment and the target value with the test are too large, the engineer should evaluate if a test can be carried out. This decision will depend on the possible sources of additional capacity that are not considered in the assessment and their estimated contribution. Such contributions depend on the type of structure; for example, in reinforced concrete slab bridges more transverse redistribution than expected could be observed, whereas girder bridges might develop less redistribution.

An inventory of the available information about the bridge structure should be made. If information is lacking, an evaluation-based decision should be taken regarding the properties that are unknown. If needed, these properties can be estimated or measured on site with destructive or nondestructive testing techniques.

Before the decision is taken whether or not a bridge is suitable for load testing, an exploratory study and/or report should be made to analyze the bridge and its state. This preliminary study and related documents, as well as the preparation of the load test (if it is decided to load test the bridge), should include the following:

- Information from the design stage of the bridge, encompassing the original design calculation reports, the design and as-built plans, and the results of material testing at the time of construction;
- Verification that the as-built information corresponds to the geometry and material parameters of the real in-situ structure;
- Evaluation reports, including periodical inspection reports, with the results of destructive and nondestructive tests carried out on the structure, and assessment reports of the structure;
- Information on changes to the structure, design reports of rehabilitation measures, and associated rehabilitation plans.
- Economical estimate of the testing project compared to other solutions, such as more detailed theoretical capacity estimations, strengthening methods, etc.

As part of the preliminary analysis of the structure, a summary of the available information should be prepared. This summary should state which information is available and which information is missing. If information is missing, measures to obtain (some of) this information can be proposed. Moreover, the summary should include a discussion of the current state of the structure based on the available inspection and assessment reports. The summary of the current state serves as a summary of the available documentation and can never replace a detailed on-site inspection of the bridge and the preparatory calculations, which are both required prior to load testing.

When crucial information about the structure is missing, these properties should be measured, estimated based on similar structures and experience, or calculated. When the properties are found, the validity of these assumptions should be discussed after the load test and in the light of the obtained measurements. When no information about the geometry of the bridge is available, the needed measures of the geometry should be determined. In addition, for concrete bridges, the amount of reinforcement and the layout of the reinforcement can be estimated and then be verified with reinforcement scanner methods.

When no information about the material properties is available, testing of the material properties can be done on samples from the bridge. To limit the number of samples that need to be collected from the bridge, it is recommended to supplement sample testing with nondestructive testing to determine material parameters. The outcome of material testing can be used to improve the initial assessment of the bridge and improve the preparations of the load test. For the post-processing stage of the test, having sufficient knowledge

 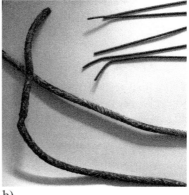

Figure 4.2 Examples of (a) concrete cores and (b) reinforcement samples from an existing concrete bridge

of the material properties will eliminate uncertainties, so that one source related to the uncertainties between the measured responses and the calculated responses can be reduced.

To determine the concrete compressive strength for concrete bridges, core samples can be drilled out (see Fig. 4.2a). For core sampling, transportation, and testing, the guidelines provided by the bridge owner and/or national codes should be followed. The assessment methodology determines the number of required samples, and is explained in the Dutch Guidelines for the Assessment of Bridges RBK, for example (Rijkswaterstaat, 2013).

To determine the yield strength and tensile strength of reinforcement steel, samples can be taken and tested in the laboratory. When the steel grade is given on the structural plans, samples (see Fig. 4.2b) can be taken to determine the average value of the yield strength and tensile strength. This additional testing can be interesting for older existing bridges in which steel grades may be used that are no longer commercially available. Testing of samples that have yielded is not recommended, as hardening will result in larger measured capacities. Similarly, for steel bridges, material samples can be taken.

For steel bridges, it can be necessary to estimate or measure the residual stresses in the members to improve the assessment. Nondestructive techniques based on ultrasound measurements (Li et al., 2016) can be used to measure the residual stresses. Additionally, hole drilling strain gage methods can be used as an indication related to the residual stresses in steel bridge members (Bathgate, 1968; Schajer, 1992). More novel methods that are currently under investigation are laser ultrasonic measurements (Zhan et al., 2017) and magnetic Barkhausen noise (Vourna et al., 2015; Samimi et al., 2016).

In some cases, very limited information about a bridge is available. In such a case, a choice can be made to use load testing to gather the required information about the structural behavior. Examples of using load testing on bridges without structural plans are available in the literature (Aguilar et al., 2015; Anay et al., 2016). In some cases, it may be cheaper and easier to carry out material testing on the structure and measure missing dimensions and to use this information as the initial input for the assessment calculations. With such calculations, it can be determined if the bridge is a suitable candidate for load testing and to select the required type of load test depending on the questions that need

78 Load Testing of Bridges

to be answered with the field test. In other cases, a "quick and easy" proof load test can be used to demonstrate that the bridge fulfills the code requirements.

4.2 Types of load tests, and which type of load test to select

In-situ load testing methods can differ significantly depending on the bridge type and project scope. It is often seen that the loading methods reflect the given national guidelines and standard methods, but that there are very strict demands to the testing time, since traffic often has to be redirected or stopped during testing. Consequently, these projects can be very costly and the in-situ test procedures have to be optimized to an extent which reduces the cost as much as possible. Depending on the goal of the load test, two different main types of load tests exist[1]:

- Diagnostic load tests (static or dynamic), mostly used for verification of the design assumptions of newly built bridges and to improve the analytical assessment of existing bridges.
- Proof load tests, loading of a structure (normally an existing structure) to a given target load (or to the load for which a stop criterion is reached) in order to verify a given response at higher load levels and demonstrate with a test that the bridge fulfills the code requirements.

These types of load tests are nondestructive tests. Additionally, failure testing (also called collapse testing or destructive testing) can be carried out on one bridge to evaluate for example a large subset of bridges or to understand the behavior up to the ultimate capacity of a certain bridge type. Failure testing includes loading magnitudes that reach the damage regimes of the bridge structure where irreversible damage occurs, as well as fully developed failure.

4.2.1 Diagnostic load tests

Diagnostic load tests (Moses et al., 1994; Russo et al., 2000; Farhey, 2005; Jauregui et al., 2010; Olaszek et al., 2014) are the most common type of load test. Diagnostic load tests can be used to check a number of in-situ properties of a given bridge (Lantsoght et al., 2017f). This bridge can be a new bridge or an existing bridge. For new bridges, diagnostic load testing is used to verify the design assumptions and analytical models used for the design. Several countries such as Italy (Veneziano et al., 1984), Switzerland (Brühwiler et al., 2012), France (Cochet et al., 2004), and Ecuador (Sanchez et al., 2018) require a diagnostic load test upon opening of all bridges or certain types of bridges. The verification of the design assumptions is carried out by comparing the measured structural response and the analytically determined response. National codes describe the limits of the allowable difference between the response of the structure in the load test and the predicted response. If the difference is larger than allowed, a logical explanation for this difference should be sought. When the difference has an influence on the design calculations, the relevant calculations should be repeated with the inclusion of the information from the field test, and reported. For existing bridges, diagnostic load tests can be used to update the analytical model of the bridge and obtain a refined assessment of the structure.

The transverse flexural distribution (Ohanian et al., 2017) can be determined, as explained in ACI 342R-16 (ACI Committee 342, 2016). The overall stiffness of the structure can be determined (Hodson et al., 2013). Testing prior to opening (Yang and Myers, 2003), over time (Myers et al., 2012), and after rehabilitation (Alkhrdaji et al., 2000) can be used to compare the behavior of the bridge measured in the field to the design or assessment calculations. When a bridge is tested prior to opening and after several years of service, the results of these load tests can be compared. The effect of deterioration and other time dependent changes can then be analyzed based on the reduction in stiffness between the newly opened bridge and the bridge after decades of service life. Dynamic tests can be used to determine the impact factor (Jiang et al., 2016) and the natural frequencies of the bridge (Frýba and Pirner, 2001).

Since diagnostic load testing goes hand in hand with analytical modeling, the response measurements (strain/stress, crack widths, or deflections) play an important role. The measured responses (often based on strains) are compared to the analytically predicted responses, and the field test results can then be used to update the analytical model. In practice, the analytical model is often a finite element model (Bridge Diagnostics Inc., 2012; Sanayei et al., 2016).

The loads that are selected for carrying out a diagnostic load test must be large enough to result in measurable responses, but are significantly smaller than the loads involved with proof load testing. Known loads such as loading vehicles can be used for this purpose.

A challenging part of the post-processing of a diagnostic load test is to identify the effect of each different contribution to the overall structural response (Barker, 2001). These contributions can be:

- The actual impact factor
- The actual dimensions
- The stiffness from nonstructural elements such as curbs and railings
- The actual lateral live-load distribution
- The effect of the flexibility of the supports
- The actual longitudinal live-load distribution
- Unintended or additional composite action in composite sections.

Not all of these effects occur at the ultimate limit state (ULS). Certain effects only occur at small load levels, for example, the interaction of some of the nonstructural elements. Therefore, it is recommended to carry out the load test with loads that can evaluate the behavior of the structure at the rating level. Other effects, such as unintended composite action and unintended support fixity, can be lost after cracking and/or at higher load levels. Therefore, it is important to analyze the sources of differences between the measured and analytical response to see which elements can be used at the ULS and can thus be used for the assessment.

4.2.2 Proof load tests

In a proof load test (Saraf and Nowak, 1998; Faber et al., 2000; Casas and Gómez, 2013; Arangjelovski et al., 2015; Lantsoght et al., 2017a, 2017c), a load representing the factored load combination prescribed by the governing code is applied to the bridge. Since the load is applied during the test and all sensors are zeroed at the beginning of the test, a proof load test evaluates the net capacity of a bridge for carrying live load. If the bridge can carry the applied target load without signs of distress, the proof load test has shown directly that the bridge fulfills the code requirements. In the past, proof load testing was carried out on bridges prior to opening to show the traveling public that this new bridge is safe for use (Bolle et al., 2011). Nowadays, proof load testing is mainly used for a direct assessment of existing bridges when a fully analytical assessment is not possible and/or lacks significant input information.

The lacking input can be related to the available information about the structure, for example, the lack of structural plans (Shenton et al., 2007) and no means to develop them, related to the structural behavior, such as transverse load redistribution in concrete slabs at higher load levels (Saraf, 1998), or related to the effect of material degradation and deterioration on the capacity of the tested bridge (Lantsoght et al., 2017c). In general, several uncertainties can be identified that can lead to an increased capacity, such as:

- Unintended composite behavior of slab-on-girder bridges (e.g. steel girder bridges with a concrete deck)
- More desirable boundary conditions
- Larger transverse redistribution than assumed during design and/or assessment
- Arching action and direct load transfer
- Compressive membrane action
- Increased concrete compressive strength as a result of the ongoing cement hydration process
- Stiffness and load-carrying capacity of nonstructural elements
- Unintended continuous behavior of simply supported girders with a continuous deck.

On the other hand, one can identify uncertainties resulting from deterioration mechanisms, material degradation, and damage that lead to a decreased capacity as well, such as:

- Corrosion-induced damage
- Alkali-silica reaction
- Environmental effects (chloride intrusion, frost/thaw, high/low temperatures, humidity, etc.)
- Leaching
- Fatigue damage
- Poor connection detailing
- Ultraviolet and ozone damage in bridges made with plastic composites.

In addition, time-dependent damage can occur to:

- Joints and bearings
- Substructures (settlements, scour, etc.)
- Girders, caused by vehicle impact
- Synergy effects which include several deterioration and damage mechanisms.

Since the loads have to represent the factored load combination, the applied loads in proof load tests are of a high magnitude. As such, the costs and risks involved with proof load testing are of a higher category than for diagnostic load testing.

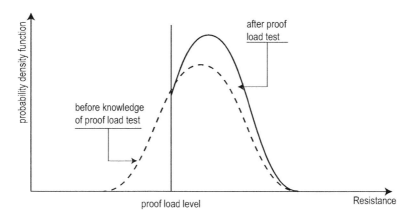

Figure 4.3 Truncation of probability density function of resistance after proof load test, based on Nowak and Tharmabala (1988)

In proof load tests, the measurements are important, as they need to be followed in real time during the experiment to verify that no irreversible damage takes place or that no collapse can occur. Threshold values for the structural responses, measured by the sensors, need to be determined prior to a proof load test. These threshold values are called the stop criteria. The stop criteria normally depend on the construction type, construction material, and expected failure mode. These stop criteria should be changed if the structural behavior is different than expected during preloading of the real in-situ structure. If a stop criterion is exceeded during a proof load test, further increasing of the load is not permitted, and the test should be terminated. It is then found that the bridge does not fulfill the requirements of the target load level. However, depending on the highest load that was achieved during the proof load test, the conclusion may be that the bridge can be used for a lower load level. Recommendations for posting of the structure can then be formulated. If the target proof load, which represents the factored load combination, is applied to the bridge without exceeding any of the stop criteria, it is shown experimentally that the bridge fulfills the code requirements. If the load is increased further than the target proof load, more information can be obtained about the behavior of the bridge at a higher load level. This practice is typically limited to research applications, however.

The maximum load that is applied in a proof load test (i.e. the target proof load) can also be used as input for a probabilistic analysis (Lin and Nowak, 1984; Rackwitz and Schrupp, 1985; Nowak and Tharmabala, 1988; Hall and Tsai, 1989; Fu and Tang, 1995). The target proof load can be used to truncate the probability density function of the resistance when the curves of load and resistance are used to determine the probability of failure. After a successful proof load test, it is known that the capacity is larger than or equal to the load effect caused by the applied load. Therefore, the probability density function of the capacity can be truncated as shown in Figure 4.3. To influence the probability of failure, the target proof load should be large enough.

4.2.3 Failure tests

The only type of test that can answer questions with regard to the ultimate capacity of a structure is a failure test (Elmont, 1913; Rösli, 1963; Jorgenson and Larson, 1976;

Figure 4.4 Application of load for collapse test and high magnitude loading of (a) Ruytenschildt Bridge in the Netherlands and (b) Rosmosevej Bridge in Denmark

Cullington et al., 1996; Haritos et al., 2000; Steinberg et al., 2011; Bagge et al., 2017b). Such testing is normally done on existing bridges that are functionally obsolete and serve to calibrate theoretical approaches as well as provide an indication regarding the actual response of a real-life structure. Often the capacity of real-life structures is higher than the predicted capacity due to rearrangement of the load path, interaction between the structural elements, and so forth. However, one collapse test is often considered as insufficient for the predictions of other similar bridge structures. Consequently, a series of similar bridges needs to be tested to answer capacity questions, which is rarely possible. Such testing provides, nevertheless, very important information concerning the responses and failure mechanisms at high loading magnitudes.

An example of this goal for a collapse test is testing of the Ruytenschildt Bridge (Lantsoght et al., 2017d) in Friesland to learn more about the behavior of reinforced concrete slab bridges (see Fig. 4.4a). In recent years in Sweden, a number of bridges have been tested to collapse (Bień et al., 2007; Enochsson et al., 2008; Puurula et al., 2008; Taljsten et al., 2008; Zou et al., 2009; Puurula et al., 2014; Bagge et al., 2015a, 2015b; Puurula et al., 2015, 2016; Bagge et al., 2017a). In Denmark, open T bridges have been studied in failure tests (see Fig. 4.4b) (Schmidt et al., 2018; Halding et al., 2017). Collapse tests, however, are very uncommon and require thorough preparation in terms of execution and safety, as well as much larger budgets than other types of field tests.

Selection of the load test type relates to the client's needs and can thus differ depending on the unique project. Most available codes and guidelines do not permit the proof load testing of fracture- and fatigue-critical steel bridges or the proof load testing of concrete bridges that are shear-, punching-, or torsion-critical, as such tests involve larger risks. However, in current practice such bridges are often found to be insufficient upon analytical assessment, and the client may wish to subject the structure to a proof load test. This application currently lies within the realm of research.

4.3 When to load test a bridge, and when not to load test

Economic estimations compared to competing bridge evaluation methods (recalculation of capacity, probabilistic evaluations, in-situ monitoring, etc.) need to be conducted before an initial step toward full-scale bridge testing can be done. After determining the goal of the full-scale test, it is necessary to see if the bridge under consideration is a good candidate for the required type of load test. An inspection should determine if there are no site-specific limitations or restrictions to carrying out the load test. Coordination with the traffic authorities is required to see if (partial) closure of the bridge is possible during the test, and for how much time this closure can be in effect.

Depending on the available time to prepare the test on site, it needs to be decided how contact and/or non-contact sensors can support the loading procedure and evaluate the expected response most efficiently. The number and type of sensors should be sufficient to gather the required information for updating the analytical model if a diagnostic load test is considered, or to monitor the bridge response and check the stop criteria if a proof load test is considered.

When a load test is considered for improving the assessment of a certain bridge, it is necessary that the analytically determined rating factor is slightly smaller than 1 or that the Unity Check is only slightly larger than 1. For proof load tests on reinforced concrete slab bridges (Lantsoght et al., 2017e), which have a large capacity for transverse redistribution, sections with a Unity Check between 1 and 1.3 are good candidates. For other bridge types, where less additional sources of capacity and redistribution are expected, sections with a Unity Check between 1 and 1.1 are good candidates for a proof load test or a diagnostic load test. For these bridges a diagnostic load test can be sufficient to determine parameters required to improve the analytical model used for the assessment, and thus refine the assessment.

When a detailed assessment of the bridge already results in Unity Checks smaller than 1 or rating factors larger than 1, a load test for assessment is not recommended. A detailed assessment can be an assessment based on a linear finite element model, for example. Different codes and countries have develop methods based on Levels of Approximation (as used in the *fib* Model Code 2010 (fib, 2012), see Fig. 4.5) for assessment of existing bridges (Brühwiler et al., 2012; Shu et al., 2015; Lantsoght et al., 2017b), where load testing (if included) is considered as the highest level of approximation. A high level of approximation means that the expected results are more precise than for the lower levels of approximation, but that the required cost and computational time are much larger.

Other situations in which load testing is not recommended is when the safety cannot be guaranteed during the load test, for example, when there is a risk of a brittle failure during a

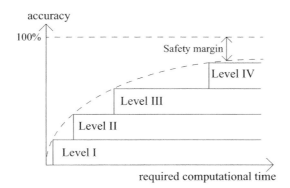

Figure 4.5 Levels of approximation

Source: Reprinted from Lantsoght et al. (2017b) with permission.

proof load test. Besides the structural safety, the safety of the personnel involved on site and the traveling public should be considered before deciding on load testing a bridge. If there are serious concerns regarding the safety, the engineers should not proceed with the load test.

4.4 Structure type considerations

4.4.1 Steel bridges

For steel bridges, measuring the structural response is straightforward when the residual stresses have been estimated. Contact sensors, such as strain gages or vibrating wire gages, can be used to directly measure the structural response in terms of strains. The type of superstructure will determine where the structural responses (often strains) need to be measured, and will thus determine the required sensor plan.

Steel bridges are good candidates for diagnostic load tests, and methods to update the analytical models with the results from the field test are well established (Barker, 2001). Often a diagnostic load test is sufficient for the assessment of steel bridges, as the isotropic material provides a structural behavior that is easier quantified compared to anisotropic concrete structures, for example.

For proof load testing of steel bridges, special attention should be paid to fracture-critical members and members with fatigue damage (often cross beams and connections). These elements need to be monitored closely during a proof load test, and relevant stop criteria should be defined prior to the test.

4.4.2 Reinforced concrete bridges

For reinforced concrete bridges, the type of structure will influence the sensor plan. For girder, tub, and box girder bridges, strains can be measured over the height of the structural elements. From these measurements, the strain profiles and thus stresses and internal forces can be derived. For other structure types, such as slab bridges, applying sensors over the height of the structural element is not possible. For those cases, sometimes only the bottom face can be instrumented see (Figs. 4.6 and 4.7b).

In a diagnostic load test, a number of strain measurements over the element height (girder, tub, or box girder) can be used to calibrate the analytical models. When such measurements cannot be taken, such as for slab bridges, the calibration of the model should be carried out based on the measurements of strains on the bottom of the cross section, or based on structural responses that are affected by a larger number of (unknown) parameters, such as deflection measurements. These measurements can be useful to evaluate transverse flexural distribution, but at times cannot cover all the unknowns required to improve the assessment.

In a proof load test, the onset of nonlinear behavior should be monitored with adequate stop criteria and sufficient sensors. When shear failures or other brittle failure modes can take place, stop criteria for these failure modes should be developed. At this moment, stop criteria for brittle failure modes in concrete are the topic of research (Schacht et al., 2016; Lantsoght et al., 2018). It has been shown that proof load testing can be used to evaluate shear-critical reinforced concrete slab bridges (Lantsoght et al., 2017c), but this type of test should only be carried out by experts until codes and guidelines that include recommendations for the testing of concrete bridges for brittle failure modes are available.

General Considerations 85

a)

b)

Figure 4.6 Monitoring with LVDTs, lasers, strain gages, acoustic emission sensors, and digital image correlation during a proof (a) and high magnitude (b) load test of a reinforced concrete slab bridge. Only the bottom face can be instrumented for these bridges

a)

b)

Figure 4.7 Instrumentation used during a test on a prestressed concrete bridge (a); monitoring preparation for OT (overturned T-section) bridge (b)

4.4.3 Prestressed concrete bridges

For prestressed concrete bridges, which typically consist of girders, tubs, or box girders, the strains can also be measured over the height of the girders with contact sensors such as strain gages or vibrating wire gages.

Diagnostic load testing of prestressed concrete bridges is especially desirable since the measured strain profiles over the height can be directly used to calibrate the analytical assessment models. It is for such bridges seen that the cracking moment is larger than in non-prestressed reinforced concrete bridges. Consequently the occurrence of cracking in a field test is typically not acceptable in prestressed bridges. An example of instrumentation for a prestressed girder bridge is shown in Figure 4.7a.

In a proof load test, the definition of stop criteria is an important part of the preparation. If the occurrence of cracks would change the overall structural behavior of the tested super-structure, then cracking cannot be allowed during the proof load test. A stop criterion should then be defined to avoid exceedance of the cracking moment in a proof load test on a prestressed concrete bridge. In the currently available codes and guidelines, such a stop criterion is not available. Moreover, (brittle) failure modes such as punching of the wheel prints through the deck should be avoided – especially if this undesirable failure mode is not related to the evaluation of the critical element during the test but is instead caused by the method of load application, when for example all load is concentrated at a single position. The stop criteria should also warn for the possibility of such a failure mode.

4.4.4 *Masonry bridges*

Masonry bridges are typically constructed as arch bridges. Their behavior is somewhat similar to the behavior of plain concrete and stone arch bridges. Therefore the stipulations in this section are valid for all masonry, stone, and plain concrete arch bridges. These bridges are complex structures, and measuring the structural response often does not lead to a straightforward interpretation of the structural behavior. The overall behavior is three-dimensional and results from the interaction between arch barrel and the spandrel walls (Fanning et al., 2005). Often, the measurements need to capture this overall behavior, and thus non-contact methods that follow the behavior of the entire structure such as radar interferometry (Mayer et al., 2012), total station, and digital image correlation (Bentz and Hoult, 2016) can be used. Contact sensors (linear variable differential transducers [LVDTs], laser triangulation sensors) can be applied to measure deflections locally and to calibrate the results of the non-contact methods. Measuring strains is typically not recommended for masonry and stone bridges, given the larger heterogeneity of the material. For plain concrete arch bridges, strain measurements can be used.

For arch bridges, it is recommended to use a series of nondestructive evaluation methods during the inspection on site as part of the preparation stage. Key input parameters required for the numerical analysis should be determined based on these nondestructive tests. The presence of internal voids, flaws, layering condition, and the mapping of non-homogeneity, moisture content, and so forth need to be evaluated and determined prior to the load test; these results should be implemented in the analytical model used for the assessment and preparation of the load test (Orban and Gutermann, 2009).

Diagnostic load testing of masonry, stone, and plain concrete arch bridges can be used to determine answers to particular questions, such as the overall stiffness of the structure or the influence of bearing restraint. However, the difficulty in the assessment lies in the modeling of these types of bridges. Their complex three-dimensional behavior often results in a large number of unknowns and the necessity of a large number of assumptions, so that the developed analytical models can be subject to discussion.

General Considerations 87

In most cases, proof load tests are required for the assessment of masonry, stone, and plain concrete arch bridges (Orban and Gutermann, 2009). With a proof load test, it can be assessed directly if the considered bridge fulfills the code requirements. Proof load tests are particularly interesting for arch bridges, as these bridges typically have a capacity that is found to be considerably larger in experiments than the capacity that is determined analytically (Orban and Gutermann, 2009).

4.4.5 *Timber bridges*

Timber bridges are characterized by the change in behavior over time caused by the change in material properties due to environmental influences on the timber elements. Timber bridges are sometimes historic truss bridges with solid timber elements, and sometimes they are new bridges with girders made of composite timber products (such as glulam), which are less subject to environmental influences.

Since the properties of timber are heterogeneous, special attention should be paid to the positions where sensors are applied. It is recommended to focus on the overall structural behavior and to measure deck deflections or timber strains during a diagnostic load test (Gentry et al., 2007), which can be used to determine the performance and the static and dynamic load distribution characteristics of the timber bridge (Wipf et al., 2000). Diagnostic load testing of timber bridges can be used to determine the distribution of loads (Wipf et al., 2000) or to evaluate the performance of rehabilitation measures (Gutkowski et al., 2001). During the preparation stage of a diagnostic load test, the condition assessment is important to determine the state of the timber bridge and the used material prior to the test. This assessment can combine visual inspections, photographic and video documentation, and moisture content measurements to determine the extent of wood deterioration (Wipf et al., 2000).

Proof load testing of timber bridges can be used to directly evaluate if a given bridge fulfills the code requirements. When the effect of material deterioration cannot be assessed, a proof load test may be required to evaluate the bridge's performance.

4.5 Safety requirements during load testing

4.5.1 *General considerations*

A crucial part in the preparation of a load test, and an element that needs to be considered prior to deciding whether a load test can be used to meet the requested goal, is assuring safety during the load test. If structural safety and the safety of personnel and the traveling public cannot be guaranteed, a load test should not be attempted.

The structural safety, and the risk of damage to the structure and collapse, should be carefully contemplated during the preparation stage of the load test. During the on-site inspection, special attention should be paid to structural and nonstructural elements that have significant deterioration and of which the performance may be subpar. The analysis prior to the test should identify which members are fatigue or fracture critical and which members could fail in a brittle manner. If such elements are present, it must be analyzed if measurements and stop criteria can be used to warn when a threshold of performance is surpassed. If it is not possible to test the bridge in a safe manner, the load test should not be attempted.

The integral safety during the load test is the responsibility of the party entrusted with the general safety considerations as agreed with the client. This party is responsible and liable for the preparation, preparatory calculations, determination of required measurements, load application, application and position of sensors, carrying out of the test, analysis and interpretation of the structural response during the test, delivering the report of the test, finalization and cleanup of the site after the test, and the safety of personnel, the structure, and the traveling public during the entire process. The responsible executing party can contract subcontractors for certain (sub)tasks of its responsibilities. When subcontractors are involved, the responsible executing party will then need to coordinate the activities with the subcontractors, arrange for meetings and communication between and with the subcontractors, communicate the safety requirements to all subcontractors, and communicate the tasks of the subcontractors to the client.

During the preparation stage, the party responsible for the safety should carry out the technical inspection. This inspection should be carried out in accordance with the governing guidelines for inspections, for example, CUR 117 (CUR, 1984) in the Netherlands. After this inspection, a report should be written which summarizes the possible dangers, risks, and possible problems that may occur during the preparation, execution, and dismantling of the test setup on site. This risk analysis report should provide an overview of the possible risks and provide solutions for these risks. Possible risks and situations that should be considered and evaluated include mechanical problems, electrical problems, electronic problems, failure or illness of personnel, and external factors (weather, local parties, press, etc.). Where required, duplicate equipment (such as backup sensors and additional computers) should be included in the required equipment that needs to be transported to the test location. This report also needs to include an overall planning of the activities on site (preparation, execution, dismantling), as well as detailed planning that identifies which tasks are consecutive, which tasks are parallel, and where the risk for delays lies. It is good practice to deliver the risk analysis and planning report at least five working days prior to the start of the activities on site. This report should be accepted by the client and owner of the bridge before activities on the site can be commenced. The report should also be distributed to all subcontractors and parties involved with the load test.

4.5.2 *Safety of personnel and traveling public*

As part of the initial considerations of the load test, an engineer responsible for the safety on site (here called the safety engineer) should be appointed. The safety engineer is responsible for the safety report of the test, which should consider the practical aspects of the execution of the load test. The report should include the means of communication that are used during the test (such as walkie-talkies). The safety requirements for during the test need to be added to this report: when can personnel (by exception) go under the bridge during the load test, and who decides after the load test that it is safe to go under the bridge again. In general, during a load test, nobody is allowed to go under the tested object except when explicit permission is given. During the test, only the executing party is allowed on the bridge. Prior to the test, a list of people that are allowed to enter the test site should be drafted. Additional people can only be granted access in exceptional situations. Nobody is allowed to enter the site without attending the safety briefing by the safety engineer. This briefing should be given in a language that can be understood and spoken by all people involved with the test – a requirement that is particularly important when the load test is carried out by an international team or in multilingual countries. If necessary, the briefing should be given in multiple languages.

Moreover, the safety engineer is responsible for the safety on site and the safety and well-being of the personnel during the period of activities on site (preparation of the test on site, execution of test, and dismantling of the setup and cleanup of the site). Examples of risks and situations that need to be evaluated by the safety engineer are the possible presence of asbestos in the structure and the presence of traffic on or next to the bridge (including shipping on rivers and canals).

The safety engineer should prepare the safety card, which contains the information of actions required in case of an accident, fire, or other type of calamity. This card should be displayed at a position where it can be seen by all personnel working on the site, it should be accessible during all activities on site and its contents should be communicated to all personnel before they start their activities on site. The safety card should include the phone numbers and addresses of the nearest hospital, general practitioners, and pharmacy, as well as the phone numbers of the emergency services, police, and firefighters. At all times, a first aid kit and at least one person with first aid certification should be present on the test site.

Besides the safety of the personnel involved with the load test, the safety of the traveling public and the safety of locals who may be curious about the load test and perhaps attempt to enter the test site should be guaranteed. An example of signaling used to show locals that they should not approach a bridge subjected to a collapse test is given in Figure 4.8a and an example of road closure in Figure 4.8b. For the safety of the traveling public, a traffic control plan and possibly a temporary detour should be developed together with the road authorities. This traffic control plan and possible detour should be communicated to the local communities with anticipation. When the interest of locals and possibly local press is expected, they should be referred to the communications expert of the road authority.

It is good practice to deliver the safety report and safety card at least five working days prior to the start of the activities on site, and these should be accepted by the client and owner of the bridge prior to the start of the activities on site. The safety report and

a)   b)

Figure 4.8 Example of signaling for the traveling public during a collapse test. Translation: do not access – weakened structure caused by testing (a). Closure of a road leading under the bridge enabling a safe working area (b)

90 Load Testing of Bridges

safety card should also be communicated to all subcontractors and parties involved with the load test.

4.5.3 Structural safety

To guarantee the structural safety during a load test, a good preparation of the test and the measurements during the test are important. The preparation should consist of assessment calculations, calculations to predict the structural behavior during the test, an evaluation of the overall stability, a verification of the substructure, and design of loading configuration paying attention to its safety. The structural behavior during the test can be predicted with hand calculations to have a first estimate of the expected responses, or by using a numerical model. For this numerical model, assumptions need to be made regarding the boundary conditions, composite action, and the modulus of elasticity of the concrete (and the influence of cracking thereon). These preparatory calculations should be combined into a report. It is good practice to deliver this report at least five work days before the start of the activities on site to the client and bridge owner, and the report should be accepted by the client before any activities can be started on site. The report should also be shared with all subcontractors and other parties involved with the load test.

Especially for proof load tests, the structural safety should be an important consideration. During a proof load test, stop criteria are used to verify if no irreversible damage occurs in the bridge due to the test. To check the stop criteria, the measured structural responses should be displayed in real time during the test and should be followed carefully. It is good practice to have at least two engineers dedicated to following the measurements. The load should be applied in increments, and after each load cycle the response measurements should be analyzed. Careful preparation and planning of the proof load test is important, and safety considerations should be thought through prior to the test to avoid damage to the structure or injury to personnel or the public. The response measurements should be analyzed by an experienced bridge engineer who makes decisions about the structural safety and the possibility to increase the applied load on the structure.

4.6 Summary and conclusions

This chapter discussed the general considerations that need to be considered prior to deciding on a load test. For that purpose, a number of questions should be explored.

1 Is all information about the structure available? If not, can this information be estimated or measured at the structure with a nondestructive or a destructive test?
2 What is the aim of the load test? Which type of load test should be selected for this purpose?
3 Are there site-specific limitations and restrictions that should be considered and that require changes to the plan for testing the bridge? Can the test be carried out in a safe manner?
4 How does the type of structure affect the choices with regard to instrumentation and possible aims of a load test?
5 How can the structural safety, and the safety of personnel and the traveling public, be guaranteed at all time during the activities on site? Who is responsible for safety?

The first question was answered by carrying out an inventory of the available information of the bridge. If important parameters are unknown, these should be measured with destructive or nondestructive testing methods.

The second question was answered by considering the definitions of the different types of load tests. Diagnostic load tests can evaluate the overall stiffness, transverse redistribution, effect of nonstructural elements on the stiffness, stiffness of the bearings, and unintended composite action between girders and the deck. When these properties are quantified, the analytical model used for the assessment of an existing bridge or the analytical model used for the design of a new bridge can be updated. A proof load test, in which a load representative of the code-prescribed factored loads is applied to the bridge, can be used to directly prove that a given bridge fulfills the code requirements. In addition, proof loading can be used to evaluate the structural response up to a given stop criterion, which is the threshold before irreversible damage occurs.

The third question is essential to the decision whether or not to load test a bridge. An inspection of the site is necessary to see what the limitations and restrictions of the site may be. Afterwards, these limitations and their effect on the position of sensors and all physical activities on site should be thought through. If the test cannot be carried out in a safe manner or poses a risk for the traveling public, a load test should not be attempted. The results of the assessment should also have identified if the Unity Check (or Rating Factor) is close enough to one for a proof load test to be possible to demonstrate sufficient capacity.

The fourth question was answered in this chapter by highlighting the particularities of different structural types: steel bridges, reinforced concrete bridges, prestressed concrete bridges, masonry bridges, and timber bridges. Depending on the structural type and material, different locations and different types for the measurements may be recommended and, for proof load testing, different stop criteria should be defined.

Finally, the fifth question was answered by discussing the different general elements of safety that need to be considered before the decision on load testing a bridge is made. Again it was stressed that a load test should not be attempted when this test poses a risk for the structural safety, the safety of the personnel on site, and the safety of the traveling public. To make sure all aspects of safety are considered, the project leader is made responsible for overall safety. The project leader will then appoint a safety engineer to be responsible for the safety during all activities on site.

Note

1 This terminology follows the AASHTO MBE (AASHTO, 2011) definitions. In the UK and Ireland, a different terminology is used (see Chapter 3).

References

ACI Committee 342 (2016) ACI 342R-16: Report on flexural live load distribution methods for evaluating existing bridges. American Concrete Institute, Farmington Hills, MI.

Aguilar, C.V., Jáuregui, D. V., Newtson, C. M., Weldon, B. D. & Cortez, T. M. (2015) Load rating a prestressed concrete double-tee beam bridge without plans by proof testing. *Transportation Research Board Annual Compendium of Papers*. Washington, DC.

92 Load Testing of Bridges

Alkhrdaji, T., Nanni, A. & Mayo, R. (2000) Upgrading Missouri transportation infrastructure: Solid reinforced-concrete decks strengthened with fiber-reinforced polymer systems. *Transportation Research Record*, 1740, 157–163.

Anay, R., Cortez, T. M., Jáuregui, D. V., ElBatanouny, M. K. & Ziehl, P. (2016) On-site acoustic-emission monitoring for assessment of a prestressed concrete double-tee-beam bridge without plans. *Journal of Performance of Constructed Facilities*, 30.

Arangjelovski, T., Gramatikov, K. & Docevska, M. (2015) Assessment of damaged timber structures using proof load test: Experience from case studies. *Construction and Building Materials*, 101, Part 2, 1271–1277.

Bagge, N., Sas, G., Nilimaa, J., Blanksvärd, T., Elfgren, L., Tu, Y. & Carolin, A. (2015a) Loading to failure of a 55 year old prestressed concrete bridge. *IABSE Workshop*. Helsinki, Finland.

Bagge, N., Shu, J., Plos, M. & Elfgren, L. (2015b) Punching capacity of a reinforced concrete bridge deck slab loaded to failure. *IABSE Conference 2015 – Structural Engineering: Providing Solutions to Global Challenges*. Geneva, Switzerland.

Bagge, N., Nilimaa, J., Puurula, A., Täljsten, B., Blanksvärd, T., Sas, G., Elfgren, L. & Carolin, A. (2017a) Full-scale tests to failure compared to assessments: Three concrete bridges. *Fib Symposium 2017*. Maastricht, The Netherlands.

Bagge, N., Popescu, C. & Elfgren, L. (2017b) Failure tests on concrete bridges: Have we learnt the lessons? *Structure and Infrastructure Engineering*, 14(3), 292–319.

Barker, M. G. (2001) Quantifying field-test behavior for rating steel girder bridges. *Journal of Bridge Engineering*, 6, 254–261.

Bathgate, R. G. (1968) Measurement of non-uniform bi-axial residual stresses by hole drilling method. *Journal of Strain*, 4(2), 20–29.

Bentz, E. C. & Hoult, N. A. (2016) Bridge model updating using distributed sensor data. *Institute of Civil Engineers: Bridge Engineering*, 170, 74–86.

Bień, J., Elfgren, L. & Olofsson, J. (2007) *Sustainable Bridges: Assessment for Future Traffic Demands and Longer Lives*. Dolnośląskie Wydawnictwo Edukacyjne, Wroclaw, Poland.

Bolle, G., Schacht, G. & Marx, S. (2011) Loading tests of existing concrete structures: Historical development and present practise. *Fib Symposium*, Prague, Czech Republic.

Bridge Diagnostics Inc. 2012. *Integrated Approach to Load Testing*. BDI, Boulder, CO. 44 pp.

Brühwiler, E., Vogel, T., Lang, T. & Luechinger, P. (2012) Swiss standards for existing structures. *Structural Engineering International*, 22, 275–280.

Casas, J. R. & Gómez, J. D. (2013) Load rating of highway bridges by proof-loading. *KSCE Journal of Civil Engineering*, 17, 556–567.

Cochet, D., Corfdir, P., Delfosse, G., Jaffre, Y., Kretz, T., Lacoste, G., Lefaucheur, D., Khac, V. L. & Prat, M. (2004) *Epreuves de chargement des ponts-routes et passerelles pietonnes*. Stra – Service d'Etudes techniques des routes et autoroutes, Bagneux-Cedex, France.

Cullington, D. W., Daly, A. F. & Hill, M. E. (1996) Assessment of reinforced concrete bridges: Collapse tests on Thurloxton underpass. *Bridge Management*, 3, 667–674.

CUR (1984) *Onderzoek naar aansprakelijkheid voor schade aan betonoppervlakken*. Civieltechnisch Centrum Uitvoering Research en Regelgeving, Gouda, the Netherlands.

Elmont (1913) Test-loading until breaking point of a 100-foot arch bridge. *Canadian Engineer*, 24, 739–744.

Enochsson, O., Puurula, A., Thun, H., Elfgren, L. & Taljsten, B. (2008) Test of a concrete bridge in Sweden: III. Ultimate capacity. In: Kohl, M. & Frangopol, D. (eds.) *IABMAS 2008*. Seoul, Korea.

Faber, M. H., Val, D. V. & Stewart, M. G. (2000) Proof load testing for bridge assessment and upgrading. *Engineering Structures*, 22, 1677–1689.

Fanning, P. J., Sobczak, L., Boothby, T. E. & Salomoni, V. (2005) Load testing and model simulations for a stone arch bridge. *Bridge Structures*, 1, 367–378.

Farhey, D. N. (2005) Bridge instrumentation and monitoring for structural diagnostics. *Structural Health Monitoring*, 4, 301–318.

FIB (2012) *Model Code 2010: Final Draft*. International Federation for Structural Concrete, Lausanne.

Frýba, L. & Pirner, M. (2001) Load tests and modal analysis of bridges. *Engineering Structures*, 23, 102–109.

Fu, G. K. & Tang, J. G. (1995) Risk-based proof-load requirements for bridge evaluation. *Journal of Structural Engineering-ASCE*, 121, 542–556.

Gentry, T. R., Wacker, J. P., Brohammer, K. N. & Wells, J. (2007) In situ materials and structural assessment of stress-laminated deck bridge treated with chromate copper arsenate. *Transportation Research Record*, 2028, 28–33.

Gutkowski, R. M., Shigidi, A.M.T., Tran, A. V. & Peterson, M. L. (2001) Field studies of strengthened timber railroad bridge. *Transportation Research Record*, 1770, 139–148.

Halding, P. S., Schmidt, J. W., Jensen, T. W. & Henriksen, A. H. (2017) Structural response of full-scale concrete bridges subjected to high load magnitudes. *SMAR 2017–4th Conference on Smart Monitoring, Assessment and Rehabilitation of Civil Structures*. Berlin, Germany.

Hall, W. B. & Tsai, M. (1989) Load testing, structural reliability and test evaluation. *Structural Safety*, 6, 285–302.

Haritos, N., Hira, A., Mendis, P., Heywood, R. & Giufre, A. (2000) Load testing to collapse limit state of Barr Creek Bridge. *Fifth International Bridge Engineering Conference, Vols 1 and 2: Bridges, Other Structures, and Hydraulics and Hydrology*. Tampa, FL.

Hodson, D. J., Barr, P. J. & Pockels, L. (2013) Live-load test comparison and load ratings of a post-tensioned box girder bridge. *Journal of Performance of Constructed Facilities*, 27, 585–593.

The Institution of Civil Engineers: National Steering Committee for the Load Testing of Bridges (1998) *Guidelines for the Supplementary Load Testing of Bridges*. The Institution of Civil Engineers, London, UK.

Jauregui, D. V., Licon-Lozano, A. & Kulkarni, K. (2010) Higher level evaluation of a reinforced concrete slab bridge. *Journal of Bridge Engineering*, 15, 172–182.

Jiang, D., Terzioglu, T., Hueste, M.B.D., Mander, J. B. & Fry, G. T. (2016) Experimental study of an in-service spread slab beam bridge. *Engineering Structures*, 127, 525–535.

Jorgenson, J. L. & Larson, W. (1976) Field testing of a reinforced concrete highway bridge to collapse. *Transportation Research Record: Journal of the Transportation Research Board*, 607, 66–71.

Lantsoght, E. O. L., Koekkoek, R., Yang, Y., Van Der Veen, C., Hordijk, D. & De Boer, A. (2017a) Proof load testing of the viaduct De Beek. *39th IABSE Symposium: Engineering the Future*. Vancouver, Canada.

Lantsoght, E. O. L., De Boer, A. & Van Der Veen, C. (2017b) Levels of Approximation for the shear assessment of reinforced concrete slab bridges. *Structural Concrete*, 18, 143–152.

Lantsoght, E. O. L., Koekkoek, R. T., Hordijk, D. A. & De Boer, A. (2017c) Towards standardization of proof load testing: Pilot test on viaduct Zijlweg. *Structure and Infrastructure Engineering*, 16.

Lantsoght, E. O. L., Van Der Veen, C., De Boer, A. & Hordijk, D. A. (2017d) Collapse test and moment capacity of the Ruytenschildt Reinforced Concrete Slab Bridge. *Structure and Infrastructure Engineering*, 13, 1130–1145.

Lantsoght, E. O. L., Van Der Veen, C., De Boer, A. & Hordijk, D. A. (2017e) Proof load testing of reinforced concrete slab bridges in the Netherlands. *Structural Concrete*, 18, 597–606.

Lantsoght, E. O. L., Van Der Veen, C., Hordijk, D. A. & De Boer, A. (2017f) State-of-the-art on load testing of concrete bridges. *Engineering Structures*, 150, 231–241.

Lantsoght, E. O. L., Van Der Veen, C. & Hordijk, D. A. (2018) Proposed stop criteria for proof load testing of concrete bridges and verification. *IALCCE 2018*. Ghent, Belgium.

Li, Z., He, J., Teng, J. & Wang, Y. (2016) Internal stress monitoring of in-service structural steel members with ultrasonic method. *Materials*, 9.

Lin, T. S. & Nowak, A. S. (1984) Proof loading and structural reliability. *Reliability Engineering*, 8, 85–100.

Mayer, L., Yanev, B., Olson, L. D. & Smyth, A. (2012) Monitoring of the Manhattan Bridge and interferometric radar systems. *IABMAS 2012*. Stresa, Italy.

Moses, F., Lebet, J. P. & Bez, R. (1994) Applications of field testing to bridge evaluation. *Journal of Structural Engineering-ASCE*, 120, 1745–1762.

Myers, J. J., Holdener, D. & Merkle, W. (2012) Load testing and load distribution of fiber reinforced, polymer strengthened bridges: Multi-year, post construction/post retrofit performance evaluation. *FRP Composites and Sustainability: Focusing on Innovation, Technology Implementation and Sustainability.* Springer, New York, NY.

Nowak, A. S. & Tharmabala, T. (1988) Bridge reliability evaluation using load tests. *Journal of Structural Engineering-ASCE*, 114, 2268–2279.

Ohanian, E., White, D. & Bell, E. S. (2017) Benefit analysis of in-place load testing for bridges. *Transportation Research Board Annual Compendium of Papers*, 14.

Olaszek, P., Lagoda, M. & Casas, J. R. (2014) Diagnostic load testing and assessment of existing bridges: Examples of application. *Structure and Infrastructure Engineering*, 10, 834–842.

Orban, Z. & Gutermann, M. (2009) Assessment of masonry arch railway bridges using non-destructive in-situ testing methods. *Engineering Structures*, 31, 2287–2298.

Puurula, A. M., Enochsson, O., Thun, H., Taljsten, B. & Elfgren, L. (2008) Test of a concrete bridge in Sweden: I. Assessment methods. In: Kohl, M. & Frangopol, D. (eds.) *IABMAS 2008*. Seoul, Korea.

Puurula, A. M., Enochsson, O., Sas, G., Blanksvärd, T., Ohlsson, U., Bernspång, L., Täljsten, B. & Elfgren, L. (2014) Loading to failure and 3D nonlinear FE modelling of a strengthened RC bridge. *Structure and Infrastructure Engineering*, 10, 1606–1619.

Puurula, A. M., Enochsson, O., Sas, G., Blanksvärd, T., Ohlsson, U., Bernspång, L., Täljsten, B., Carolin, A., Paulsson, B. & Elfgren, L. (2015) Assessment of the strengthening of an RC railway bridge with CFRP utilizing a full-scale failure test and finite-element analysis. *Journal of Structural Engineering*, 141, D4014008:1–11.

Puurula, A. M., Enochsson, O., Sas, G., Blanksvärd, T., Ohlsson, U., Bernspång, L., Taljsten, B. & Elfgren, L. (2016) 3D non-linear FE analysis of a full-scale test to failure of a RC railway bridge strengthened with carbon fibre bars. *19th IABSE Congress Stockholm 2016: Challenges in Design and Construction of an Innovative and Sustainable Built Environment.* Stockholm.

Rackwitz, R. & Schrupp, K. (1985) Quality-control, proof testing and structural reliability. *Structural Safety*, 2, 239–244.

Rijkswaterstaat (2013) Guidelines assessment bridges: Assessment of structural safety of an existing bridge at reconstruction, usage and disapproval (in Dutch), RTD 1006:2013 1.1.

Rösli (1963) "Die Versuche an der Glattbrücke in Opfikon" Bericht Nr. 192, Report. Eidgenössische Materialprüfungs- und Versuchanstalt Für Industrie, Bauweses und Gewerbe, Dübendorf, Dübendorf, December 1963.

Russo, F. M., Wipf, T. J. & Klaiber, F. W. (2000) Diagnostic load tests of a prestressed concrete bridge damaged by overheight vehicle impact. *Transportation Research Record*, 1696, 103–110.

Samimi, A. A., Krause, T. W. & Clapham, L. (2016) Multi-parameter evaluation of magnetic Barkhausen noise in carbon steel. *Journal of Nondestructive Evaluation*, 35.

Sanayei, M., Reiff, A. J., Brenner, B. R. & Imbaro, G. R. (2016) Load rating of a fully instrumented bridge: Comparison of LRFR approaches. *Journal of Performance of Constructed Facilities*, 2.

Sanchez, T. A., Bonifaz, J. & Robalino, A. (2018) Bridge load testing in Ecuador: Case studies. *IALCCE 2018*. Ghent, Belgium.

Saraf, V. K. (1998) Evaluation of existing RC slab bridges. *Journal of Performance of Constructed Facilities*, 12, 20–24.

Saraf, V. K. & Nowak, A. S. (1998) Proof load testing of deteriorated steel girder bridges. *Journal of Bridge Engineering*, 3.

Schacht, G., Bolle, G., Curbach, M. & Marx, S. (2016) Experimental evaluation of the shear bearing safety (in German). *Beton – und Stahlbetonbau*, 111, 343–354.

Schajer, G. S. (1992) Non-uniform residual stress measurements by the hole drilling method. *Journal of Strain*, 28(1), 19–22. https://doi.org/10.1111/j.1475-1305.1992.tb00784

Schmidt, J. W., Halding, P. S., Jensen, T. W. & Engelund, S. (2018) High magnitude loading of concrete bridges. *ACI Special Publication*, Evaluation of concrete bridge behaviour through load testing: International perspectives. SP-323–9, 9.1–9.20.

Shenton, H. W., Chajes, M. J. & Huang, J. (2007) Load rating of bridges without plans: Newark, Delaware: Department of civil and environmental engineering, university of Delaware; Department of civil engineering, Widener University.

Shu, J., Plos, M., Zandi, K. & Lundgren, K. (2015) A multi-level structural assessment proposal for reinforced concrete bridge deck slabs. *Nordic Concrete Research*, 53–56.

Steinberg, E., Miller, R., Nims, D. & Nims, D. (2011) Structural evaluation of LIC-310-0396 and FAY-35-17-6.82 box beams with advanced strand deterioration. Athens, OH.

Taljsten, B., Bergstrom, M., Nordin, H., Enochsson, O. & Elfgren, L. (2008) Test of a concrete bridge in Sweden: II. CFRP strengthening and structural health monitoring. In: Kohl, M. & Frangopol, D. (eds.) *IABMAS 2008*. Seoul, Korea.

Veneziano, D., Galeota, D. & Giammatteo, M. M. (1984) Analysis of bridge proof-load data I: Model and statistical procedures. *Structural Safety*, 2, 91–104.

Vourna, P., Ktena, A., Tsakiridis, P. E. & Hristoforou, E. (2015) A novel approach of accurately evaluating residual stress and microstructure of welded electrical steels. *NDT & E International*, 71, 33–42.

Wipf, T. J., Ritter, M. A. & Wood, D. L. (2000) Evaluation and field load testing of timber railroad bridge. *Transportation Research Record*, 1696, 323–333.

Yang, Y. & Myers, J. J. (2003) Live-load test results of Missouri's first high-performance concrete superstructure bridge. *Transportation Research Record*, 1845, 96–103.

Zhan, Y., Liu, C., Kong, X. & Lin, Z. (2017) Experiment and numerical simulation for laser ultrasonic measurement of residual stress. *Ultrasonics*, 73, 271–276.

Zou, Z. Q., Enochsson, O., He, G. J. & Elfgren, L. (2009) Finite element analysis of small span reinforced concrete trough railway bridge. *Advances in Concrete and Structures*, 400–402, 645–650, 978.

Chapter 5

Preparation of Load Tests

Eva O. L. Lantsoght and Jacob W. Schmidt

Abstract

This chapter discusses the aspects related to the preparation of load tests, regardless of the chosen type of load test. After determination of the test objectives, the first step should be a technical inspection of the bridge and bridge site. With this information, the preparatory calculations (assessment for existing bridges and expected behavior during the test) can be carried out. Once the analytical results are available, the practical aspects of testing can be prepared: planning, required personnel, method for applying the load, considerations regarding traffic control and safety, and the development of the sensor and data acquisition plan. It is good practice to summarize all preparatory aspects in a preparation report and provide this information to the client/owner and all parties involved with the load test.

5.1   Introduction

This chapter gives an overview of the steps that are carried out to prepare for a load test, regardless of the type of load test that is selected to meet the objectives of the test. These general considerations include formulating the specific objectives of the test, carrying out a technical inspection of the bridge, carrying out preliminary calculations, and developing the practical aspects of the test.

It is good practice to report the preparation of the load test in a preparation report. The first item that should be included in this report are the test objectives. In a first step, the test objectives were explored in the previous chapter from the point of view to decide whether or not a load test should be carried out. Once it is decided to carry out a load test, the test objectives should be clearly stated and agreed upon with the bridge owner and/or client. These objectives then form the basis for the preparation of the load test: which information about the bridge is needed, which calculations should be carried out, and how should the load be applied. The responses should be measured in such a way that the required information to meet the test objectives can be collected during the field test.

Since the available time on site during a load test is usually limited, a good preparation is necessary to streamline all activities on site. For this purpose, a technical inspection of the bridge and bridge site should be carried out before any other preparation steps are done. For an existing bridge, the inspection should focus on two aspects: the condition of the bridge and possible site limitations. The condition of the bridge should be documented with photographs. Maps of deterioration, cracks, and other forms of damage should be developed based on the inspection. The effects of deterioration should then also be taken into

DOI: https://doi.org/10.1201/9780429265426

account for the preparatory calculations. The structure should be inspected for changes with respect to the available plans, such as widening or a different lane layout, and special attention to the joints and bearings should be paid. Second, the bridge site should be investigated. Limitations with regard to the application of sensors, application of the load, transport of the load to and from the site, and possible hazardous situations should be evaluated and reported. When a load test is carried out on a new bridge, the technical inspection should focus on possible details that deviate from the design plans, and possible restrictions on the site for carrying out a load test.

The preliminary calculations for an existing bridge serve two purposes: carrying out the assessment calculations based on the Unity Check or Rating Factor, and predicting the behavior of the bridge during the test. For a new bridge the design calculations are available, and the analytical models may be readily available. In that case, only the expected structural responses for the test load should be taken from the available model if a finite element model was used for the design. The model can be updated with the tested properties of the materials that were used for construction. For existing bridges, the assessment calculations may be available in an evaluation report of the structure. These calculations should be checked, and where the technical inspection identified changes to the structural system or deterioration, these effects should be considered in the assessment calculations. The second type of calculations, which are used to predict the behavior of the bridge during the test and the expected magnitude of the structural responses, should be carried out based on average material parameters and by omitting all load and resistance factors. Since load testing often has the purpose to reduce the uncertainties with regard to the behavior of the tested bridge, these exploratory calculations can only give an indication of the expected behavior. To cover these uncertainties, a range of values or situations (such as comparing the results with and without composite action) can be considered in the preparatory calculations.

Once the goals of the test are defined, the bridge is inspected, and its behavior is estimated with preliminary calculations, the practical preparations for the test can be done to ensure efficiency at the test site. An overall planning of the activities on site should be made, as well as a detailed planning which identifies who will be carrying out which task, when, and with which equipment. A second element is determining how the load will be applied and where. The load can be applied with dead loads, heavy trucks, or a test frame and hydraulic jacks. The method of load application and the position of the load (single position or driving path) depends on the test goals. The safety of the personnel, bridge, and traveling public should be ensured during all activities on site. For this purpose, the possible hazardous situations for the personnel should be evaluated, preparatory calculations are necessary, and traffic signaling or detours should be developed. Finally, a plan for the application of sensors and data visualization during the test (if required) should be developed. This plan should fulfill the goals during the test (e.g. evaluation of the stop criteria during a proof load test), as well as the goals of the post-processing (e.g. updating of the developed finite element model with the field observations).

5.2 Determination of test objectives

Before determining if a load test is the right way to address the open questions with regard to the structure, the test objectives should be clearly stated. Depending on the test objectives, the

Preparation of Load Tests 99

Table 5.1 Examples of goals that can be achieved with different types of load tests, for new and existing bridges

	Diagnostic load test	Proof load test
New bridge	Verify design assumptions (stiffness, deflection, load distribution)	Demonstrate, prior to opening, that bridge can carry code-prescribed loads (uncommon nowadays)
Existing bridge	Verify structural behavior (stiffness, deflection, load distribution, composite action, rotation capacity at supports, continuity at supports). Verify behavior after rehabilitation/strengthening.	Demonstrate that bridge can carry code-prescribed loads (when loads have changed over time, material deterioration or degradation occurs, uncertainties of behavior at higher load levels are large)

type of load test can be selected. These elements have been discussed in general terms in Chapter 4. A summary of possible objectives for load tests for new and existing bridges is given in Table 5.1.

Determining the test objectives is a critical step at the initial preparation of the load test. Since the steps required to prepare and execute the load test and to analyze its results afterwards depend on the goals of the test, defining the test objectives is of the utmost importance. The executing party and owner and/or client should agree on the test objectives early on when a load test is considered for the bridge under study and write these objectives down in a memo that should be communicated with all parties involved.

Besides determining whether or not load testing is recommended and which load test should be used, the test objectives are also required for the type, position, and range of the sensors that are applied during the load test. In addition, the test goals determine the sampling rate and type of the data acquisition system as well as the data processing and visualization. For example, for a proof load test, immediate data processing and visualization is required so that the stop criteria can be evaluated.

It is good practice to summarize all assumptions, calculations, and decisions that were made prior to the test in a preparation report. This report should state the test objectives and sketch how these objectives will be met during and after the load test.

5.3 Bridge inspection

5.3.1 *Inspection results*

The first step during the preparation of a load test is the technical inspection of the bridge and the bridge site. A detailed reference that can be consulted on how to conduct a technical inspection is the *Bridge Inspector's Reference Manual* (Ryan et al., 2012). The bridge plans available in the archives should be studied prior to the inspection and taken to the bridge site so that changes with respect to the available plans can be noted and, where possible, measured. Possible alterations to the bridge that may not be shown on the drawings are:

*   Widening of the bridge
*   Strengthening projects
*   Actual geometry and reinforcement placing
*   Sound barriers

Figure 5.1 Example of frozen bearing

- Changes to the lane layout
- Changes to the width of the shoulder at the edges of the lanes and between the driving directions
- Changes to the thickness of the wearing surface or asphalt layer.

The joints and bearings should be inspected so that it can be evaluated if restraint of deformations (e.g. caused by changes in temperatures) can occur. Frozen bearings (see Fig. 5.1) can be identified by the following observations: bending, buckling, improper alignment of members, or cracks in the bearing seat. Pitting, section loss, deterioration, and the buildup of debris at the bearing can result in frozen bearings as well. If restraint of deformations occurs and results in additional stresses on the cross sections, this load effect should be considered in the assessment calculations and preparation of the test. Since these observations depend on temperature, the temperature during the inspection should be measured and reported (Ryan et al., 2012). If the temperature is above the design temperature, the bearing should be in its expanded position. If the temperature is below the design temperature, the bearing should be in its contracted position. Unless otherwise noted, the design temperature is 18°C or 68°F.

During the inspection of concrete bridges, a map of the cracks and deterioration should be drawn, along with the related cause. This map should represent the length, direction, and width of the cracks to scale for the accessible faces of the structure. For example, for slab bridges, the crack map should include the bottom face, side faces, and the top face if no wearing surface is provided. For girder bridges, the bottom and side faces of the girders should be presented, as well as the bottom of the deck, and when no wearing surface

Preparation of Load Tests   101

is applied, the top face of the deck. The width of cracks with a width $w \geq 0.15$ mm (0.006 in.) should be indicated on the map of cracks. Additionally, regions with material damage and degradation should be marked on the drawing with the map of cracks. Examples are positions of delamination and where rebar corrosion can be observed. A possible method for developing the map of damages is to mark the cracks with a marker and to make photographs of the accessible faces of the structure. For reference, a grid can be drawn on the faces. The photographs can then be compiled in photo editing software. Special attention should be paid to the scale, which should be the same in all photographs, and possible effects of wide-angle lenses and other sources of distortion of the photographs. Once all photographs are combined to represent the entire face of the bridge that is studied, the marked cracks can be drawn by hand over the photograph in the editing software and saved as a separate layer. When the layer with the photographs is disabled, the drawing of the crack map (including other forms of damage where relevant) remains. An example of an inspection map which depicts damages is shown in Figure 5.2.

During the inspection of steel bridges, the position of signs of corrosion, fatigue damage, and fracture-critical details should be identified and represented on a drawing of the occurring damage. This drawing should show all accessible faces of the structure. For box and tub girders, the inspection and drawing of the occurring damage should include the inside of the box or tub.

During the inspection of timber bridges, all positions with material damage and degradation should be identified and represented on a drawing. This drawing should show all identified damage, represent it on scale, and cover all accessible faces of the structure.

During the inspection of masonry bridges, a map of cracks should be developed in a similar way as the map of cracks for concrete bridges. Additionally, where the mortar of the joints is degraded and where bricks are missing, this damage should be included on the drawing that represents the condition of the structure. Other forms of observed material degradation and damage should be added as well. The drawing should cover all accessible faces of the structure.

During the inspection of bridges using fiber reinforced plastics (FRPs), special attention should be paid to damages and material degradation. In particular, delamination due to concentrated load exposure and adhesive joint locations should be checked in combination with durability deterioration caused by ultraviolet light, ozone, temperature, and humidity. A drawing representing the positions, type, and severity of the observed damage for all accessible faces of the structure should be prepared.

5.3.2 *Limitations of testing site*

Besides the inspection of the bridge to be tested, the bridge site and its accessibility should be checked during the inspection. Possible obstructions for the measurements, the load application, and the access of personnel to the site should be identified. Examples of limitations to the test site include:

* Elevated sidewalks on the bridge, which makes measuring and marking sensor and load positions more difficult;
* Restrictions to the access to the bridge structure, or limited space (or height) under the bridge or at the side faces;
* Restrictions related to the access to the bridge site itself;

102 Load Testing of Bridges

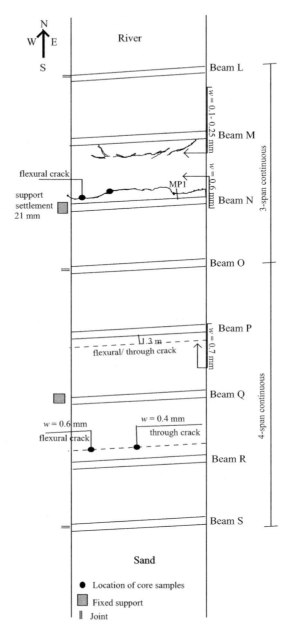

Figure 5.2 Example of map of damages. Conversion: 1 mm = 0.04 in., 1 m = 3.3 ft
Source: Reprinted with permission from ASCE from Lantsoght et al. (2016).

- Restrictions caused by roadway or waterway traffic under or on (parts of) the bridge that may not be halted, or only when a special exception is granted for a short amount of time;
- Obstructions that complicate the application of sensors to the bridge.

Figure 5.3 Proof load test on viaduct where one lane of traffic (right side of photograph) remains open for traffic (Fennis et al., 2014; Koekkoek et al., 2015)

The restrictions regarding roadway or waterway traffic that cannot be halted significantly impact the load test. Especially if the bridge cannot be closed for traffic during the load test (see Fig. 5.3 for an example), the effect of the passing traffic on the measured structural responses cannot be ignored. This situation is not desirable but is sometimes necessary because of the accessibility to certain towns or dwellings. When the bridge cannot be closed for traffic during the test, the safety aspects need to be considered in even larger detail. At all times, crossing the bridge should be safe for the traveling public. When the load is applied through trucks, the truck drivers should be alerted about the passing of traffic on, for example, the remaining lanes of the bridge. When a proof load test is carried out, the structural responses should be followed in detail, the stop criteria should be checked meticulously, and at no point should the load magnitude cause danger to the traveling public in terms of performance of the bridge, presence of large amounts of counterweights, or possible collapse of the structure. In general, it is not recommended to allow traffic on a bridge during a proof load test. If closing is not possible, a temporary closing when the largest load levels are applied should be considered.

These limitations regarding access to the bridge structure and the test site need to be considered when choices are proposed during the preparation stage. They will influence the safety plan, sensor plan, loading protocol, and possibly the type of sensors that are selected as well as the way in which the load is applied.

5.4 Preliminary calculations and development of finite element model

5.4.1 Development of finite element model

An important part of the preparation of a load test is the development of a (linear) finite element model of the possible test bridge. Not all codes (e.g. the *Manual for Bridge Evaluation* [AASHTO, 2016]) require that a finite element model is developed prior to the load test. In fact, depending on the objectives of the load test, it is not always desirable to develop a finite element model. In some cases, simplified analytical calculations are sufficient and keep load testing an attractive and reasonably economic option. The codes and guidelines that do not require a finite element model mention that the field test can be used to update the analytical model that is used for rating the structure. The analytical model can be of varying levels of detail and complexity. For new bridges, finite element models are often available as part of the design calculations. For existing bridges, this model serves the following purposes:

- It is used for the assessment (or rating) of the bridge prior to the test. Sometimes, the assessment based on a refined finite element model will show that the bridge fulfills the code requirements, and it may then be decided that a field test is not necessary.
- It will be used to identify the loading positions and the critical positions for which the structural response should be monitored during the test. The magnitude of the expected response is necessary to require the type and measurement range of the sensors.
- It can be used to identify overall structural behavior, such as stiffness and transverse load distribution, prior to the test. This behavior can then be compared to the measured responses during the test. Additionally, developing a finite element model gives the engineer a better understanding of the behavior and details of the bridge. Prior to the load test, the model can be used to evaluate the influence of certain factors: the stiffness of nonstructural members such as barriers and parapets, unintentional restraint at the bearings, and other factors that typically are not considered during assessment. To evaluate the importance of these factors, for example the effect of the nonstructural members, two models can be developed: a model with and a model without the nonstructural members. The responses from these two models can then be compared to evaluate if the influence of the studied factor is significant. Similarly, for composite bridges, two models can be developed: one in which no composite action occurs and one with the composite cross section. The responses can then be compared and can be used to evaluate the structural behavior during the field test (Zhou et al., 2007).
- If the preparations are carried out according to different Levels of Approximation as described in the *fib* Model Code for concrete structures (fib, 2012), the results of hand calculations can be compared with the results in the model. For example, for transverse distribution, the distribution factors from codes such as the AASHTO LRFD code (AASHTO, 2015) can be used to have a first idea. Additionally, the method of Guyon-Massonet can be used as a first approximation, or the recommendations from ACI 342R-16 (ACI Committee 342, 2016) for concrete bridges. The results from a simplified analytical method can be compared with the results from the finite element model.
- After a diagnostic load test, the model should be updated based on the responses measured in the field and used for the assessment (or rating) of the bridge after the test.

When the finite element model is used to identify the critical loading positions and the magnitude of the load in a proof load test that corresponds to the factored live-load combination, a model with shell elements is sufficient (Lantsoght et al., 2017a). If structural responses such as strains over the height of the cross section should be determined prior to the test, for example to prepare the stop criteria, solid elements are necessary.

For concrete bridges, in the linear finite element model the uncracked cross-sectional stiffness can be used together with a Poisson ratio of 0.15. Guidance for the modeling with nonlinear finite elements is given in guidelines such as the Dutch guideline (Rijkswaterstaat, 2017). When a concrete bridge is susceptible to a brittle failure mode, or when a steel bridge has fracture- or fatigue-critical details or possible structural stability issues, a nonlinear finite element model is recommended to analyze the structure in more detail and compare the responses from the model during the test with the measured responses. This nonlinear finite element model will also give a more precise estimate of the responses in concrete bridges after cracking and redistribution occurs.

The loads that are applied in the finite element model are the loads that are used for the assessment of the bridge. For a proof load test, the load combination with these loads should result in the same sectional moment or force (depending on the studied failure mode) as the applied load during the test. The following loads are typically modeled:

- Self-weight
- Superimposed dead load
- Live load
- Loads or effects such as temperature changes that lead to stresses on the cross section as a result of a restraint of deformation, or such as support settlements (if any) that change the distribution of sectional moments and forces.

The self-weight is modeled based on the load resulting from the material density. If a simplified model is used, in which the geometry is simplified, then the equivalent load of the omitted parts should be added as an external load.

For bridges with a superimposed dead load resulting from an asphalt layer, the resulting load can be determined based on a volume load of 23 kN/m^3 (0.15 kip/ft^3). The thickness of the asphalt layer can be taken from the structural plans. However, if additional layers have been added over time, the value in the plans may be unconservative. For that case, it is recommended to determine the thickness based on drilled core samples or by using a nondestructive test method.

The live loads that are applied on the bridge depend on the considered code. The live-load model typically consists of distributed and concentrated loads, for example as defined in AASHTO LRFD (AASHTO, 2015) and NEN-EN 1991–2:2003 (CEN, 2003). The distributed loads can be distributed lane loads, pedestrian loads, and distributed loads on the remaining area. The concentrated loads can be design trucks or design tandems placed in the bridge lanes. The notional lane width that is prescribed in the codes should be used; this lane width and the resulting layout can be different from the actual lane layout of the structure. For the design trucks or tandems, the position should be sought that results in the largest load effect. This position should give the most unfavorable case, which is governing for assessment.

When shell elements are used in the finite element model, the distribution of the concentrated wheel prints over the layer of asphalt and to the middle of the cross section should be

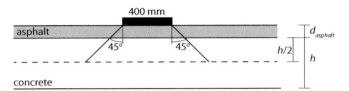

Figure 5.4 Distribution of wheel print to center of cross section, applied to a concrete bridge, showing the tire contact area from NEN-EN 1991–2:2003. Conversion: 1 mm = 0.04 in.

considered. An approximation for finding the resulting wheel print at the middle of the cross section is to use a vertical distribution under 45°, as shown in Figure 5.4.

5.4.2 Assessment calculations

For new bridges, the design calculations are available, and the preparation before the load test is limited to reading out the expected structural responses from the analytical model used for the design. For existing bridges, two types of preliminary calculations should be carried out: assessment calculations to evaluate if the bridge fulfills the code requirements, and calculations to predict the behavior during the test and the bridge's maximum capacity. The assessment calculations are based on load and resistance factors and characteristic material properties, whereas the calculations to predict the behavior are based on average values for the material properties and take all load and resistance factors equal to one.

The assessment calculations can be carried out with a combination of finite element models, hand calculations, or spreadsheets that automate hand calculations. These tools should be developed as part of the preparation stage of the load test. After the test, these tools can be reused to develop the improved assessment of the bridge, taking into account the field measurements.

The assessment calculations are carried out based on a Unity Check or a Rating Factor. The Unity Check is the ratio of the factored load effect as caused by the load combination to the factored resistance:

$$UC = \frac{\text{Load effect due to factored load combination}}{\text{Factored capacity}} \quad (5.1)$$

If the Unity Check is larger than one, it is concluded that the bridge does not fulfill the code requirements. Unity Checks can be calculated based on simple hand calculations, which are typically more conservative than the Unity Checks that result from using refined finite element models (Shu et al., 2015; Lantsoght et al., 2017c).

The Rating Factor (AASHTO, 2016) gives the available capacity for live loads. If the Rating Factor is smaller than one, the available capacity is insufficient. Just as for the Unity Checks, different approaches can be used, from fast conservative hand calculations to more time-consuming refined finite element models. The Rating Factor *RF* for LRFR is calculated as follows according to the *Manual for Bridge Evaluation* (AASHTO,

2016), equation 6A.4.2.1–1:

$$RF = \frac{C - (\gamma_{DC})(DC) - (\gamma_{DW})(DW) \pm (\gamma_P)(P)}{(\gamma_{LL})(LL + IM)} \tag{5.2}$$

with, for the Strength Limit States:

$$C = \varphi_c \varphi_s \varphi R_n \quad \text{with} \quad \varphi_c \varphi_s \geq 0.85 \tag{5.3}$$

and for the serviceability limit state:

$$C = f_R \tag{5.4}$$

where:
RF = rating factor
C = capacity
f_R = allowable stress specified in the LRFD Code (AASHTO, 2015)
R_n = nominal member resistance (as inspected)
DC = dead load effect due to structural components and attachments
DW = dead load effect due to wearing surface and utilities
P = permanent loads other than dead loads
LL = live-load effect
IM = dynamic load allowance
γ_{DC} = LRFD load factor for structural components and attachments
γ_{DW} = LRFD load factor for wearing surfaces and utilities
γ_P = LRFD load factor for permanent loads other than dead load = 1.0
γ_{LL} = evaluation live-load factor
ϕ_c = condition factor
ϕ_s = system factor
ϕ = LRFD resistance factor

The load factors can be found in table 6A.4.2.2–1 in the *Manual for Bridge Evaluation*. The condition factor depends on the observed deterioration and is given in table 6A.4.2.3–1 of the *Manual for Bridge Evaluation*. The system factor considers redundancy in the structural system and is given in table 6A.4.2.4–1 of the *Manual for Bridge Evaluation*. A bridge with less redundancy is more failure critical, which is represented by a lower rating.

For concrete bridges, if the load effect is determined based on a finite element calculation, the effect can be averaged over a certain width in the transverse direction. For bending moment in reinforced concrete slab bridges, the peak can be averaged over 3 m (9.8 ft) in the transverse direction, or another width that corresponds to local practice (Lantsoght et al., 2017a). For shear in reinforced concrete slab bridges, the peak shear stress can be averaged over $4d_l$, with d_l the effective depth to the longitudinal reinforcement (Lantsoght et al., 2017b). An example is shown in Figure 5.5. For other types of structures and materials, the local rules of thumb can be followed, or a sensitivity study can be carried out to identify a suitable transverse distribution of the peak of the load effect.

A steel bridge is considered fracture critical when it contains fracture-critical details. A concrete bridge is considered shear critical when the Unity Check for shear is larger than for bending moment, or when the Rating Factor for shear is smaller than for bending moment. Special precautions should be taken for proof load testing of shear-critical or

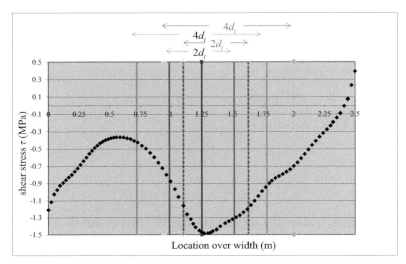

Figure 5.5 Example of distribution of shear stresses in transverse direction in reinforced concrete slab tested in the laboratory with seven bearings for the support. Due to small imperfections, the slab did not rest on all bearings at the beginning of the test, resulting in an asymmetric stress profile. Conversion: 1 m = 3.3 ft, 1 MPa = 145 psi

Source: Adapted from Lantsoght et al. (2017b).

fracture-critical structures as high loads are involved. For diagnostic load testing, the applied loads are smaller and the involved risk is smaller, but the fact that the structure is fracture or shear critical should be taken into account during the preparation stage, and critical responses should be monitored during the test.

All information from the technical inspection should be taken into account for the assessment calculations. If material parameters have been determined from sample tests, the resulting characteristic values can be used. If section losses have occurred, the reduced areas and moments of inertia should be used for the assessment.

When assessment calculations are available with the documentation of the bridge under study, the assumptions made for these calculations should be verified. It is not sufficient to simply report the resulting Unity Check or Rating Factor from the available reports. The assumptions that were made, for example, with regard to section losses or restraint of bearings, should be reported, and it should be evaluated if these assumptions are still valid or if new assessment calculations should be added.

When in the available documentation with reported Unity Checks or Rating Factors the assumptions are not given, the assessment calculations should be repeated. It is of the utmost importance to report and document the assumptions that lie at the basis of the available assessment calculations. When these calculations and load tests are used to make official decisions with regard to posting, strengthening, or closing an existing bridge, all steps that lie at the basis of this decision should be reported and available for reference in the future.

5.4.3 *Estimation of bridge behavior during load test*

The second type of calculations that need to be carried out prior to a load test are the calculations that are used to estimate the bridge behavior during the load test. These calculations should be based on average (measured) material parameters and capacity expressions without the resistance factor. The calculations should represent the actual situation of the bridge as closely as possible. For new bridges, the average material properties can be used as input in the analytical models used for the design, and the resulting expected response can be obtained. For existing bridges, section losses, reduction of the capacity caused by material damage or degradation, and other factors that may have been identified during the technical inspection should be taken into account. When additional load-carrying mechanisms that can increase the capacity of the structure are expected, the influence of these mechanisms can be explored through sensitivity analyses. Moreover, the calculations should identify the expected failure mechanism and associated maximum load.

For truss bridges, the maximum load should be determined by evaluating the tensile and compressive strength of the critical truss member. The maximum load in compression should consider the structural stability of the member.

For arch bridges, the maximum compression in the arch should be determined based on the compressive strength as well as the structural stability of the arch. If the bridge is a tied arch, the maximum capacity of the deck should be determined by studying the bending moment capacity, shear capacity, and tensile capacity. For the bending moment capacity, the interaction with the occurring tension should be considered. For the shear capacity of a concrete deck, the reduction in capacity caused by the applied tension should be considered.

For steel girder bridges, in general the capacity in bending, shear, and normal force should be calculated, and all stability-related failure modes should be verified, to identify the most likely failure mode and its associated maximum load. The effect of fatigue and resulting reduction of the capacity should be considered.

For concrete bridges, the capacity in bending moment, and shear and punching should be verified. If elements are subjected to normal forces, the interaction of these forces to bending moment, shear and punching should be considered. For slender compression members, the maximum load taking into account structural stability issues should be determined. For concrete girder and slab bridges, the bending moment capacity can be determined for a cross section at midspan and a cross section over the support. If the height of the cross section is variable and/or the reinforcement layout changes throughout the span, a representative number of sections should be checked. For the studied sections, it is recommended to determine the following capacities without safety factors on forces and materials:

- Ultimate bending moment resistance based on average material properties;
- Moment-curvature diagram based on average material properties;
- Maximum applied load from vehicle or test tandem that causes yielding of the tension reinforcement;
- Maximum applied load from vehicle or test tandem that causes flexural failure;
- Shear resistance based on average material properties;
- Maximum applied load from vehicle or test tandem that causes shear failure;
- Punching shear resistance based on average material properties;
- Maximum applied load from vehicle or test tandem that causes a punching shear failure;
- Load-displacement diagram based on average material properties, which can be used for field test verification;
- Initiation of the first cracks based on the cracking moment;
- Thresholds related to the stop criteria, for proof load tests for failure modes and bridge types where no codified stop criteria are available.

5.4.4 Shear capacity considerations

Concrete shear failure is often described as a brittle and unwarned failure mode where only limited deformations occur before initiation of the critical shear crack. Consequently, special attention and thus considerations with regard to this failure mechanism should be undertaken before loading of the bridge. This means that the shear capacity has to be evaluated as accurately as possible. If mechanisms are available that can increase the shear capacity as compared to the code-prescribed capacity, the effect of these mechanisms should be studied. For example, if plain bars are used, or if transverse redistribution can occur as in slab bridges, the expected increase in the shear capacity should be considered and the maximum load without and with the capacity-increasing effect should be reported.

For the punching shear capacity, the critical position of the loading vehicle or test tandem should be considered that results in the lowest punching shear capacity. The difference between the punching capacity with three or four sides needs to be explored, both for a single wheel print, two wheel prints (Fig. 5.6), and the entire tandem. If a loading tandem is placed in the first lane, the punching perimeter with four sides should be compared to the punching perimeter with three sides (see Fig. 5.7), and the smallest perimeter length should be used for the calculations. For slabs with a small thickness, such as the deck slabs of girder bridges, punching shear can be the governing failure mode. For these cases, the effect of compressive membrane action should be explored and the maximum loads without and with this capacity-increasing effect should be reported (Amir et al., 2016).

If models are available to evaluate the expected capacity of a certain bridge type, but these models are not reported in the codes, then the code-prescribed capacities as well as the capacity determined from the specific model should be determined, reported, and compared. For example, for reinforced concrete slab bridges, a plasticity-based model (the Extended Strip Model [Lantsoght et al., 2017d]) can be used to determine the maximum load that is expected to cause failure, considering the interaction between two-way flexure and one-way shear.

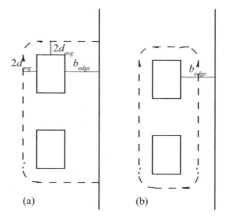

Figure 5.6 Punching of two wheels: (a) perimeter with three sides; (b) perimeter with four sides. d_{avg} is the average of the effective depths to the x- and y-direction flexural reinforcement, and b_{edge} is the edge distance

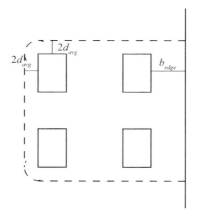

Figure 5.7 Punching of the entire tandem, showing perimeter with three sides. d_{avg} is the average of the effective depths to the x- and y-direction flexural reinforcement, and b_{edge} is the edge distance

If discussion arises during the preparation for a proof load test with regard to the maximum load that the bridge can carry and the expected failure mode, nonlinear finite element models can be used to explore the effect of different assumptions (Rijkswaterstaat, 2017). However, it must be remarked that often the goal of a field test is to remove certain uncertainties that cannot be directly covered by analytical models. Whether or not it is cost-effective to develop a nonlinear finite element model as part of the preparation depends on the test objectives and the type of load test under preparation. After the test, this model can then be updated with the field observations, where extrapolations to higher load levels can be the outcome (Lantsoght et al., 2018).

The used theoretical evaluations should provide a support to the identification of the critical areas of the structure and serve as a method to estimate placing of the monitoring equipment as well as support safety measures when planning the bridge test.

5.5 Planning and preparation of load test

5.5.1 *Planning*

To prepare for a successful execution of a load test, it is important to plan the on-site activities beforehand, as shown in Figure 5.8. All steps that form part of the load test on site, such as applying the sensors, preparing the load, marking positions or driving paths, the actual load test, and dismantling all equipment, should be considered in this detailed planning. Examples of actions that can be included in the detailed planning can be:

- Collecting of sensors, wires, logging gear, and other required equipment.
- Verification of the bridge site closing time or related closing and rerouting schedule.
- Transportation of sensors, other required equipment, and personnel to the test site.
- If no scaffolding is constructed, it should be determined for which operations a manlift is necessary, which then will help to determine the required number of manlifts on site.

112 Load Testing of Bridges

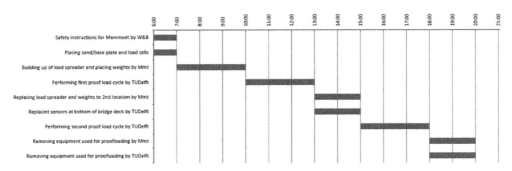

Figure 5.8 Example of time schedule for planning, showing one day

- Marking a grid on the structural members for reference. This grid may already be available from the detailed technical inspection, where it was used for drawing the map of damage and deterioration.
- Marking the positions of the sensors.
- Marking the critical positions of the wheel prints when jacks are used to apply the load, or the driving path when loading trucks are used to apply the load.
- Applying the sensors to their positions required during the test.
- Testing if the sensors are functioning correctly.
- Delivering the required loading equipment to the site, building the setup for applying the load, and applying the loading to its position as used during the test, when a system with jacks and counterweights is used.
- Coordination of bringing trucks, drivers, and weights to the test location, and weighing the trucks, when loading trucks are used.
- Execution of the load test according to a prescribed loading protocol.
- If another span is tested, changing the sensors to this span. If a test setup is built to apply the load, this setup should be moved to the next span that is to be tested as well.
- Removal of the loading equipment from the tested span and from the bridge site.
- Removal of the applied sensors.
- Transportation of sensors, other equipment, and personnel away from the test site.

It should be identified who will be responsible for which task and which equipment is necessary. All potentially dangerous activities should be carried out by at least two individuals. Examples of such activities include handling of loading equipment, handling of ballast, applying measurement equipment from an elevated surface such as a ladder or in a manlift, hoisting, lifting, climbing, and activities close to water or moving traffic. The detailed planning is necessary for streamlining the activities on site, and should be discussed with all personnel that will be involved with these actions. This detailed planning can only be developed after the technical inspection so that site limitations or changes to the structure can be accounted for.

An overall planning, outlining the larger actions that need to be carried out on site, should be prepared and included in the preparation report. In the general preparation for the load test, it must also be discussed with the owner of the bridge and/or client when a preparation report including the planning is expected, for how much time the bridge can be available for testing, and how much time after the test the final report should be delivered. It is good practice to deliver the preparation report including the planning to the bridge owner and/or client no later than five working days before the start of the activities on site. This report should also be communicated to all parties involved and should be accepted by all parties involved with the execution of the test.

5.5.2 *Personnel requirements*

Depending on the governing codes, there may be requirements stated with regard to the personnel. For example, ACI 437.2M-13 (ACI Committee 437, 2013) requires that a licensed professional be responsible for the preparation of the load test and for decision-making on site during the test. The *Manual for Bridge Rating through Load Testing* (NCHRP, 1998) stipulates that a qualified bridge engineer should be responsible for the planning and execution of the load test, and requires that the engineer has experience in testing and instrumentation and field investigations, and possesses adequate knowledge of bridge structural behavior. In addition, this manual requires that adequate staff be available to perform the load test, to provide traffic control during the test, and to assist in evaluating the results.

In most countries, current practice for the execution of load tests requires that the planning be developed by an engineer. During proof load tests, the stop criteria need to be checked by the responsible engineer based on real-time measurements (Fig. 5.9). The engineer needs to make the decision with regard to loading to the next load level, or should decide that further loading could result in permanent damage to the structure and that the test should be terminated prior to reaching the target proof load.

In addition to the requirements with regard to the responsible engineer, it is good practice to have one individual on site who is a certified safety engineer and one individual who is trained in first aid. As mentioned in Section 5.5.1, for all potentially dangerous activities sufficient personnel should be present to be able to work safely. The required number of personnel and their task division should be identified in the detailed planning. For the safety of the traveling public, personnel should be available to guide the traffic, or a detour should be determined and signposted.

5.5.3 *Loading requirements*

The loading setup relates to the type of load test that will be carried out and its objectives. If a diagnostic load test will be carried out, the applied load should result in measurable responses to be used for comparison to and updating of the analytical model. If a proof load test will be carried out, the applied load should be representative of the factored load combination, so that it can be experimentally shown that the bridge fulfills the code requirements.

For a diagnostic load test, the magnitude of the load that results in measurable responses can only be determined after the technical inspection and preliminary calculations have been carried out. These preliminary calculations and the developed analytical model are necessary to determine the expected response under a certain magnitude of applied load. If the behavior is expected to change for different load levels (e.g. as a result of unintended composite action), the applied load should be representative of the heaviest service load. This load level is then required to ensure that the measured responses and the updating

Figure 5.9 Measurement engineers following measurements in real time during a proof load test
Source: Photo by M. Roossen, used with permission.

of the analytical model are in line with the responses at the load level that governs for rating and assessment.

For a proof load test, the required target proof load and resulting loading requirements can also only be determined after the technical inspection and preliminary calculations. The analytical element model that is developed as part of the preliminary calculations will be used to identify the target proof load. This target proof load should represent the factored live load on the bridge, and the equivalence between the loads can be determined based on sectional moments or forces.

Depending on the type of test, different methods can be used for applying the load. The most common methods are:

- The use of dead weights that are applied to the bridge directly (Fig. 5.10a).
- The use of a loading frame or other type of test setup and hydraulic jacks (Fig. 5.10b).
- The use of test vehicles: dump trucks or specifically designed trucks for load tests (Fig. 5.11).

Examples of the use of dead weights in a load tested building are shown in Figures 5.12 and 5.13. Dead loads are more commonly used for load testing of buildings than for bridges.

Preparation of Load Tests 115

a) b)

Figure 5.10 Load application methods: (a) dead weights on bridge and (b) application of loading rig with hydraulic jacks

Figure 5.11 Load application with dump trucks

When a loading vehicle is used, the choice of the vehicle (in terms of axle layout and load capacity) should be based on the following considerations:

- The loading vehicle should be representative of the load used for the design of the bridge (for a new bridge) or the load used for the assessment of the bridge (for an existing bridge).
- If the bridge should be rated for permit vehicles, the loading vehicle should be representative of this load and vehicle type.
- The loading vehicle should be similar to the heavy vehicles that are expected to use the bridge (for new bridges) or that are using or will be using the bridge (for existing bridges) in terms of weight and axle configuration.
- The loading vehicle should be able to induce the critical state of stress in the element that needs to be verified or evaluated.
- The loading vehicle should also be selected based on its availability within a reasonable distance from the testing site and based on practical considerations regarding the execution of the load application and associated costs.

Different controlling failure modes, as may be tested for in a proof load test, may require different vehicles. An example, showing a special load testing vehicle, is given in Figure 5.14.

116 Load Testing of Bridges

Figure 5.12 Example of application of dead weight with water
Source: Photograph by S. Camino. Printed with permission.

The loading requirements for proof load testing are stricter, as higher load levels are applied during proof load testing. The requirements then have as their goal, to ensure a safe execution of the load test. To maintain structural safety during a proof load test, it is imperative that the load can be offloaded rapidly when large deformations occur (e.g. when a stop criterion is exceeded). As such, the application of dead load alone is not recommended for proof load tests on bridges, as this loading method may result in collapse if the structure is weakened due to undesirable structural damage as well as large deformations. An added disadvantage of the use of dead load is that when deflections occur in the tested structure, arching action can develop in the loads, resulting in effectively lower loading in the structure as the load is carried through the arch to the supports. This disadvantage can be mitigated by using water in soft basins as a load application method. When a system with hydraulic jacks and a loading frame is used, see Figures 5.10b and 5.15, large deformations will result in the deactivation of the hydraulic jacks, so that the structure is not loaded anymore and the load of the applied counterweights is carried directly to the supports.

For proof load testing, a cyclic loading protocol with a number of load cycles for each load level is recommended, so that the stop criteria can be checked after each load cycle, and so that the linearity and reproducibility of the results can be checked after each load

Preparation of Load Tests 117

Figure 5.13 Example of application of dead weight with cement bags
Source: Photograph by S. Camino. Printed with permission.

cycle. Therefore, a requirement to the loading method is that the loading protocol can be applied in the test. This protocol can be achieved with loading vehicles (Steffens et al., 2001; Bretschneider et al., 2012) and a loading frame with hydraulic jacks, see Figure 5.15. Moreover, to make sure that all loading cycles at the same load level are comparable, it is necessary that the applied loading is executed at a prescribed loading speed. This requirement can be achieved by using loading vehicles driving at a constant (crawl) speed or by using hydraulic jacks with a prescribed loading speed.

5.5.4 Traffic control and safety

For all activities on site, all aspects of safety should be considered and checked for by the responsible safety engineer. All possibly hazardous situations and possible problems during the on-site activities should be reported in a risk analysis as part of the preparatory report. This analysis can only be carried out after the technical inspection of the bridge and bridge site, during which attention should be paid to the potentially hazardous situations. Possible

Figure 5.14 Example of application of load with load testing vehicle
Source: Photograph by D. Hordijk. Printed with permission.

problems that can occur during the activities on site should be considered, and a solution or backup plan should be thought through before the test. Possible problems include mechanical failures, electrical and electronical failures, sickness of personnel, and external conditions such as bad weather.

The specific tasks of the safety engineer depend on national practice and codes. In the Netherlands, the safety engineer is responsible for the safety card, the safety briefing, inspection of on-site safety, and preparation of a report regarding safety for all parties involved prior to the test. The safety card gives a brief overview of the actions that need to be taken in case of an accident, fire, or other calamity, and contains the phone numbers and addresses of the emergency services, police, fire station, nearest hospital, nearest doctor, and nearest pharmacist. The safety briefing is required for all personnel prior to starting their on-site activities to review the basic safety principles and dangers related to working on site. For some projects, a safety certificate is required for all personnel involved.

During all on-site activities, the safety of the traveling public and possible local spectators should be safeguarded. The responsibility for the safety of the traveling public and local spectators lies with the parties executing the load test, not with the traveling public and local spectators themselves. A traffic control plan and, if the bridge is closed during the load test, a detour should be developed together with the local road authorities. An example of a temporary traffic situation during a load test is shown in Figure 5.16.

The last aspect of safety that should be ensured during all on-site activities is structural safety. For this purpose, thorough preparations of the load test are recommended together with adequate instrumentation of the structure. Since load tests are sometimes used to

Figure 5.15 Use of counterweights, steel spreader beam, and a single hydraulic jack

answer questions with regard to behavior of the structure, thorough preparations prior to the test can sometimes only give an indication of the expected behavior. For this reason, varying possible effects (restraint at bearings, unintended composite action, etc.) in the finite element models during the preparation stage can give a possible range of expected responses. To make sure the responses during the test lie within the expected range, or to find an explanation for responses that lie outside of this range, the responses should be measured. Extensive finite element models prior to load tests are not always cost-effective, and the decision with regard to modeling depends on the bridge type and test objectives. When such calculations are not carried out, the measurements become even more important. The sensor output should be followed and evaluated during the load test. For proof load testing, the measured responses also serve the purpose of verifying the stop criteria. As such, the development of the sensor plan is a part of the preparations for the test and the preparations to ensure safety during the test.

5.5.5 Measurements and sensor plan

An important part of the preparation for a load test is the development of the sensor plan, which should show the position, type (Ettouney and Alampalli, 2012), and range of the selected sensors and give insight in the data collection scheme by including the data logger type, sampling rates, test controls, and data transmission means, as well as the

120 Load Testing of Bridges

Figure 5.16 Overview of traffic situation during a load test. From left to right: one lane open for traffic, one lane used for load testing, temporary bike bridge

Source: Photograph by S. Fennis. Printed with permission.

mounting and wiring details, where relevant. An example sensor plan (not showing the data collection system and wiring details) is given in Figure 5.17. The preparation report should also include a detailed list explaining the requirements per sensor and a justification for its inclusion in the sensor plan.

The first step in developing a sensor plan is to select the suitable sensors. The required range of the sensors should be selected based on the preliminary calculations. For materials with time-dependent behavior such as concrete, the effect of time-dependent behavior on the responses should be considered. For concrete, the structural response also differs depending on whether the cross section is uncracked or cracked. An extra margin for the structural responses should be taken into consideration so that no sensor runs out of its measurement range when larger structural responses than expected occur during the test. Moreover, the accuracy of the sensor or chosen measurement technique should be sufficient for the load test. Where small variations may be critical, the selected sensor should be able to capture these responses. Additionally, the sampling rate of the sensor or selected measurement technique (when this sensor or technique is digital) should be selected, taking the loading speed and expected speed of change in responses in consideration (i.e. for dynamic load testing, a faster sampling speed is required than for a static load test with a monotonic loading protocol during which the load is kept constant for 24 hours). All sensors should be calibrated before each load test, and their correct conversion factors should be updated in the data logging and visualization software. Finally, all deployed sensors should be suitable for field testing and outdoor conditions, and their operation should be affected as little as possible by (changes in) environmental conditions (monitoring equipment is often developed for laboratory conditions). Examples of limitations to sensors in field conditions include difficulties with laser sensors during rain (the rain drops reflect the laser beam), as well as challenges

Preparation of Load Tests 121

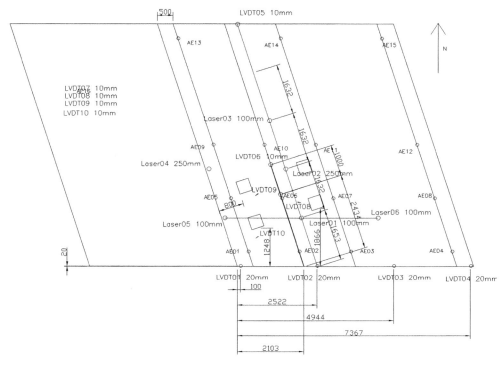

Figure 5.17 Example of sensor plan, showing position of loading tandem, position and range of lasers and LVDTs to determine deflection profiles, acoustic emissions sensors and additional LVDTs for monitoring crack width of cracks that are selected on site

Source: Modified from Lantsoght et al. (2017e).

using photogrammetry and digital image correlation (DIC) calibration when the camera lens fogs over or when there is limited daylight. Besides the possible effects of temperature and humidity, which will be discussed later in this section, the proper functioning of the sensor should depend as little as possible on the environmental conditions. Any external influence in the sensor results should be justified and quantified in regards to the precision before test initiation.

A requirement for the data acquisition equipment is that its sampling rate should correspond to the loading speed and expected speed of change in responses. When the structural responses should be followed in real time during the experiment, analog to digital conversion and data analysis software should be present and running during the experiment. For proof load testing, the structural responses must always be followed in real time so that the stop criteria can be checked.

The sensor plan should be developed in such a way that the critical structural responses can be followed during the load test and that the required information about the structural behavior can be gathered during the test. The sensor plan should be developed keeping in mind the requirements for monitoring during the test and the required information to meet the objectives of the test for the analysis afterwards. For a diagnostic load, the measurements are followed during the test to check if the structural responses are within the expected range, and the responses are used after the test to update the analytical model that was used for design or assessment of the bridge. For a proof load test, the measurements are followed during the test to check if the structural responses are within the expected range and to verify if no irreversible damage occurs. The latter criterion is expressed based on the stop criteria. These stop criteria need to be determined prior to the proof load test, and the acceptable threshold values, which are determined based on the expected failure mode and preliminary calculations, should be agreed upon prior to the test.

Parameters that are typically measured during a load test are:

Local measurements

- The linear strains at locations of interest can be measured. For example, for concrete bridges the strain on the bottom of the cross section can be followed for the verification of the stop criteria. For steel girder bridges the longitudinal strain on the bottom surface at midspan can be monitored to follow the maximum strains when loading trucks drive over the bridge; the strains can be determined over the height of the girder to find the position of the neutral axis and evaluate if unintended composite action occurs; or strains in the transverse direction can be used to find the transverse distribution.

Exterior and global measurements

- The deflections at a number of positions or at the critical position can be measured. When only the critical position is measured, the load-displacement diagram can be analyzed, which gives an idea of the overall stiffness of the structure and may indicate the onset of nonlinear behavior. When the deflections at a number of positions are measured, the deflection profiles in the longitudinal and/or transverse directions can be developed. These profiles give an indication of the overall structural behavior, onset of nonlinear behavior, (re)distribution, settlements, and so forth.
- The surface deformations can be monitored to estimate strains over a certain length or surface or to cover cracking over a certain length or surface.
- The rotations at the supports can be measured to get an idea of the real boundary conditions at the supports. As such, these measurements can give an idea of the restraint magnitude occurring in joints and/or bearings.
- For concrete bridges, existing cracks can be monitored during the test and the opening and closing of the crack during the test can be measured. If a selected critical crack is monitored, there may be discussion about the selection of this crack. A solution to this problem is the use of non-contact measurements that can follow all cracks in a pre-determined area.
- In dynamic load tests, it is necessary to measure accelerations.
- The reference strains outside the loaded member/span can be monitored to measure the effect of temperature and humidity and find the net structural response.
- The compression of the supports for bridges with elastomeric bearings should be measured to find the net deflections.
- The settlements of the substructure need to be monitored when significant settlements are expected and/or a stop criterion regarding the substructure is defined.

- The compression and expansion at the joints needs to be monitored when limited space in the joints is observed during the technical inspection.
- The environmental conditions during the test should be measured: the ambient temperature, wind speed (if relevant), and humidity during the test. Additionally, the surface temperature of the bridge at relevant positions can be measured.

Interior measurements

- Acoustic emissions (see Fig. 5.18) can be monitored during load testing to obtain information about changes in structural behavior before these changes can be observed with the naked eye, such as for concrete cracking.
- Optical fibers provide a promising strain measurement method. These sensors can be cast into the structure during concrete casting or adhesively bonded into a surface slit on existing bridges. They can cover larger distances than traditional localized strain measurements.

When critical situations or elements were determined during the inspection as part of the preparatory stage, the corresponding structural response should be measured during the load test, and the acceptable threshold values should be determined prior to the load test. Depending on the goals of the load test and the type of critical situation or element, the responses should be monitored in real time during the test or checked visually during the test at certain time intervals.

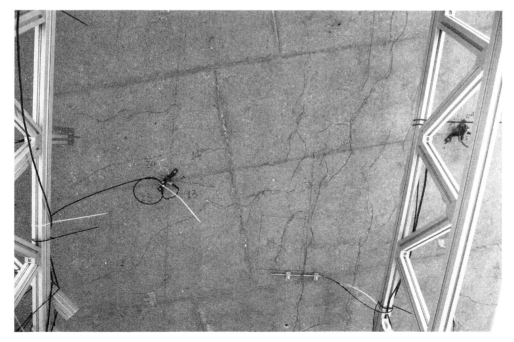

Figure 5.18 Application of acoustic emissions sensors
Source: Photograph by Y. Yang, used with permission.

124 Load Testing of Bridges

The effect of the ambient conditions on the sensors during a load test should be taken into account. Changes in temperature and humidity affect the measured structural responses in two ways:

1   The measured structural response includes the structural response due to the changes in temperature and humidity.
2   The sensor may behave differently when the ambient conditions change, which can result in unwanted or erroneous measurements.

For the first case, the net structural response due to the applied load should be filtered out and used for the analysis of the test. The structural response due to changes in temperature and humidity should be subtracted from the measured output. For this purpose, sensors can be applied outside of the tested region so that only the effect of changes in temperature and humidity be measured. An example of such a reference sensor, as part of a sensor plan, is shown in Figure 5.19. Another solution would be taking measurements for a certain amount of time prior to the load application to follow the structural responses as a function of the changing ambient conditions (Zhou et al., 2007). However, if the bridge is open to traffic

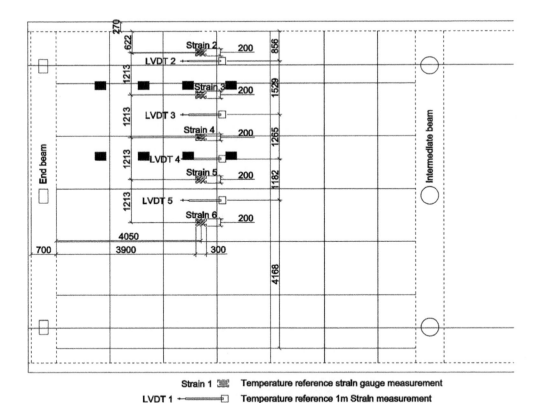

Figure 5.19 Example sensor plan, indicating LVDT1 as a reference strain measurement to compensate for the effects of temperature and humidity on the strain measurements

Preparation of Load Tests 125

and traffic is busy throughout the day, a clean registration of the baseline response will be difficult. For the second case, the inherent error in the sensor due to changes in temperature and humidity, the provided documentation of the sensor fabricator should be consulted. For strain gages in particular, temperature-compensating gages may be used.

The magnitude of the effect of temperature and humidity on the structural members that are tested and on the sensors depends on the conditions of the load test. If the test is short in duration and carried out during the night, the effect will be smaller than when the test duration is longer and carried out during a sunny day, when the temperatures rise steadily.

When all sensors are selected, the sensor plan can be developed. Since the particularities for developing a sensor plan for a diagnostic load test and a proof load test are quite different, this section focuses on the general requirements to the measurements and sensors, whereas Part III gives guidance for the development of a sensor plan for a diagnostic load test and Part I of Vol. 13 for a proof load test. The sensor plan can only be developed after the technical inspection of the bridge and its location. Site limitations or restrictions to the site access may prohibit the application of certain sensor types or positions. In some cases where access is severely restricted, only non-contact sensors may be applied.

In general terms, a sensor plan contains or can contain the following information:

- Type, position, and range of all sensors that will be applied;
- Additional sensors with type and range of which the position will be applied in the field (e.g. for monitoring crack width in concrete bridges);
- Number, type, and range of backup sensors that need to be taken to the field;
- The required elements for data acquisition and real-time monitoring: data logger (and list identifying which sensor is connected to which canal), amplifier, analog-to-digital converter, and data visualization software;
- The properties of the different data acquisition and visualization elements, including the sampling rates;
- A list with calibration values of all sensors and the date of the most recent calibration;
- Details of sensor mounting for contact sensors;
- Wiring details for wired sensors.

5.6 Summary and conclusions

This chapter discussed the preparatory steps that are required prior to a load test to ensure the safety during the test and an efficient use of time on site. It is good practice to report the observations, choices, and calculations that were carried out during the preparation stage in a report that should be communicated to the bridge owner and/or client prior to the start of the on-site activities, as well as to all parties involved.

Typically, the preparatory report summarizes the following elements:

- Statement of the test objectives, and a discussion of how these objectives will be met with the load test.
- Observations from the technical inspection of the bridge, documented with photographs and summarized in maps of deterioration and damage. If there are limitations on the test site with regard to the application of sensors or the load, and when possibly

126 Load Testing of Bridges

hazardous situations are observed, these should be reported as well. Critical structural elements or conditions observed during the inspection that should be monitored during the test should be discussed and documented as well.

- Preliminary calculations to prepare for the test: a summary of the design calculations for new bridges and assessment calculations for existing bridges, and a prediction of the expected structural responses during the test.
- An overview of the decisions with regard to the practical aspects during the load test: planning, personnel requirements, loading requirements, regulations with regard to safety, traffic control, sensor plan, and data acquisition and visualization requirements.

The test objectives determine in the first place if a load test is the correct method for meeting the objectives. Once it is decided that a load test is the right means to meet the objectives, and the type of load test is determined, the specific goals for the test should be identified. These goals are the basis for determining how the test will be carried out, what will be measured, and which post-processing and analysis is required.

Before any calculations and planning can be done, a technical visit to and inspection of the bridge and the bridge site should be carried out. This inspection is the first step in preparing for the test. The condition of the bridge should be evaluated and reported, and the site limitations that may affect the way in which the load test is executed should be documented and considered for the following calculations.

To prepare for the load test, preliminary calculations should be carried out. Depending on the test objectives, uncertainties, and available time, a finite element model of the bridge can be used for these calculations. It should be evaluated if a finite element model is cost-effective for the project. For new bridges, such models can be available from the design calculations. For existing bridges, such models can be built taking into account the material deterioration (resulting in section losses, for example) and restraints at the supports. The assessment calculations can then be carried out based on these models. Then, the expected structural response during the experiment should be determined by using average (measured) material parameters and omitting all load and resistance factors. Where available, methods that have been developed based on laboratory experiments can be used to give a refined prediction of the expected behavior. When the uncertainties on the structure are large, the structural response for ranges of parameters with uncertainties can be explored. The expected behavior is required to determine the positions where the structural responses should be measured during the test and the expected magnitude of the response, so that a suitable sensor type can be selected.

Once the condition of the bridge and site are known and the expected behavior is calculated, the practical preparations can be carried out. A detailed planning of all activities on site is necessary to streamline the activities. The required personnel, their qualifications, and assigned tasks during the test should be considered. Depending on the objectives of the test, the loading requirements can be determined. The method of load application should be determined (dead load, trucks, hydraulic jacks in a loading frame), and the critical position or wheel path during the test should be determined and reported on drawings. The safety of the bridge, personnel, and traveling public should be considered during the preparation stage. Where necessary, detours should be coordinated with the local road authority. Based on the goals of the test and the expected responses, the adequate sensor type, range, and sampling rate, sensor location, wiring, mounting, data acquisition, and data visualization (when required) should be selected.

After this selection, a drawing showing these details, called the sensor plan, should be developed and added to the report.

References

AASHTO (2015) *AASHTO LRFD Bridge Design Specifications, 7th Edition with 2015 Interim Specifications*. American Association of State Highway and Transportation Officials, Washington, DC.

AASHTO (2016) *The Manual for Bridge Evaluation with 2016 Interim Revisions*. American Association of State Highway and Transportation Officials, Washington, DC.

ACI Committee 342 (2016) ACI 342R-16: Report on flexural live load distribution methods for evaluating existing bridges. American Concrete Institute, Farmington Hills, MI.

ACI Committee 437 (2013) Code Requirements for Load Testing of Existing Concrete Structures (ACI 437.2M-13) and Commentary. Farmington Hills, MI.

Amir, S., Van Der Veen, C., Walraven, J. C. & De Boer, A. (2016) Experiments on punching shear behavior of prestressed concrete bridge decks. *ACI Structural Journal*, 113, 627–636.

Bretschneider, N., Fiedler, L., Kapphahn, G. & Slowik, V. (2012) Technical possibilities for load tests of concrete and masonry bridges. *Bautechnik*, 89, 102–110 (in German).

CEN (2003) *Eurocode 1: Actions on Structures: Part 2: Traffic Loads on Bridges, NEN-EN 1991–2:2003*. Comité Européen de Normalisation, Brussels, Belgium.

Ettouney, M. & Alampalli, S. (2012) *Infrastructure Health in Civil Engineering*. CRC Press, Boca Raton, FL.

Fennis, S.A.A.M., Van Hemert, P., Hordijk, D. & De Boer, A. (2014) Proof loading Vlijmen-Oost: Research on assessment method for existing structures (in Dutch). *Cement*, 5, 40–45.

FIB (2012) *Model Code 2010: Final Draft*. International Federation for Structural Concrete, Lausanne.

Koekkoek, R. T., Yang, Y., Fennis, S.A.A.M. & Hordijk, D. A. (2015) Assessment of viaduct vlijmen oost by proof loading. Stevin Report Nr. 25.5–15–10. Delft University of Technology, Delft, the Netherlands.

Lantsoght, E. O. L., Van Der Veen, C., Walraven, J. C. & De Boer, A. (2016) Case study on aggregate interlock capacity for the shear assessment of cracked reinforced-concrete bridge cross sections. *Journal of Bridge Engineering*, 21, 04016004–1–10.

Lantsoght, E. O. L., Van Der Veen, C., Hordijk, D. & De Boer, A. (2017a) Recommendations for proof load testing of reinforced concrete slab bridges. *39th IABSE Symposium: Engineering the Future*. Vancouver, Canada.

Lantsoght, E. O. L., De Boer, A. & Van Der Veen, C. (2017b) Distribution of peak shear stress in finite element models of reinforced concrete slabs. *Engineering Structures*, 148, 571–583.

Lantsoght, E. O. L., De Boer, A. & Van Der Veen, C. (2017c) Levels of approximation for the shear assessment of reinforced concrete slab bridges. *Structural Concrete*, 18, 143–152.

Lantsoght, E. O. L., Van Der Veen, C., De Boer, A. & Alexander, S.D.B. (2017d) Extended strip model for slabs under concentrated loads. *ACI Structural Journal*, 114, 565–574.

Lantsoght, E. O. L., Van Der Veen, C., De Boer, A. & Hordijk, D. A. (2017e) Collapse test and moment capacity of the Ruytenschildt Reinforced Concrete Slab Bridge *Structure and Infrastructure Engineering*, 13, 1130–1145.

Lantsoght, E. O. L., De Boer, A., Van Der Veen, C. & Hordijk, D. A. (2018) Modelling of the proof load test on viaduct De Beek. In: *Euro-C, Computational Modelling of Concrete and Concrete Structures*. Bad Hofgastein, Austria.

NCHRP (1998) Manual for Bridge Rating through Load Testing. Washington, DC.

Rijkswaterstaat (2017) *Guidelines for Nonlinear Finite Element Analysis of Concrete Structures*. Rijkswaterstaat, Utrecht, the Netherlands.

Ryan, T. W., Mann, E.E.Z., Chill, M. & Ott, B. T. (2012) *Bridge Inspector's Reference Manual*. Federal Highway Administration, U.S. Department of Transportation, Washington, DC.

Shu, J., Plos, M., Zandi, K. & Lundgren, K. (2015) A multi-level structural assessment proposal for reinforced concrete bridge deck slabs. *Nordic Concrete Research*, 53–56.

Steffens, K., Opitz, H., Quade, J. & Schwesinger, P. (2001) The loading truck BELFA for loading tests on concrete bridges and sewers (in German). *Bautechnik*, 78, 391–397.

Zhou, Y. E., Cupples, T. H., Brown, R. & Hwang, G. (2007) Re-evaluation of weight limits for six bridges in Montgomery County, Maryland. *International Bridge Conference IBC 2007*. Pittsburgh, PA.

Chapter 6

General Considerations for the Execution of Load Tests

Eva O. L. Lantsoght and Jacob W. Schmidt

Abstract

This chapter discusses the aspects related to the execution of load tests regardless of the chosen type of load test. The main elements required for the execution of the load test are the equipment for applying the load and the equipment for measuring and displaying (if required) the structural responses. This chapter reviews the commonly used equipment for applying the loading and discusses all aspects related to the measurements. The next topic is the practical aspects related to the execution. This topic deals with communication on site during the load test and important safety aspects during a load test.

6.1 Introduction

This chapter introduces the general aspects that all load tests have in common for their execution. Particularities for diagnostic load testing are discussed in Part III and for proof load testing in Part I of Vol. 13. The determination of the target load during the test is different for diagnostic and proof load testing and is discussed in the relevant chapters. This chapter will in particular discuss possible methods for applying the load and the equipment necessary for the presented methods. Depending on the magnitude of the target load and bridge geometry, some load applications will be more suitable than others. It is therefore the aim of this chapter to qualitatively describe which load methods are more suitable for larger load magnitudes.

A second section in this chapter contains the general considerations related to the measurement equipment. The development of the sensor plan is closely related to the goals of the load test and will be different for diagnostic load testing and proof load testing. Therefore, a special focus is on a discussion of the general requirements that the equipment (sensors and data acquisition and visualization equipment) should fulfill for use in a load test.

A final section of this chapter deals with the practical aspects related to the execution of a load test. During the load test, the communication between all parties involved and present in the field is important. The safety during the execution of the load test should also be considered at all times. The safety of the traveling public and the parties involved with the execution of the load test is discussed. The structural safety needs to be monitored during the load test, and the attention that needs to be paid to the structural responses during a load test depends on the magnitude of the load (and as such, on the type of load test). For diagnostic load testing, a verification of the linearity of the responses and comparison to the analytically predicted responses may be sufficient, whereas for proof load testing the structural

DOI: https://doi.org/10.1201/9780429265426

130 Load Testing of Bridges

safety requires checking the stop criteria. These aspects are discussed further in Part III for diagnostic load tests and in Part I of Vol. 13 for proof load tests.

It is good practice to keep the general considerations for the execution of load tests in mind during the preparation stage of the load test. As such, the testing engineers can prepare a planning for the execution to ensure a smooth process in the field.

6.2 Loading equipment

Depending on the goals of the load test, several options are available to apply the load. For static load tests, dead loads, loading vehicles, and a system with hydraulic jacks can be used. For dynamic load tests, vehicles, a group of pedestrians (for pedestrian bridges) or an external source of excitation can be used. The elements important for the preparation are discussed in Chapter 5.

Dead loads are only suitable for static load tests. However, dead loads are not commonly used for bridge load testing, as applying these loads is time-consuming, and arching can occur in the loads which makes the loading of the bridge ineffective. Additionally, the loading method is difficult to control if deformations rapidly increase during testing and could lead to collapse. Consequently, high magnitudes of dead loads do not have an inherent safety against collapse. For proof load testing, the use of a system with hydraulic jacks is to be preferred as high static and cyclic loads can be applied in a controlled manner. Hydraulic jacks can however be combined with dead loading in a desirable way when high magnitude loading is applied as shown in Chapter 5. Dead load, using distributed loading for static load tests, is typically used for proof load testing of buildings, for load testing of the sidewalks of bridges, and for load testing of pedestrian bridges.

Load testing of bridges is mostly done by utilizing loading vehicles (Zhou et al., 2007). Loading vehicles can be used for diagnostic and proof load testing. The advantage of using loading vehicles is that they can easily be placed at different positions and in different configurations, and that the test can easily be extended to include dynamic testing. The disadvantage is that the maximum load that can be applied is limited to what a vehicle can carry. Some special testing vehicles with a larger load capacity (Steffens et al., 2001) have been developed (see Fig. 6.1), and in some cases military vehicles and tanks (Varela-Ortiz et al., 2013) have been used.

The required weight and number of test vehicles needs to be determined in function of the goals of the load test. For a diagnostic load test, the applied loading vehicles should cause load effects in the structure and/or critical structural element that correspond to service load levels. It is important for diagnostic load tests that the weight is large enough and represents the highest service load, so that structural behavior that may change depending on the loading level can be correctly evaluated (e.g. transverse distribution and unintended composite action). Figure 6.2 shows an example of using multiple trucks to carry out a diagnostic load test prior to opening a steel box girder bridge built with the incremental launching technique (Bonifaz et al., 2018).

For a proof load test, the applied loading vehicles should cause load effects in the structure and/or critical structural element that correspond to the effect of the factored loads used in design or assessment. For proof load testing, the applied weight is of the utmost importance, as the magnitude of the target proof load is the measure that is used to directly assess the bridge with the field test. The determination of the weight and number of test vehicles

Figure 6.1 Example of special loading vehicle
Source: Photograph by D. A. Hordijk, used with permission.

thus results from the analysis calculations carried out as part of the preparation stage. It is good practice that this analysis considers the code-prescribed loads used for design (for new bridges) or assessment (for existing bridges), as well as special permit vehicles and other heavy vehicles that are using the bridge or that will use the bridge. Such an analysis during the preparation stage also leads to an indication of how the load compares to the maximum expected live load on the considered bridge.

When vehicles are used for dynamic testing, the structural response can be measured for different driving speeds to determine the dynamic amplification. To determine the damping after impact, a bump can be placed on the bridge deck and the structural response from driving over the bump can be determined.

The use of hydraulic jacks for load testing is limited to proof load tests where loading vehicles cannot provide the axle layout or target load required to investigate the critical stress state and/or failure mode in a given structural element or the bridge structure. A system with hydraulic jacks has the preference for proof load testing of buildings under cyclic loads. For failure testing and research purposes, a system with hydraulic jacks is the preferred approach since such load is controllable and well defined and can reach high magnitudes.

To determine the structural response of pedestrian bridges under moving live loads, a group of pedestrians can be used. These pedestrians are asked to stand on the bridge (static load test), to walk over the bridge, and to jog over the bridge (dynamic load test). Such tests typically are carried out on new bridges prior to opening (Ministerio de Fomento-Direccion General de Carreteras, 1999; Beben and Anigacz, 2014). In some cases, groups of cyclists are also used.

Figure 6.2 Load testing of a new bridge prior to opening with trucks: Villorita Bridge, Quito, Ecuador

Source: Photograph by T. A. Sanchez. Used with permission.

When the natural frequencies of a bridge need to be determined, an external source of excitation can be applied to the bridge to force vibration (Frýba and Pirner, 2001). As such, the forced and free vibration shapes can be determined. For unusual bridge structures and long-span bridges, the frequencies and modes of vibration should be compared to the analytically determined values (Cunha et al., 2016).

6.3 Measurement equipment

6.3.1 Measurement requirements

Typically, structural design and assessment approaches use sectional moments and forces and second-order effect evaluations as a means to evaluate stress states of the structural elements. These values are of great importance when verifying and evaluating a response of the tested structure. In-situ testing can be extremely challenging due to difficulties in assessing the structure, environmental exposure, short time for testing, and so forth. It is difficult to measure sectional moments and forces, and second-order effects directly during a load test. Reaction forces can be measured when load cells are installed at the supports, which may be an option for particular new bridges but would require challenging jacking operations for existing bridges.

Structural responses (strains, deflections, and in concrete bridges, cracking) are typically measured and compared to calculated responses. The preparation stage of a load test includes preliminary calculations to predict the expected structural response during the test for the determined sectional moments and forces. This analysis should be carried out

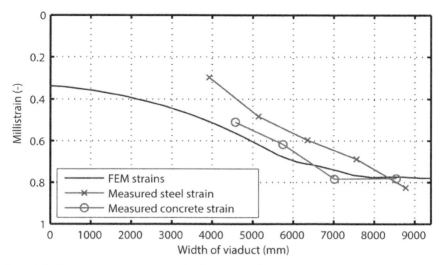

Figure 6.3 Comparison between results from linear finite element model and strains measured during field test. Conversion: 1 mm = 0.04 in.

based on measured average material properties, taking the observed damage and deterioration from the technical inspection into account, and is usually carried out with a linear finite element model (see also Chapter 5). The output from the finite element model can also be used for comparison to the field measurements after the test (see Fig. 6.3). Based on the expected structural response and considering the goals of the load test, the sensor plan can be developed (see also Chapter 5). The expected responses can be used for verification and/or to develop a stop criterion prior to a proof load test.

Since all measurements are zeroed at the beginning of the load test, only the response caused by the applied load is measured. The response under the permanent loads should be calculated and added to the measured response for evaluation and assessment. For newly designed bridges, the response due to the applied load can be directly compared to the expected response based on the analytical models that were used for the design. An example of the loaded area and measured deformations using digital image correlation is shown in Figure 6.4 which indicates the load separation and surface deformations (Schmidt et al., 2018).

6.3.2 Data acquisition and visualization equipment

A data acquisition and visualization system containing amplifiers, data loggers, analog-to-digital converters, and software for data visualization is necessary when the structural responses need to be monitored in real time during the load test. Therefore, the choice for the type and complexity of the data acquisition and visualization equipment should be made based on the type of load test and its goals. For example, for proof load testing, when stop criteria need to be evaluated in real time, real-time data visualization (see Fig. 6.5) is of the utmost importance, and all important parameters and plots showing the relations between the parameters should be preprogrammed prior to the test.

134 Load Testing of Bridges

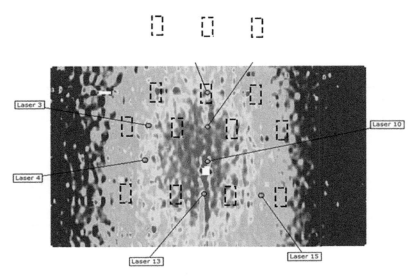

Figure 6.4 Example of the loaded area and measured deformations

For load testing, the data acquisition and visualization system should be suitable for outdoor conditions. Sensitive equipment should be protected against environmental effects. A military-grade computer can be useful for running the data visualization software. The data acquisition equipment can be built into robust weatherproof boxes, stored in robust weatherproof boxes (as shown in Fig. 6.6), or adequately covered. The operation of the data acquisition and visualization system should be independent of the weather conditions and should be guaranteed under conditions such as rain, wind, humidity, and so forth.

The type of load test and its goals drive the sampling rate of the sensors as well as the sampling rate of the data acquisition equipment. When cycles of loading are used to different magnitudes (e.g. in a proof load test when the load is applied with hydraulic jacks), the structural responses during loading and unloading should be captured, which requires a higher sampling rate. During a dynamic load test, the speed of loading and unloading may be even faster, requiring an even higher sampling rate. Conversely, during a monotonic loading protocol when the load is kept constant for a long period of time (e.g. when the standard monotonic loading protocol for proof load testing of concrete buildings according to ACI 437.2M-13 [ACI Committee 437, 2013] is used), a lower sampling rate is sufficient.

The data visualization software requires as input the calibration factors of all sensors. These factors should be determined by calibration prior to the load test. Before the beginning of the planned load test, a test run should be done to check the proper functioning of all sensors and the data acquisition and visualization system. The output should be checked to see if the sensors are registering responses, and if these responses fall within the range of expected values. If this is not the case, an explanation for these observations should be sought and remedied where necessary.

In order to visualize the relation between the structural responses and the applied load or the position of the load, the data acquisition system should gather the output of the applied

General Considerations for Test Execution 135

Figure 6.5 Real-time data visualization as part of a data acquisition and visualization system

sensors as well as the information from the position or magnitude of the load. These values should then be processed together in the data visualization software.

6.3.3 Sensors

A large number of sensors exists (both contact and non-contact sensors). Since the possibilities and tools rapidly change, this section only describes the requirements for the sensors. A recent overview of available sensors can be found in Ettouney and Alampalli (2012). Depending on the goals of the load test, tried and tested sensors (such as strain gages) or cutting-edge measurement techniques can be used. Often, a combination of different

136 Load Testing of Bridges

Figure 6.6 Use of weatherproof boxes for data acquisition system

sensors is used, and when new technologies are explored, the same structural response is often also measured with tried and tested sensors for comparison and evaluation. Examples of using new sensor types and measurement techniques for the practice of load testing are given in Part III of Vol. 13.

In addition to the structural responses that are monitored during the load test and which are discussed in Chapter 5, the applied load can be measured. Whether or not this measurement is required depends on the goals of the load test. When vehicles are used, the position of the load can be measured with trackers and used to compare to the position of the structural responses. The magnitude of the load is then determined before each run of the vehicle or group of vehicles by weighing the trucks. When hydraulic jacks are used to apply a cyclic loading protocol, the magnitude of the load can be measured with load cells. The position of the load is then determined prior to the test and is a single position, and it does not need to be monitored during the load test.

6.3.4 Interpretation of measurements during load test

The importance and required extent of the interpretation of measurements during load tests depends on the goals of the load test. For some types of diagnostic load testing, no real-time visualization of the measurements is required, and therefore no interpretation of these measurements is done. For proof load testing, on the other hand, a crucial aspect for a safe execution of the load test is the stop criteria. These thresholds need to be evaluated during the proof load test. The topic of stop criteria is further discussed in Part I of Vol. 13.

6.4 Practical aspects of execution

6.4.1 Communication

For a safe and timely execution of a load test, communication between all parties involved is crucial. Prior to the load test, all parties involved in the field should meet and talk through the practical details of the test. The workflow processes during the test should be determined in this meeting, written down, distributed to all parties involved, and be visible to all parties during the load test. These descriptions should include the responsible person for the different decisions during the load test and the lines of communication during the load test. During the execution of the load test, all parties involved should be able to communicate, for example by using walkie-talkies and mobile phones (see Fig. 6.7).

When the load is applied with trucks, the required driving path and speed should be clearly communicated to the drivers. When the load is applied with dead loads or hydraulic jacks, the responsible engineer should give the permission to increase the load prior to each load step. For a proof load test, the responsible engineer makes this decision based on the analysis of the measurements and stop criteria.

6.4.2 Safety

Safety is of crucial importance during a load test. In Chapter 4, all considerations that need to be addressed before deciding if a load test can be carried out are summarized.

Figure 6.7 Means of communication during load test

138 Load Testing of Bridges

In Chapter 5, the elements that need to be considered during the preparation stage are discussed. In this chapter, the important safety considerations related to the execution stage of the load test are highlighted.

Safety considerations are an important part of the preparation stage, and all elements of the execution should be thought through so that the load test can be carried out as safely as possible. All information on what to do in case of accidents or calamities should be communicated to all personnel that will carry out activities on site, and the address and phone number of emergency services, the nearest pharmacy, doctor, and hospital, firefighters, and police should be clearly displayed. A list with names of personnel that are allowed on site from all parties involved can be made to ensure no unauthorized people access the site. Upon arrival to the test site personnel should register with the safety engineer, and upon leaving the test site they should also register the time of leaving the site to have a log of the personnel present on site.

For a safe execution of the load test, the following considerations should be taken into account:

- All personnel involved on site should receive a safety briefing before starting any activities.
- All personnel involved on site should be wearing the correct personal protective equipment.
- Sufficient light for night activities should be provided.
- Safe access to the bridge structure and the locations where sensors are applied should be provided.
- At all times, a person with a first aid qualification should be on site.
- A short briefing every morning before the start of activities to revise the main safety concerns and discuss the actions for the day is recommended.
- After the load test, an evaluation meeting in which the safety and execution procedures are discussed is recommended.

At all times, the minimum required personal protective equipment includes safety shoes, hard hat, and reflective jacket. A life vest is necessary for activities close to water (see Fig. 6.8). Safety goggles and hearing protection should be available and used for specific tasks.

Safe access to the bridge structure and sensor locations can include a pontoon for activities on bridges over water, or scaffolding. These temporary structures should be checked every day by the safety engineer.

Safe access to the bridge location is also important, and the access routes to the bridge should be communicated with all parties involved. Parking should be available on site. If the bridge is closed for traffic (see Fig. 6.9), all signposting for the detour should be in place before the activities begin on site and before access to the bridges is restricted.

6.5 Summary and conclusions

This chapter summarizes general considerations for the execution of load tests, regardless the type of load test. Particularities for diagnostic load tests and proof load tests are discussed in Part III and Part I of Vol. 13, respectively.

General Considerations for Test Execution 139

Figure 6.8 Safety shoes, life jacket, hard hat, and protective clothing for load test

Figure 6.9 Signposting for detour during load test
Source: Excerpt from photograph by D. A. Hordijk. Used with permission.

The first general element for all load tests is that load is applied. Different methods and types of equipment for applying the load are available. In most cases, loading vehicles (trucks or a special vehicle) are used. Other options include using dead load, hydraulic jacks, a group of pedestrians, or an exciter (for modal analysis). The choice for a method and type of equipment depends on the goals of the load test.

A second general element for all load tests is that the structural response caused by the applied loads is measured. These responses can be monitored in real time or analyzed after the load test, depending on the type of load test and the objectives of the load test.

Finally, this chapter discusses practical aspects during the load test. The most important elements here are communication and workflow lines, and safety. While again these aspects depend on the particularities of the load test, some best practices and general considerations are discussed in this chapter.

References

ACI Committee 437 (2013) Code Requirements for Load Testing of Existing Concrete Structures (ACI 437.2M-13) and Commentary. Farmington Hills, MI.

Beben, D. & Anigacz, W. (2014) Interferometric radar application for dynamic testing of bridge structures. *IABMAS 2014*. Foz de Iguacu, Brazil.

Bonifaz, J., Zaruma, S., Robalino, A. & Sanchez, T. A. (2018) Bridge diagnostic load testing in Ecuador: Case studies. *IALCCE 2018*. Ghent, Belgium.

Cunha, A., Caetano, E., Magalhães, F. & Moutinho, C. (2016) Dynamic identification and continuous dynamic monitoring of bridges. In: Bittencourt, T., Frangopol, D. M. & Beck, A. (eds.) *Maintenance, Monitoring, Safety, Risk and Resilience of Bridges and Bridge Networks*. IABMAS 2016, Foz do Iguacu, Brazil.

Ettouney, M. & Alampalli, S. (2012) *Infrastructure Health in Civil Engineering*. CRC Press, Boca Raton, FL.

Frýba, L. & Pirner, M. (2001) Load tests and modal analysis of bridges. *Engineering Structures*, 23, 102–109.

Ministerio De Fomento – Direccion General De Carreteras (1999) *Recomendaciones para la realización de pruebas de carga de recepción en puentes de carretera*. Ministerio de Fomento, Madrid, Spain.

Schmidt, J. W., Halding, P. S., Jensen, T. W. & Engelund, S. (2018) High Magnitude Loading of Concrete Bridges, *ACI Special Publication*, Evaluation of concrete bridge behaviour through load testing: International perspectives. SP-323–9, 9.1–9.20.

Steffens, K., Opitz, H., Quade, J. & Schwesinger, P. (2001) The loading truck BELFA for loading tests on concrete bridges and sewers (in German). *Bautechnik*, 78, 391–397.

Varela-Ortiz, W., Cintrón, C.Y.L., Velázquez, G. I. & Stanton, T. R. (2013) Load testing and GPR assessment for concrete bridges on military installations. *Construction and Building Materials*, 38, 1255–1269.

Zhou, Y. E., Cupples, T. H., Brown, R. & Hwang, G. (2007) Re-evaluation of weight limits for six bridges in Montgomery County, Maryland. *International Bridge Conference IBC 2007*. Pittsburgh, PA, USA.

Chapter 7

Post-Processing and Bridge Assessment

Eva O. L. Lantsoght and Jacob W. Schmidt

Abstract

This chapter discusses the aspects related to processing the results of a load test after the test. The way in which the data are processed depends on the goals of the test. As such, the report that summarizes the preparation, execution, and post-processing of the load test should clearly state the goal of the load test, how the test addressed this goal, and what can be concluded based on an analysis of the test results. Typical elements of the post-processing stage include discussing the applied load, the measured structural responses, and then evaluating the bridge based on the results of the load test.

7.1 Introduction

The final step in a load test is the post-processing of data obtained during testing, and using this data to verify the bridge response and thus fulfill the aims identified before testing. This step includes developing the deliverables from the load test (often a technical report) which includes an assessment of the tested bridge.

For a diagnostic load test, a comparison between the analytically predicted responses and the measured responses can be used to update and improve the analytical model, so that an improved rating or assessment of the bridge results. After a diagnostic load test with a specific goal, such as verification of stress transfer, deflections, transverse distribution, unintended composite action, and so forth, the data obtained during the field test should be analyzed to address this specific goal.

After a successful proof load test in which the bridge has carried the target load without exceeding any stop criteria, it can be directly concluded that the bridge can carry the code-prescribed loads. In this case, the post-processing stage is limited to discussing the way in which the target load was determined, showing placing and description of the loading, presenting the graphs of the measured responses in a report, demonstrating that no stop criteria were exceeded, and writing down the conclusion that the bridge can carry the predefined target loads. Additionally, a proof load test can be used to verify bridge response assessments based on extensive analytical models. The proof load test then serves to confirm the assumptions regarding material modeling, geometries, structural element interactions, load separation, boundary conditions, and so forth. For bridges with material degradation and deterioration, the proof load test can be used to evaluate the deterioration model assumed in the analytical model.

DOI: https://doi.org/10.1201/9780429265426

142 Load Testing of Bridges

In all cases, it is necessary to develop a report after a load test for future reference and bridge management decisions. When calibrating a finite element model based on field test results is one of the goals, the improved model should be submitted with the report so that this improved model is available for future decisions (e.g. checking if a superload can be allowed passage over the bridge).

7.2 Post-processing of measurement data

7.2.1 Applied load

A first element that should be calculated if not directly available and reported is the magnitude of the total applied load. When the load is applied with vehicles, the weight of these vehicles is measured on site. The weight per axle and distance between axles should be included in the report of the load test. Depending on the goals of the load test, the position of the vehicles may be measured as well. Post-processing of these measurements could include developing plots of the loading protocol in terms of position of the load versus time. If different vehicles and configurations of these vehicles are used, resulting in a number of loading scenarios, then these scenarios should be documented and the report of the load test should contain such information.

When hydraulic jacks are used, the applied load should be measured by using, for example, load cells. If the same load should be applied on all wheel prints, the measurements that show that this goal was met during the test should be post-processed and shown in the report (see Fig. 7.1). If the load is applied to represent a truck with different axle loads, the measurements that show that this goal was met during the test should be post-processed and shown in the report. Additionally, the sum of the externally applied load should be made. The total applied load is the sum of all separate loads (on the separate wheel prints), as well as the weight of the jacks and other elements that are used for load application. The weight of the elements used for load application should be provided by the engineers responsible for the loading procedure. The position of the wheel prints on site should be measured carefully so that it corresponds to the sensor plan as close as possible, and any changes from the originally determined positions should be explained, reported, and changed on the final drawing of the sensor plan for the report.

7.2.2 Verification of measurement data

After a load test, all measurements should be post-processed. A first step here is to develop plots of all measured structural responses versus time or versus load and to check these plots for anomalies in the data. If such anomalies are observed, an explanation should be sought. Possible explanations include a sensor running out of its available measurement range, effects of wind on the structure on which sensors are mounted, and so forth. Special attention should be paid to the measurement data that are used to meet the goals of the load test. Using two independent measurement methods for the same output could be one of the means to understand discrepancies related to monitoring and responses.

The data should also be checked based on the expected structural response determined analytically prior to the test. The cause of significant differences between prediction and measurements should be determined as well as possible. Additionally the symmetry, linearity, and reproducibility of the data should be checked. Symmetry of the measured

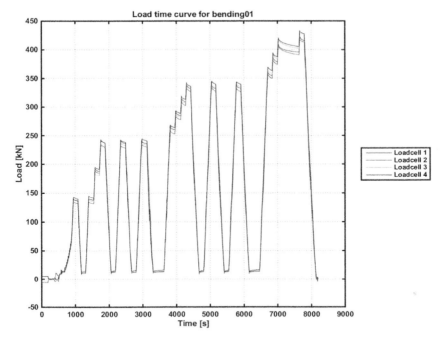

Figure 7.1 Comparison between four load cells on which the same load should be measured. Conversion: 1 kN = 0.225 kip

structural responses means that for symmetry of geometry or load application, similar responses should be measured. Linearity of the measured structural responses means that for equal increases in the load, equal increases in the structural response should be measured. Reproducibility of the measurements means that when a load case is repeated, similar structural responses should be measured. When these requirements are not met, possible explanations could be:

- Differences in the structural responses as a result of the influence of changes in temperature and humidity;
- Sensor malfunction;
- Imperfections in the structure;
- Small misalignments with the method of load and sensor application or on the load path when vehicles are used.

The engineer should also address the quality of the data based on the amount of noise in the measurements.

7.2.3 Correction for support deformations

When elastomeric bearings (Fig. 7.2) are used, these bearings can be compressed during the load test. This deformation should be measured during the load test. Consequently, the measured deflections of the superstructure can be corrected for the support deformations to find the net deflection of the superstructure.

Figure 7.2 Example of elastomeric bridge bearings in laboratory conditions

Additionally, for bridges on soft soils, the settlement of the supports during the load test should be measured, and the measured deflections of the superstructure should be compensated with the support settlement to find the net deflections of the superstructure. The plots of the measurement data that should be included in the report of the load test should show the deflection measurements corrected for the effect of the support deformations.

7.2.4 Correction for influence of temperature and humidity

The influence of temperature and humidity on the measured structural responses can be important, as discussed in Chapter 5. Temperature and humidity can influence both the structural response as well as the sensor function and sensitivity. The latter effects can sometimes be mitigated with correction factors that are provided by the manufacturer of the sensor. The first effect can be mitigated by measuring the structural response caused by only temperature and humidity changes at a position that is not affected by the applied load (see Fig. 7.3) or by using "zero load" cases. The measurements that are affected by the influence of temperature and humidity should be corrected for these effects. The discussions in the report of the load test should address the effect of temperature and humidity, and the plots of measurements data in the report of the load test should be corrected for the effect of temperature and humidity.

7.2.5 Reporting of measurements

Once the structural responses are corrected for the influence of support deformations and the influence of temperature and humidity, the plots of the measurements can be developed

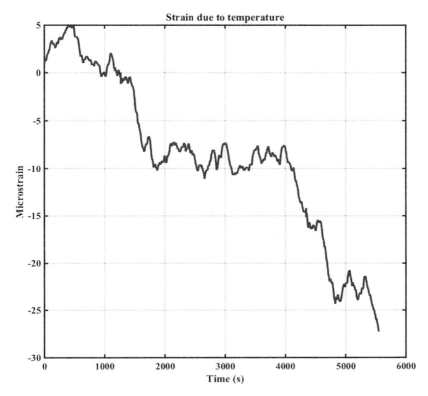

Figure 7.3 Strain measured at position not influenced by the applied load: strain development over time due to changes in temperature and humidity

for the report. The measurements that are included in the final report of the load test depend on the goals of the load test. The structural responses necessary to address the testing aims should be included in the report, and the verification of these aims should be based on the observed structural responses.

Depending on the goals of the load test and type of load test, the following measurements can be included in the report:

- Monitoring plan, describing the location of each monitoring device and justification for the use (see also Chapter 5, which describes the development of the sensor plan).
- Loading scheme: the measured loading scheme should be similar to the planned loading scheme, and the report should address deviations from the plan, if any. Depending on the method of load application, the measured loading scheme can include the measured weights of the vehicles, the measured positions of the vehicles versus time during the load test, or the measured applied load on each jack when a system with hydraulic jacks is used. For more information on the required data for the report regarding the applied load, see Section 7.2.1.
- Load-displacement diagram at representative positions: the displacement is often the vertical deflection, but measurements in the horizontal direction can be used to verify out-of-plane movements of bridge elements, supports, and so forth. If cycles of load are

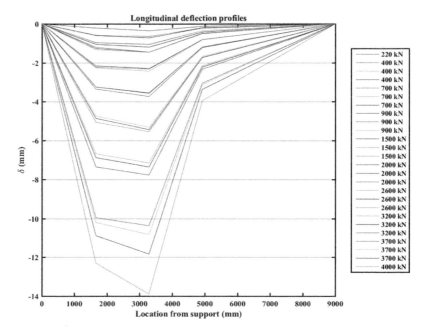

Figure 7.4 Longitudinal deflection profiles for different load levels. Conversion: 1 kN = 0.225 kip, 1 mm = 0.04 in.

applied, the envelope as well as the full load-displacement diagram should be added to the report.
- Plots of deflections: if lines of deflection sensors are applied in the longitudinal and transverse direction, the resulting deflection plots at selected magnitudes of the load can be used to show the linearity of the overall structural response. Figure 7.4 depicts an example of a longitudinal deflection plot at different load steps.
- Strain profiles: if strains are measured over the height of girders, the resulting strain profiles should be drawn at selected magnitudes of the load. The position of the neutral axis should be determined for selected magnitudes of the load.
- Strain measurements: besides using the strain measurements to derive strain profiles, these measurements can also be reported as a function of the time and as a function of the applied load. These plots can be used to evaluate the linearity and reproducibility of the data, and to verify stop criteria for proof load tests.
- Crack width measurements: for concrete bridges, existing cracks and related opening can be monitored during a load test, or if a proof load test induces new crack development to the structure, these cracks can be monitored and followed as a part of the stop criterion evaluation. The post-processing of the data of crack width measurements can include plots of the crack width versus time, crack width versus load, drawings of the damage to the bridge before and after the load test, and the comparison of the measured crack widths to stop or acceptance criteria for proof load tests.

For proof load tests, the report should restate the stop and acceptance criteria that were selected prior to the load test. The derived plots and measured structural responses can then be used to report that the stop criteria are not exceeded and to verify the acceptance criteria. Since the stop criteria are verified during the proof load test, their main importance is during the test. The post-processed data should be used to make a second check of the stop criteria. Acceptance criteria are verified after a proof load test to demonstrate that the structural behavior is within acceptable limits.

If a stop criterion is exceeded during a proof load test prior to reaching the target proof load, the outcome of the proof load test is that the bridge cannot carry the code-prescribed loads. Depending on the load level at which the stop criterion is exceeded, the bridge may fulfill the requirements for a lower safety level or for reduced traffic loads. Since the decision for posting this bridge then hinges on the stop criterion and the load at which it was exceeded, for such a case it is important to discuss this stop criterion and the measurements that were analyzed for this stop criterion in the report. This discussion should include the effect of the support deformations and influence of temperature and humidity. If it turns out that the corrected data show that the stop criterion was not exceeded, a retest may be necessary. For cases where a stop criterion is exceeded during the proof load, this step of post-processing should be done already in the field, so that (time permitting) a retest can be done.

7.3 Updating finite element model with measurement data

The goal of a diagnostic load test can be to have a better understanding of the overall behavior of the tested bridge. A way to address this goal is by updating a finite element model with measurement data from a field test so that the improved model can be used for the assessment. The details of these procedures are included in Part III. Barker (2001) identified the sources of differences between the measurements and the finite element model for diagnostic load tests on steel bridges as follows:

- Frozen bearings, resulting in a restraint of deformations at the supports, which led to moments over the supports that were not included in the model;
- Differences in the longitudinal distribution of bending moments;
- Differences in the transverse distribution of bending moments;
- The stiffness of nonstructural elements such as barriers, curbs, and railings;
- The actual impact factor;
- The actual dimensions;
- Unintended or additional composite action.

In addition to the above discrepancies related to steel bridges, some additional differences could be the case for concrete bridges (Nanni et al., 1999), (Alkhrdaji et al., 1998), (Goodpasture and Burdette, 1973), (Schmidt et al., 2018):

- Actual influence of the steel reinforcement strain hardening magnitude;
- Actual load distribution in the concrete bridge deck;
- Effect of compressive and tensile membrane action;
- Time-dependent effects on the concrete properties;
- Confinement from the restraint of the bridge deck.

The differences between the dimensions in the finite element model and the actual dimensions should be limited after the visual inspection on site during the preparation stage of the load test. If there are doubts regarding certain dimensions, these dimensions can be measured on site. For concrete cross-sections, the difference between considering the cross section as cracked or as uncracked should be evaluated.

Similarly, for proof load testing, the field measurements can be used to update the finite element model that was used for the preparation of the load test. The improved model can be used for future load ratings of the bridge, or to evaluate its ability to carry a superload.

Depending on the goals of the load test and the available measurements, the strain and deflection profiles from the measurements can be compared to those predicted prior to the field test with a finite element model. If parametric studies were done by changing the uncertain properties in the finite element model, then the measured response should be compared to the predicted range of responses. This comparison can then be used to estimate the uncertain properties where these cannot be determined directly from the measurements during the load test. These insights can then be used to improve the available finite element model. The existing finite element model can be improved by using mathematical optimization functions. However, it is important for the bridge engineer who will evaluate the bridge and the field test to keep in mind the sources of the differences between the model and the measurements, so that the engineer can understand mechanisms behind the optimization of the model and the overall behavior of the bridge.

One element that requires further research is how finite element models can be used when only one span can be load tested. For bridges crossing the highway, it may not be permitted to carry out a proof load test on spans that are directly above the highway, as the risks involved are too large (Lantsoght et al., 2017, 2018) and lane closures or complete closure of the highway would be necessary for safety reasons. If a span is proof loaded that is not the critical span, then the proof load test can only evaluate the tested span. A possible way to assess the critical span after the proof load test is by updating the finite element model based on the measurements from the load test and then using the updated model to assess the critical span. Future research should explore this method and use probabilistic methods to quantify the uncertainties on such approaches.

7.4 Bridge assessment

For existing bridges, load tests can be used for assessment. The method for assessment depends on the governing codes and guidelines. The way in which the information from the load test is used for an assessment depends on the type of load test. For diagnostic load tests, the AASHTO *Manual for Bridge Evaluation* (AASHTO, 2016) presents a simple method based on the ratio between predicted and measured responses to update the rating of the considered structural elements, as well as a method to determine the target proof load so that the rating factor of the considered structural elements becomes larger than or equal to one.

In the assessment of the tested bridge after the load test, the differences between the applied loads and the loads required for assessment should be considered. In Europe, the load models from the Eurocode NEN-EN 1991–2:2003 (CEN, 2003) that are used for assessment cannot directly be translated into a certain truck type. In the Americas as well as in some European countries, the assessment uses actual truck types. For such cases, the relation between the applied load and the load required for the rating is clear. For assessment using the NEN-EN 1991–2:2003 loads, an intermediate step is necessary. At this moment, this intermediate step is the use of equivalent sectional forces and moments. Future research should address this gap and embed load testing into the codes for assessment.

When an updated finite element model is used for the assessment after a load test, the bridge engineer should address questions about the applicability and extrapolation of the test results and updated model to higher load levels corresponding to the ultimate limit state (ULS) and future ratings (Bridge Diagnostics Inc., 2012). If unintended composite action or boundary conditions (e.g. frozen bearings) are the source for differences between the responses in the model and the measured responses, these changes to the structure may not be valid for all load levels. For high loads, the unintended composite action may be lost, so it would not be conservative to take this positive effect into account for a rating at the ULS. If future maintenance activities include the replacement of bearings, the effect of this change on the assessment should be considered and discussed in the report of the load test.

7.5 Formulation of recommendations for maintenance or operation

The final step in using the results from a load test is making a decision regarding the future operation of the bridge. The bridge owner is responsible for this decision. This decision could be to keep the bridge in operation as it is, post the bridge, strengthen the bridge, or demolish (and possibly replace) the bridge. The owner is also responsible for the permit loads. The report of the load test can only include a recommendation for the future operation of the bridge based on the outcome of the test. If certain elements of the bridge behavior are still uncertain after the load test, the bridge engineer can recommend further (material) testing, more advanced calculations based on more complex models, additional site inspections possibly amplified with nondestructive testing techniques, or long-term monitoring of the structure.

7.6 Recommendations for reporting of load tests

After a load test, the report of the test should contain all relevant information related to test outcome as well as the resulting recommendations.

The report should contain the following information about the bridge:

- Name, location, year of construction
- Overview photograph of the bridge
- Type of structure.

The following information from the preparation stage of the load test should be included:

- Overview of the available information: plans, original calculations, inspection reports, reports of material testing, etc.;
- Summary of the technical inspection carried out as part of the preparation of the load test;
- Results of assessment before the load test and models used to assess the bridge for an existing bridge, or summary of design assumptions and models used to design the bridge for a new bridge;

150   Load Testing of Bridges

- Thresholds related to the load test, such as target load and stop criteria (for proof load tests) and how the loading procedure will address these;
- Loading protocol (load paths and configurations if vehicles are used, or loading scheme when hydraulic jacks are used);
- Sensor plan with a link to the limiting thresholds of the load test;
- Expected structural responses, and when relevant, the expected capacity of the critical sections;
- Safety considerations.

From the execution of the load test, the report should include the following information:

- Date and time of the load test
- Weather conditions during the load test
- Personnel on site
- Most important observations during the load test (the log of all observations can be added to the report as an appendix).

The report should also include the following post-processing results:

- The actual loading protocol (and differences with the planned protocol should be addressed);
- Plots of the relevant measurements, with a discussion of the verification of the measurements (see Section 7.2.2) and derivation of properties that follow from the measurements (location of member neutral axis, resulting stresses where strains are measured, etc.);
- Verification of stop and acceptance criteria when relevant (proof load tests);
- Comparison between predicted responses and measured responses, and updating of the analytical models, when relevant (typically diagnostic load tests);
- Parameter studies related to updating the finite element model or recommended actions if the outcome of the theoretical models differs from the test result of the real bridge structure and the parameter studies cannot identify the cause of these differences.

The final recommendations in the report can address the following elements:

- Assumptions used for the assessment of the bridge after the load test based on the experiences from inspections before the bridge test, identified critical areas during load testing, and information gained from the field test;
- Results of the assessment of the bridge after the load test;
- Recommendations for maintenance of the bridge, when relevant;
- Recommendations for posting or permit loads of the bridge, where relevant.

7.7   Summary and conclusions

The final step in the project of a load test is to post-process the data gathered during the load test and combine all relevant information into a report about the load test. The way in which the data are analyzed depends on the goals of the load test, the type of load test, and whether the tested structure is a new or existing bridge. For a diagnostic load test on a new bridge prior to opening, it may be sufficient to show that the differences between the measured and predicted responses are within acceptable limits. For a proof load test on an existing bridge, it can be sufficient to show that the bridge can carry the target load and thus fulfills the code-prescribed loads and load combination. For cases where the conclusion does not follow directly from the load test, all assumptions and calculation procedures should be discussed.

The report should include the relevant structural responses measured during the load test as well as the applied load. When the load is applied with vehicles, the magnitude of the load should be mentioned in the report, as well as the loading paths and combinations of trucks used for different scenarios. When the load is applied with hydraulic jacks, the measured forces should be reported. A next step should be to discuss the quality of the measurement data in terms of reproducibility, symmetry, and linearity, and in terms of noise on the measurements. These measured structural responses should be corrected for the effect of support deflections and the effects of temperature and humidity. The report should also include a comparison between the measured structural responses and the predicted responses and should address the differences. All relevant measurements should be presented in a visual way after developing the most relevant graphs of the data.

In some cases, the goals of the load test cannot be directly met by analyzing the measurement data. For these cases, the finite element model that was used to prepare the load test can be updated with the measured structural responses, and through this process the sources for the differences between the analytically determined responses and the measured responses can be identified. The updated model can then be used to improve the assessment of the bridge, by including only the effect of the mechanisms that act at the ULS.

The bridge engineer can develop recommendations for the maintenance and decision-making for the tested bridge, but the responsibility for the operation decisions lies with the bridge owner. Based on the technical inspection, the field test, and the analysis of the field test data, the load test report should include an improved assessment of a bridge when the tested bridge is an existing bridge. For new bridges, the field test data can be used to identify differences between the actual structural behavior and the assumptions used during design. These differences should then be considered in future assessments of the bridge.

References

AASHTO (2016) *The Manual for Bridge Evaluation with 2016 Interim Revisions.* American Association of State Highway and Transportation Officials, Washington, DC.

Alkhrdaji, T., Nanni, A., Chen, G. & Barker, M. (1998) Destructive and Non-Destructive Testing of Bridge J857, Phelps County, Mo. Volume I: Strengthening and Testing to Failure of Bridge Decks (No. R11, C-5-34301). University of Missouri-Columbia, University of Missouri-Rolla, Center for Infrastructure Engineering Studies (CIES).

Barker, M. G. (2001) Quantifying field-test behavior for rating steel girder bridges. *Journal of Bridge Engineering*, 6, 254–261.

Bridge Diagnostics Inc. (2012) *Integrated Approach to Load Testing.* BDI, Boulder, CO.

CEN (2003) *Eurocode 1: Actions on Structures: Part 2: Traffic Loads on Bridges, NEN-EN 1991–2:2003.* Comité Européen de Normalisation, Brussels, Belgium.

Goodpasture, D. W. & Burdette, E. G. (1973) Full scale tests to failure of four highway bridges. *Bulletin 643.American Railway Engineering Association*. Lanham, MD. pp. 454–472.

Lantsoght, E. O. L., Koekkoek, R. T., Van der Veen, C., Hordijk, D. A. & de Boer, A. (2017) Pilot proof-load test on viaduct De Beek: Case study. *Journal of Bridge Engineering*, 22, 05017014.

Lantsoght, E. O. L., De Boer, A., Van Der Veen, C. & Hordijk, D. A. (2018) Modelling of the proof load test on viaduct De Beek. *Computational Modelling of Concrete and Concrete Structures Euro-C 2018*. Bad Hofgastein, Austria.

Nanni, A., Alkhrdaji, T., Chen, G., Baker, M., Yang, X. & Mayo, R. (1999) Overview of testing to failure program of a highway bridge strengthened with FRP composites. In: Dolan, C. W., Rizkalla, S. H. & Nanni, A. (eds.) *4th International Symposium FRP Reinforcement Concrete Structures (FRPRCS4)*. SP-188. American Concrete Institute International, Farmington Hills, MI. pp. 69–80.

Schmidt, J. W., Halding, P. S., Jensen, T. W. & Engelund, S. (2018) High magnitude loading of concrete bridges, ACI technical publication, Evaluation of concrete bridge behaviour through load testing: International perspectives. SP-323–9, 9.1–9.20.

Part III

Diagnostic Load Testing of Bridges

Chapter 8

Methodology for Diagnostic Load Testing

Eva O. L. Lantsoght, Jonathan Bonifaz, Telmo Andres Sanchez and Devin K. Harris

Abstract

This chapter deals with the methodology for diagnostic load testing. All aspects of diagnostic load testing that are shared with other load testing methods have been discussed in Part II. In this chapter, the particularities of diagnostic load testing of new and existing bridges are discussed. These elements include loading procedures, monitoring behavior during the test, reviewing test data, calibrating analytical models, and evaluating the test results.

8.1 Introduction

Load testing can be used to serve a number of purposes in bridge evaluation and is often referred to as either diagnostic or proof load testing (Lantsoght et al., 2017b). For new bridges, diagnostic load tests can demonstrate that the bridge behaves in the same way as it was designed (Kirkpatrick et al., 1984a, 1984b; McGrath et al., 1995; Lawver et al., 2000; Barnes et al., 2003; Yang and Myers, 2003; Ferrand et al., 2005; Konda et al., 2007; Harris et al., 2008; Au et al., 2013; Hernandez and Myers, 2015; Harris et al., 2016; Taylor et al., 2016; Bonifaz et al., 2018; Hernandez and Myers, 2018a). For this purpose, the measured responses during a field test are compared to the predicted responses obtained from the analytical model that was used for the design.

For existing bridges, diagnostic load tests can be used to evaluate certain aspects of the structural behavior and/or to calibrate the analytical model used for the assessment with field data in order to get an improved rating of the structure (Aktan et al., 1992; Saraf, 1998; Al-Mahaidi et al., 2000; Velázquez et al., 2000; Jáuregui and Barr, 2004; Hag-Elsafi and Kunin, 2006; Mordak and Manko, 2008; Jeffrey et al., 2009; Hernandez and Myers, 2016; Sanayei et al., 2016; Hernandez and Myers, 2018b).

For both new and existing bridges, the aspects of structural behavior that can be determined in a diagnostic load testing include (Barker, 2001):

- The actual impact factor or dynamic amplification factor (Hernandez and Myers, 2018b);
- The stiffness of the total system including nonstructural elements that provide stiffness, such as curbs and railings (Nilimaa et al., 2015);

DOI: https://doi.org/10.1201/9780429265426

156 Load Testing of Bridges

- The internal system behavior of complex structural components (Harris et al., 2016);
- The actual stiffness after material deterioration;
- The transverse live-load distribution (Arockiasamy and Amer, 1997b, 1997a, 1998; Amer et al., 1999; Jones, 2011; ACI Committee 342, 2016; Ohanian et al., 2017);
- Bearing restraint effects;
- The actual longitudinal live-load distribution;
- Load-bearing mechanisms that typically are not taken into account, such as arching action (Taylor et al., 2007);
- Unintended or additional composite action;
- Effects of skew.

When these quantities are measured in the field and used for an improved assessment of an existing bridge, it is important to evaluate if it is conservative to assume that the quantity under consideration also acts at the load levels for which the bridge is assessed. For example, during a diagnostic load test unintended composite action between steel girders and the concrete deck may be observed. However, at larger load levels this composite action may be lost. Therefore, it is generally not rational to include effects in an assessment such as unintended composite action or the restraint effects of frozen bearings that may be overcome at larger load levels.

Diagnostic load tests can also be carried out on bridges after strengthening (Alkhrdaji et al., 2000; Russo et al., 2000; Bell and Sipple, 2009; Myers et al., 2012; Au et al., 2013). To evaluate directly the effectiveness of strengthening measures, a diagnostic load test before and after applying strengthening can be performed and the relevant structural responses can be compared.

Since the goal of a diagnostic load test is to have a more realistic estimate of the actual properties of the bridge, diagnostic load testing can be carried out at lower load levels than the capacity of the bridge system. The applied load during the test should result in measurable responses that are appropriate to describe the operational state (i.e. the expected service response) but lower than that which would cause any permanent damage. As such, the required load during the test depends on the structural response under consideration and the intended behavior under evaluation, and may also depend on the accuracy of the available sensors. The requirements for implementation of diagnostic testing varies across the globe, with some government agencies requiring testing for initial acceptance of a new bridge, while others utilize testing primarily for evaluating structures after years of service. For example, Spanish guidelines (Ministerio de Fomento: Direccion General de Carreteras, 1999) (see Part I) stipulate that for diagnostic load tests prior to opening of new bridges, the required loads must be representative of the design service load levels with a return period of 5 years. This load should cause around 60% of the ultimate limit state (ULS) load effect and should never exceed 70% of the load effect at the ULS. These values can be taken as a reference during the preparation of a diagnostic load test. When service load levels are used for diagnostic load tests and no nonlinear behavior is observed at these load levels, the same linear behavior is assumed for the loads the bridge needs to be designed or assessed for. On the contrary to this type of prescribed diagnostic testing, testing in the United States is often conducted on an ad hoc basis to understand operational performance of existing bridges.

8.2 Preparation of diagnostic load tests

8.2.1 New bridge diagnostic testing

In general terms, the preparation of a diagnostic load test follows the steps described in Part II, Chapters 4 and 5. For a new bridge, an inspection is often recommended to identify differences between the design and as-built conditions. When relevant, the analytical model used for the design can be altered to reflect the as-built conditions. Then the load paths and load levels for the diagnostic load test allow for exploration with the analytical model using load paths or load positions that result in the largest structural responses in the relevant structural members, or load paths that can be applied to have a range of responses required to meet the test objectives. As an example, the Los Pajaros Bridge (Ponton et al., 2016; LaViolette et al., 2017, Robalino and Sanchez, 2017, Sanchez et al., 2017; Bonifaz et al., 2018; Sanchez et al., 2018), a three-span steel bridge with cross section shown in Figure 8.1, was diagnostically load tested following construction using the seven load cases shown in Figures 8.2–8.8. The material from the diagnostic load tests prior to opening of the Los Pajaros Bridge and the Villorita Bridge are used in this chapter to illustrate the general concepts explained here. Full examples of diagnostic load tests detailing the preparation, execution, and post-processing of a load testing project can be found in the following chapters. After determining the load paths, the target load level should be

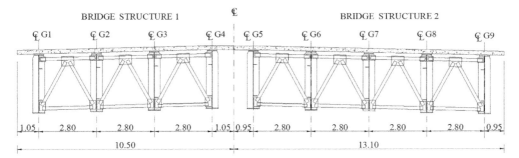

Figure 8.1 Cross section of Los Pajaros Bridge, showing the two separate structures. Units: m. Conversion: 1 m = 3.3 ft

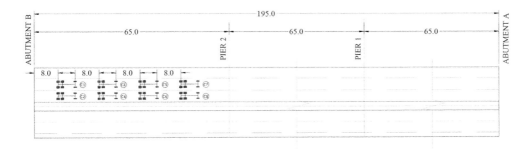

Figure 8.2 Load case 1 for Los Pajaros Bridge, Quito, Ecuador. Units: m. Conversion: 1 m = 3.3 ft

158 Load Testing of Bridges

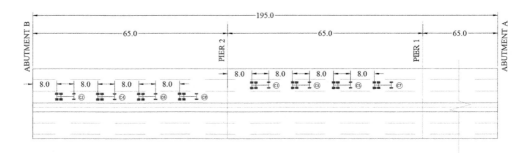

Figure 8.3 Load case 2 for Los Pajaros Bridge, Quito, Ecuador. Units: m. Conversion: 1 m = 3.3 ft

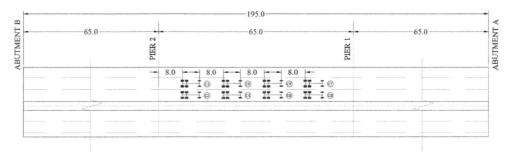

Figure 8.4 Load case 3 for Los Pajaros Bridge, Quito, Ecuador. Units: m. Conversion: 1 m = 3.3 ft

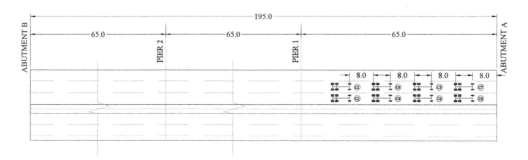

Figure 8.5 Load case 4 for Los Pajaros Bridge, Quito, Ecuador. Units: m. Conversion: 1 m = 3.3 ft

decided upon by verifying which number of loading vehicles results in measurable responses for the structural response under study and the available sensors.

For example, the Los Pajaros Bridge was designed using the AASHTO LRFD code (AASHTO, 2015); however, the bridge was part of the bridge network in Ecuador, which utilized the Spanish loading recommendation (Ministerio de Fomento: Direccion General de Carreteras, 1999) of 60% of the design load. For this test, the target load for the diagnostic load test was estimated as follows. The bridge is designed for the combination of the distributed lane load and one HL-93 truck, so the total design load can be calculated as follows:

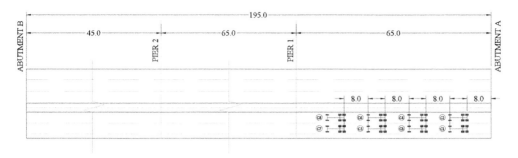

Figure 8.6 Load case 5 for Los Pajaros Bridge, Quito, Ecuador. Units: m. Conversion: 1 m = 3.3 ft

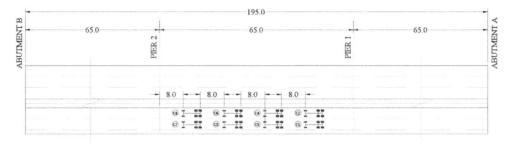

Figure 8.7 Load case 6 for Los Pajaros Bridge, Quito, Ecuador. Units: m. Conversion: 1 m = 3.3 ft

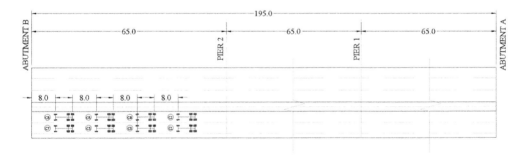

Figure 8.8 Load case 7 for Los Pajaros Bridge, Quito, Ecuador. Units: m. Conversion: 1 m = 3.3 ft

P_{1lane} = 9.36 kN/m × 195 m + 320 kN = 2145 kN (482 kip). For two lanes, P_{ULS} = 4290 kN (964 kip). Using 60% of the design load is then P_{target} = 2574 kN (579 kip). Each load case (see Figs. 8.2–8.8) consists of eight trucks of 294 kN (66 kip) each, resulting in a total load of 2352 kN (529 kip), which is close to the first estimate based on 60% of the design load. The trucks have axle distances of 1.6 m (5.2 ft) and 6 m (19.7 ft) and a distance between wheels in the transverse direction of 1.8 m (5.9 ft) (see Fig. 8.9). With the critical loading positions determined as discussed previously, and the total load and load per truck determined, the load cases (Figs. 8.2–8.8) are fully defined:

- Load Case 1: Convoy of eight trucks in lanes 1 and 2 in span 1, center of convoy of trucks at midspan (0.54 L) for bridge structure 2.

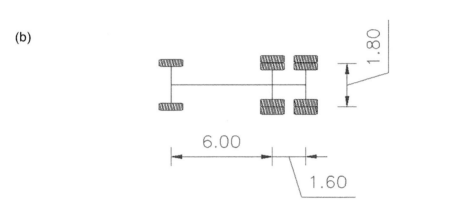

Figure 8.9 Configuration of trucks used for load test on Los Pajaros Bridge, Quito, Ecuador: (a) elevation; (b) plan view. Units: m. Conversion: 1 m = 3.3 ft

- Load Case 2: Four trucks in lane 1 of span 1 and four trucks in lane 2 of span 2, center of trucks at midspan (0.54 L) of the respective spans.
- Load Case 3: Convoy of eight trucks in lanes 1 and 2 in span 2, center of convoy of trucks at midspan (0.54 L) for bridge structure 2.
- Load Case 4: Convoy of eight trucks in lanes 1 and 2 in span 3, center of convoy of trucks at midspan (0.46 L) for bridge structure 2.
- Load Case 5: Convoy of eight trucks in lanes 1 and 2 in span 3, center of convoy of trucks at midspan (0.46 L) for bridge structure 1.
- Load Case 6: Convoy of eight trucks in lanes 1 and 2 in span 2, center of convoy of trucks at midspan (0.54 L) for bridge structure 1.
- Load Case 7: Convoy of eight trucks in lanes 1 and 2 of span 1, center of convoy of trucks at midspan (0.54 L) for bridge structure 1.

The deflections for each load case were predicted using the linear finite element model that was used for the design. Figure 8.10 shows the graph of the predicted deflections of girder 6 for load case 4 on the Los Pajaros Bridge.

After determining the load conditions for the test, the sensor plan can be developed. The most basic measurement consists of measuring deflections with a total station. This

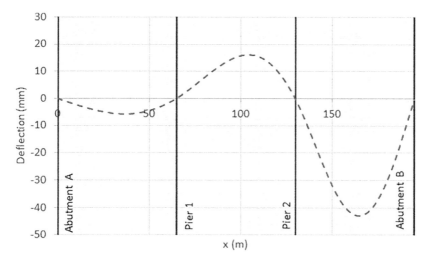

Figure 8.10 Predicted deflections for load case 4 on girder 6 of Los Pajaros Bridge. Conversion: 1 mm = 0.04 in., 1 m = 3.3 ft

measurement is often sufficient for diagnostic load tests prior to opening new bridges; however, other continuous sensor measurements (i.e. strain, displacement, rotation, and acceleration sensors) can be deployed depending on the intended measurement detail and resolution required. In addition, planning for the on-site activities should be developed as well as a safety plan.

8.2.2 Existing bridge diagnostic testing

For an existing bridge, the first step in a diagnostic test also includes a technical inspection. These inspections typically serve the purpose of identifying differences between the available plans (if any) and the actual conditions, identifying site limitations, and assessing the feasibility of the instrumentation plan. Changes to the (lane) layout, damage, deterioration, and material degradation should be reported and shown on updated drawings and damage maps. If an analytical rating prior to the test is not available, the necessary calculations should be done prior to the test. If the required input parameters regarding material properties and geometry are not known, they can be measured in the field as appropriate for the bridge type and material characteristics. For example:

- The geometry of the structure or even local features can be measured using basic surveying tools such as tape measure or total station (Hernandez and Myers, 2018b), but can also be estimated using emerging technologies such as a laser scanner or lidar unit.
- Material properties can often be determined through material sampling and/or non-destructive testing (Orban and Gutermann, 2009), when appropriate.
- For concrete bridges, the position and amount of reinforcement can be estimated with a rebar scanner or ground penetrating radar.

When the diagnostic test will be used to update the analytical model after the test to improve the rating of the bridge, the analytical model should be available prior to the test, either from previous rating calculations or as part of the preparation stage (Catbas et al., 2004; Jauregui et al., 2010; Hodson et al., 2013). If material deterioration or damage is observed during the inspection stage, these conditions need to be taken into account in the analytical model. To calculate the capacity for determining the Rating Factor (ratio of factored resistance for live load to factored live-load demand) or Unity Check (ratio of factored load effect to design capacity as used in Europe), the effect of material deterioration and degradation also needs to be considered. For example, the effect of section loss in corroded steel members (Lim et al., 2016) or the reduction in shear capacity for concrete members suffering from alkali-silica reaction (den Uijl and Kaptijn, 2004) should be taken into account in both the analytical model and diagnostic test results. Finally, the load path and target load are determined as described previously for new bridges. The remaining preparation steps include the development of the sensor plan, planning of the on-site activities, and safety plan.

8.3 Procedures for the execution of diagnostic load testing

8.3.1 Loading methods

The loading methods available for load testing are described in Part II of this book as dead loads, loading vehicles, and systems with hydraulic jacks. In some cases, impact hammers/weights or electromagnetic shakers are often used for vibration testing, but this type of testing is beyond the scope of this chapter. For diagnostic load testing, the most common loading method is the use of loading vehicles since the required loads are often representative of the typical service loads experienced by bridges. Additionally, these vehicular live loads are often more efficient with respect to deployment than static dead loads or loads applied via hydraulic jacks. Examples of using loading vehicles are given in Figure 8.11 for a steel girder bridge. Often in practice, various combinations of vehicle loading configurations are used depending on the intended loading magnitude

Figure 8.11 Diagnostic load test with loading vehicles on the Los Pajaros Bridge, Quito, Ecuador

required and design response observation. Another key advantage of using vehicular loading is that they can serve for both static and dynamic testing. Trucks can move at crawl speed or be placed at critical positions for a certain amount of time for static testing. Additionally for dynamic testing, trucks traveling at different speeds can be used to determine the dynamic amplification factor.

For static load tests, load cases are determined (see Figs. 8.2–8.8). For diagnostic load testing at crawl speed or for moving trucks at a predetermined speed, load paths are determined. To couple the position of the loading vehicles to the measured structural responses, the position of the load should be known at all times. For this purposes, automated vehicle position trackers can be attached to a truck tire or photogrammetry methods can be used. When photogrammetry methods are used and the vehicles are moving, the number of photographs that are taken per minute should be sufficiently large for the speed at which the vehicles are moving to capture the relevant information.

The advantage of the low load levels that are used in diagnostic load tests is that these tests can be carried out in a short amount of time (maximum one day). Diagnostic load tests can be carried out on one or more lanes of a bridge while other lanes remain in service, since typically there is no danger for the traveling public because of the low loads involved. To speed up the execution of a diagnostic load test, a standardized user interface that does not require programming (or only limited programming) can be combined with self-identifying sensors and a wireless sensor plan. Commercial solutions are available for standard diagnostic load testing procedures.

8.3.2 Monitoring bridge behavior during test

When a system with hydraulic jacks is used for loading the bridge, the output of the devices that measure the applied force (typically load cells) can be combined with the output of the applied sensors and processed in software to monitor the structural responses in real time. When loading vehicles are used, the output of the vehicle trackers can be processed together with the measured structural responses for real-time evaluation. When photogrammetry methods are used to identify the truck positions, the data can be processed and visualized after each loading case. When there are concerns about the linearity of the bridge's behavior at the load levels used during the diagnostic load test, real-time data visualization and interpretation should be possible for safety reasons. For other cases, real-time data visualization is not required, and an on-site check of the measured responses after the load test is sufficient.

It is common practice to compare the maximum deflection measured during a loading scenario to the analytically determined maximum deflection. Additionally, the residual deflection after the test is measured to see if the deflection returned to zero (see Fig. 8.12). The measured residual deflection should be corrected for the effects of temperature and humidity to find the structural response due to the applied loading only. The influence of temperature and humidity on sensors and the measured structural responses is discussed in Part II.

Communication during the test between the test engineer and the operator of the load (when using hydraulic jacks) or the truck drivers (when using loading vehicles) is important and should be constant during the test. Mobile phones or walkie-talkies can be used for this communication. The test engineer should be able to immediately communicate with the

164 Load Testing of Bridges

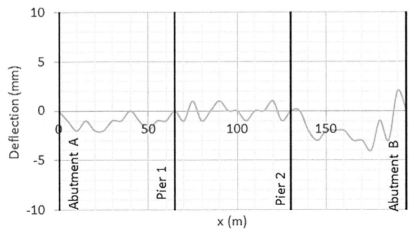

Figure 8.12 Results of measured deflections after unloading of Los Pajaros Bridge, girder 6. Conversion: 1 mm = 0.04 in., 1 m = 3.3 ft

operator of the load or truck drivers when the measured structural responses show unexpected structural behavior and possible dangerous situations.

8.4 Processing diagnostic load testing results

8.4.1 On-site validation and review of test data

After each load case, the measured data should be reviewed (see Fig. 8.13). For static tests, longitudinal and transverse deflection plots can be generated. For dynamic tests, the data can be reviewed by plotting the time and load histories of the sensors. This review serves the purpose of detecting sensor malfunctioning and detecting possible dangerous behavior of the structure during the test. If the review shows sensor malfunctioning, the sensor can be replaced and the load case can be repeated. If the review shows indications of possible dangerous behavior, the testing should be interrupted and the structural behavior should be further investigated. In these cases, the critical elements should be inspected if possible and if it can be done in a safe way.

The data need to be reviewed for symmetry, repeatability, and linearity. For symmetric load configurations, the resulting structural responses should be symmetric. Figure 8.14 shows an example of a load test where this requirement is fulfilled. For repeated load cases and load levels, the measured structural responses should be the same. Finally, when increasing load levels are used, the measured structural responses should increase linearly with the increase in load.

When all load is removed, the residual deformations should be small (see Fig. 8.12). When reviewing the data for symmetry, repeatability, and linearity, the environmental effects on the measured structural responses should be filtered out. Additionally, depending on the construction material, there may be time-dependent effects in the material that influence the measured structural responses. The time-dependent effects can be assessed by

Methodology for Diagnostic Load Testing 165

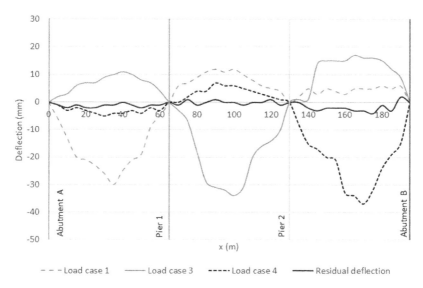

Figure 8.13 Review of different load cases for girder 6 of Los Pajaros Bridge. Conversion: 1 mm = 0.04 in., 1 m = 3.3 ft

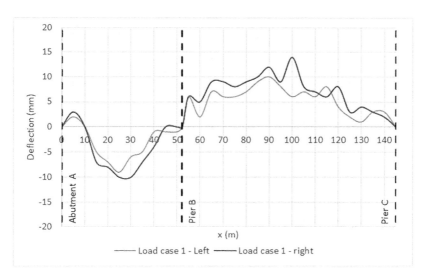

Figure 8.14 Symmetry of load cases left and right for Villorita Bridge in Quito, Ecuador. Conversion: 1 mm = 0.04 in., 1 m = 3.3 ft

reporting the residual deformations after unloading. Where necessary and interesting, the measurements can remain activated for up to 24 hours to measure the reduction of the residual deformations as a function of the time. Time-dependent effects may need to be considered in concrete, timber, masonry, and plastic composite bridges.

Possible problems with the sensor output are large amounts of noise when the structural responses are small relative to the measurement range of the sensor, a shift in the

166 Load Testing of Bridges

measurements caused by hitting of the sensor or wires, constant output when a sensor is outside of its range, and noise effects when the transducer attachment is not functioning properly or when electromagnetic radiation interferes with the measurements.

8.4.2 Processing and reporting test data

After the diagnostic load test, the measured responses should be corrected for additional influences. For example, strain measurements may need to be compensated for the effects of temperature and humidity, and deflection measurements may need to be corrected with support deflections to find the net deflection of the structure for comparison to the analytical models. The effect of temperature and humidity on sensors and structural responses is discussed in Part II. However, since diagnostic load tests typically are carried out during a short amount of time and using either moving vehicles or keeping vehicles for a short amount of time in one position, the time of a load cycle is typically short and these environmental factors are expected to have minimal impact on the results of any given loading repetition, but could impact responses across loading repetitions.

The raw data should be evaluated for issues such as a sensor running out of its measurement range and spikes in the measurements that have no structural explanation and are erroneous. Where necessary, the data should be processed by applying a noise filter (Lin et al., 2018).

Then the processed time histories should be compared to the values obtained from the analytical model (Harris et al., 2015). This initial comparison of the field output and the analytical model can be used to identify major differences between the assumptions in the analytical model and the actual bridge behavior. Based on this comparison, the test engineer may be able to identify which parameters need to be adjusted in the analytical model.

After processing of the data, the following plots – depending on the structural responses that were measured during the test and that are necessary to meet the goals of the test – can be generated for inclusion in the test report:

- The sensor output as a function of the time for each sensor (similar responses can be combined in one figure);
- The sensor output as a function of the applied load for each sensor (similar responses can be combined in one figure);
- The applied loading schemes (plot of time versus load) when hydraulic jacks are used, or the load levels and load configurations when loading vehicles are used (drawings of positions and table of measured weights of the vehicles);
- Longitudinal and transverse deflection profiles for the different load levels or load configurations that were applied;
- Strain profiles measured over the height of girders or at critical locations.

8.4.3 Verification of structural responses for new bridges

For new bridges, the structural responses obtained during the test and reported as discussed in Section 8.4.2 should be compared to the predicted structural responses from the analytical model. It is often common practice to illustrate the predicted relative to the tested structural responses; these comparisons serve as a mechanism to evaluate the validity of the analytical model used for the design of the bridge and hence the overall performance of the bridge. Possible outcomes derived from this comparison are further discussed in Section 8.5.1. Based on this comparison and the evaluation of the tested and predicted responses, the test engineer will be able to give guidance to an owner on whether or not it is safe to open the new bridge to traffic.

8.4.4 Calibration of analytical model for existing bridges

For any bridge, the general assumption is that the analytical model represents an approximation of the structural behavior and the experimental measurements serve as the ground truth. The convergence of these two mechanisms can be achieved through model calibration, which aims to improve a model's approximation by updating with better representations of in-situ conditions (constitutive properties, boundary condition, composite behavior, secondary member contribution, deterioration, etc.). In the literature, there are numerous studies that explore this calibration; however, a generalized and systematic approach presented by Barker (2001) is considered herein. This method was developed for bridges with steel-concrete composite sections but can be easily adapted for other building material or sections. This method is summarized here and follows a number of steps:

1. Inspection of the bridge to quantify the actual dimensions and dead loads, and resulting stresses due to dead load σ_D. This step is carried out during the preparation stage of the load test.
2. Calculate the experimental impact factor resulting from the diagnostic load test I_{mE}.
3. Calculate the experimental distribution factors DF_E.
4. Determine the bearing restraint moments M_{BR}.
5. Remove axial stresses from the stress profiles for steel members.
6. Calculate the total measured moment M_T.
7. Remove the bearing restraint moments M_{BR} from the total measured moment M_T to find the elastic moments M_E.
8. Calculate the experimental section moduli S_E.
9. Calculate the elastic longitudinal adjustment moments M_{LE}.
10. Determine the experimental rating factor.

Details on how to calculate the experimental rating factor according to Barker's method are given in the Appendix. Another, more general approach is the method described in the AASHTO *Manual for Bridge Evaluation* (AASHTO, 2016) and the *Manual for Bridge Rating through Load Testing* (NCHRP, 1998), where the rating factor is updated based on the ratio of tested to predicted strain and corrected for the frequency of inspection, the ability of the test team to identify difference between the tested and predicted responses, and the presence of fatigue- or fracture-critical features. The reader can find this approach presented and discussed in Chapter 3.

When a finite element model is used for the assessment of an existing bridge, it can be calibrated with the responses measured in the field (Bridge Diagnostics Inc., 2012). The methods presented in this section can be used together with ratings based on a finite element model. When the calibration is carried out with a finite element model, the different properties that are mentioned in the section can be varied until a combination of parameters is found that best represents the field results. This calibration is carried out by applying the trucks used in the field test on the finite element model and reading out the structural responses due to static live-load effects only (Hernandez and Myers, 2016; Hernandez, 2018). For easy comparison between the model and the test results, the finite element model should have nodes at the positions of the measurements. Mathematically, the best representation can be expressed in terms of the smallest error (e.g. based on a least-squares approach [Bridge Diagnostics Inc., 2012]) between the finite element model and the field test results. In practice, however, the optimization of the finite element model is an iterative procedure in which optimization methods are combined with engineering judgment. The test engineer here needs to identify which parameters should be optimized and what are the reasonable limits for these values.

When nonlinear structural responses are observed during the diagnostic load test, linear finite element models are not recommended for the assessment, and the step to nonlinear modeling should be made or the loads on the bridge should be limited to the linear range.

8.5 Evaluation of diagnostic load testing results

8.5.1 *Evaluation of results for new bridges*

Depending on the governing code or guidelines, for new bridges it is sufficient to show with a diagnostic load test that the bridge behaves as designed. Often the field measurements will show that the design assumptions are conservative. If the measured structural responses exceed the expected responses (by between 10% and 20%, depending on the governing code and the construction material; see Chapter 3), the cause for this difference needs to be identified. In that case, it may be necessary to update the analytical model and verify the design calculations. Retesting may be necessary. In summary, for new bridges, three outcomes are possible after a diagnostic load test:

1   The predicted structural response is larger than the measured structural response. The analytical model used for the design is thus on the conservative side, and no further actions are necessary.
2   The predicted structural response is more or less equal to the measured structural response. The analytical model used for the design is a good representation of the structural behavior. No further actions are necessary.
3   The predicted structural response is smaller than the measured structural response. The analytical model is not conservative. The test engineer, owner, and designer should meet to decide on further actions.

For the first case, when it is found that the analytical model is conservative, the owner may request that the designer update the analytical models used for the design to have a model that predicts the new bridge as closely as possible. This model can then be used in the future for assessments and can be updated to take into account changes to the structure or material deterioration/degradation. These changes could be based on material sampling, nondestructive testing, and future diagnostic load tests. The methods for updating an analytical model after a diagnostic load test are discussed in Section 8.4.4 and how to use the

Methodology for Diagnostic Load Testing 169

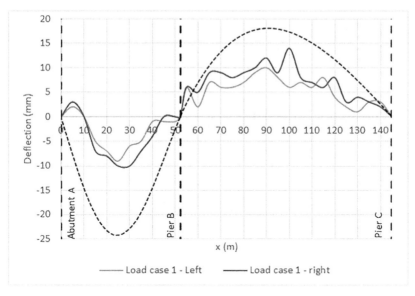

Figure 8.15 Example of difference between analytically determined maximum deflection and measured maximum deflection for flexible structure. Conversion: 1 mm = 0.04 in., 1 m = 3.3 ft

updated model for an assessment is discussed in Section 8.5.2. Figure 8.15 shows an example of a case where the analytical model of a rather flexible bridge is on the conservative side.

The second case is the ideal case, in which the analytical model closely predicts the structural responses measured during the test. It is good practice to transfer this analytical model from the designer to the bridge owner or the responsible party for the operation, management, and maintenance of the bridge so that the model can be used for future assessments and in combination with future load tests. Figures 8.16 and 8.17 show the results for maximum sagging and maximum hogging deflection, respectively, for girder 6 of the Los Pajaros Bridge. For this case, the measured deflection of 37 mm (1.5 in.) was 86% of the predicted deflection of 43 mm (1.7 in.) for the maximum sagging deflection. For the maximum hogging deflection, the measured deflection of 17 mm (0.7 in.) was 106% of the predicted deflection of 16 mm (0.6 in.).

The third case, in which the analytical model predicts smaller structural responses than the responses measured in the field, is typically addressed by limits to the ratio of tested to predicted response in the existing codes and guidelines (see Chapter 3). Where no such limits are available, they should be agreed upon prior to the diagnostic load test in accordance between the owner, test engineer, and designer. When these limits are exceeded, the following actions or a combination thereof can be proposed:

- Visual inspection directly after the diagnostic load test, to identify possible signs of distress;
- On-site inspection of the bearings;
- Material sampling or nondestructive testing of the bridge's materials;

170 Load Testing of Bridges

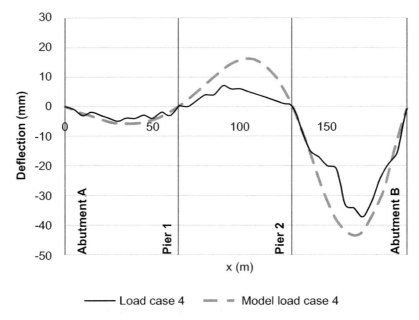

Figure 8.16 Comparison between analytically determined deflections and predicted deflections for girder 6 of Los Pajaros Bridge for the case that causes maximum sagging deflection. Conversion: 1 mm = 0.04 in., 1 m = 3.3 ft

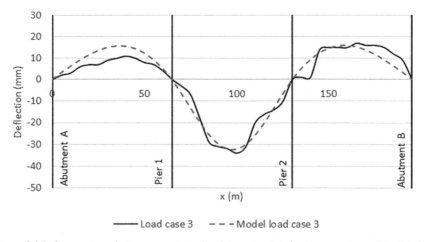

Figure 8.17 Comparison between analytically determined deflections and predicted deflections for girder 6 of Los Pajaros Bridge for the case that causes maximum hogging deflection. Conversion: 1 mm = 0.04 in., 1 m = 3.3 ft

- Re-evaluation of the analytical model by the designer or by a third party;
- Repeat diagnostic load test.

The first step is often a visual inspection to identify possible signs of distress that may have resulted in the larger than expected structural responses. If no signs of distress are observed, the bearings and/or material properties can be checked. If the bridge is in an acceptable condition and the materials and bearings are as designed, then the analytical model should be re-evaluated, either by the designer or by a third party. If errors in the analytical model are found, a full assessment of the safety of the bridge should be carried out before the bridge can be deemed safe to open to traffic. A new analytical model may need to be developed, and a repeat diagnostic load test may be necessary.

8.5.2 *Improved assessment for existing bridges*

For existing bridges, the responses measured during the field test can be used (depending on the goals of the test) to calibrate the analytical models used for the original rating of the bridge, as discussed in Section 8.4.2. It is then determined which of the mechanisms act at the ULS, and only these contributions are included in the model for rating. The updated model is then used to determine the corresponding rating, which reflects the actual performance of the bridge at a given moment during its service life (Wipf et al., 2000). Typically, the loads applied during a diagnostic load test are not the loads required for the assessment. Therefore, in the updated analytical model, the trucks applied during the diagnostic load test are removed and the loads required for the assessment are applied. Different scenarios, such as inventory and operating rating levels according to the *Manual for Bridge Evaluation* (AASHTO, 2016) in North America, superloads that may need to pass the bridge, or different load levels calibrated for different reliability indices as used in the Netherlands (Rijkswaterstaat, 2013), can then be evaluated with the updated model. For these assessment calculations, the critical loading position in the longitudinal and transverse direction should be determined for each load model applied to the analytical model. Depending on the lane layout of the bridge, the assessment should include design trucks or tandems in all lanes and pattern loading across the spans.

As for all load tests, and as highlighted in Part II, the influence of the diagnostic load test on the assessment lies on the side of the live-load effects. When a rating factor is calculated, the diagnostic load test serves to update the load effect due to the live loads. When a Unity Check is calculated, the diagnostic load test serves to update the load effect due to the considered load combination. The diagnostic load test gives no information about the capacity of the considered section for the failure mode of interest. For the assessment calculations, the capacity is determined based on the governing code equations, and the effect of material degradation and section losses should be taken into account. In special cases, more elaborate models for the capacity can be used so that additional sources of capacity that are not included in the simplified code equations can be taken into account. An example is the use of the extended strip model for the determination of the maximum load on reinforced concrete slab bridges (Lantsoght et al., 2017a).

When the different relevant scenarios are evaluated with the updated model and the resulting Rating Factors (in North America) or Unity Checks (in Europe) are obtained, recommendations can be given. If the Rating Factors for all relevant rating levels are larger than one or if the Unity Checks for the relevant load levels are smaller than one, no actions are necessary. If the Rating Factors are smaller than one or the Unity Checks are larger than one, the following recommendations can be given:

- Carrying out a proof load test (if it is expected that such a test could demonstrate adequate performance at higher load levels)
- Posting of the bridge (NCHRP, 2014)

172 Load Testing of Bridges

- Strengthening the bridge
- Demolishing and replacing the bridge.

For such bridges, permits for superloads should be carefully evaluated when requested. The engineer responsible for the diagnostic load test can only give a recommendation to the bridge owner; the final management decision lies with the bridge owner.

8.6 Summary and conclusions

This chapter describes the general methodology that can be followed for diagnostic load tests on new and existing bridges. For new bridges, diagnostic load tests can be required prior to opening the bridge. For existing bridges, diagnostic load tests can be used to have a better understanding of the overall structural behavior and/or to improve the analytically determined assessment of the bridge.

For diagnostic load tests, the preparation stage involves the determination of the required magnitude of the load during the test, the loading positions (for static load tests) and/or the load paths (for dynamic load tests). The applied load should be large enough to result in measurable structural responses. The preparation stage also involves the development of the sensor plan, planning of on-site activities, and safety plan.

During the load test, and depending on the type of load test, it may be necessary to follow the responses in real time, or after each load case or load path. Communication between the load operators or truck drivers and the testing engineers is important during the execution stage.

After the load test, a first on-site check of the measured structural responses is required. The quality of the data should be checked. The measured structural behavior should be checked in terms of repeatability, symmetry, and linearity of the measured responses. Then a first comparison to the analytically determined responses should be done. If there are doubts regarding the structural behavior or if possible sensor malfunctioning is identified, a load case or load path can be repeated at that moment. The complete post-processing of the measurement data happens in the office after finishing all on-site activities. This post-processing includes correcting the data for environmental effects, finding net deflections where necessary, and filtering out erroneous measurement results. These processed data will then be shown in the measurement report and can be used to calibrate the available analytical models. For new bridges, such calibration may not be required. For existing bridges, the calibrated analytical model can be used to obtain a more realistic assessment, taking into account the actual bridge behavior measured in the field.

References

AASHTO (2015) *AASHTO LRFD Bridge Design Specifications*, 7th edition with 2015 Interim Specifications. American Association of State Highway and Transportation Officials, Washington, DC.

AASHTO (2016) *The Manual for Bridge Evaluation with 2016 Interim Revisions*. American Association of State Highway and Transportation Officials, Washington, DC.

ACI Committee 342 (2016) ACI 342R-16: Report on flexural live load distribution methods for evaluating existing bridges. American Concrete Institute, Farmington Hills, MI.

Aktan, A. E., Zwick, M., Miller, R. & Shahrooz, B. (1992) Nondestructive and destructive testing of decommissioned reinforced concrete slab highway bridge and associated analytical studies. *Transportation Research Record: Journal of the Transportation Research Board*, 1371, 142–153.

Alkhrdaji, T., Nanni, A. & Mayo, R. (2000) Upgrading Missouri transportation infrastructure: Solid reinforced-concrete decks strengthened with fiber-reinforced polymer systems. *Transportation Research Record*, 1740, 157–163.

Al-Mahaidi, R., Taplin, G. & Giufre, A. (2000) Load distribution and shear strength evaluation of an old concrete T-beam bridge. *Transportation Research Record*, 1696, 52–62.

Amer, A., Arockiasamy, M. & Shahawy, M. (1999) Load distribution of existing solid slab bridges based on field tests. *Journal of Bridge Engineering*, 4, 189–193.

American Association of State Highway and Transportation Officials (2011) *The Manual for Bridge Evaluation*. American Association of State Highway and Transportation Officials, Washington, DC.

Arockiasamy, M. & Amer, A. (1997a) *Load Distribution on Highway Bridges Based on Field Test Data: Phase I*. Center for Infrastructure and Constructed Facilities, Florida Atlantic University, Boca Raton, FL.

Arockiasamy, M. & Amer, A. (1997b) *Load Distribution on Highway Bridges Based on Field Test Data: Phase II*. Center for Infrastructure and Constructed Facilities, Florida Atlantic University, Boca Raton, FL.

Arockiasamy, M. & Amer, A. (1998) *Load Distribution on Highway Bridges Based on Field Test Data: Phase III*. Center for Infrastructure and Constructed Facilities, Florida Atlantic University, Boca Raton, FL.

Au, A., Lam, C., Au, J. & Tharmabala, B. (2013) Eliminating deck joints using debonded link slabs: Research and field tests in Ontario. *Journal of Bridge Engineering*, 18, 768–778.

Barker, M. G. (2001) Quantifying field-test behavior for rating steel girder bridges. *Journal of Bridge Engineering*, 6, 254–261.

Barnes, R. W., Stallings, J. M. & Porter, P. W. (2003) Live-load response of Alabama's high-performance concrete bridge. *Transportation Research Record*, 1845, 115–124.

Bell, E. S. & Sipple, J. D. (2009) Chapter 27: Special topics studies for baseline structural modeling for condition assessment of in-service bridges. In: Mahmoud, K. (ed.) *Safety and Reliability of Bridge Structures*. CRC Press, Boca Raton, FL.

Bonifaz, J., Zaruma, S., Robalino, A. & Sanchez, T.A. (2018) Bridge diagnostic load testing in Ecuador: Case studies. *IALCCE 2018*. Ghent, Belgium.

Bridge Diagnostics Inc. (2012) *Integrated Approach to Load Testing*. BDI, Boulder, CO.

Catbas, F. N., Ciloglu, S. K. & Aktan, A. E. (2004) Strategies for load rating of infrastructure populations: A case study on T-beam bridges. *Structure and Infrastructure Engineering*, 1, 221–238.

den Uijl, J. A. & Kaptijn, N. (2004) Structural consequences of ASR: An example on shear capacity. *Heron*, 47, 1–13.

Ferrand, D., Nowak, A. S. & Szerszen, M. M. (2005) Field test and finite element analysis of isotropic bridge deck. *Transportation Research Record*, CD 11-S, 153–158.

Hag-Elsafi, O. & Kunin, J. (2006) Load testing for bridge rating: Dean's Mill Road over Hannacrois Creek. Transportation Research and Development Bureau, New York State Department of Transportation, Albany, NY.

Harris, D. K., Cousins, T., Murray, T. M. & Sotelino, E. D. (2008) Field investigation of a sandwich plate system bridge deck. *Journal of Performance of Constructed Facilities*, 22, 305–315.

Harris, D. K., Gheitasi, A. & Civitillo, J. M. (2015) Field testing and numerical modeling of a hybrid composite beam bridge in Virginia. *European Bridge Conference*. Edinburgh, UK.

Harris, D. K., Civitillo, J. M. & Gheitasi, A. (2016) Performance and behavior of hybrid composite beam bridge in Virginia: Live load testing. *Journal of Bridge Engineering*, 21, 04016022.

Hernandez, E. S. (2018) *Service Response and Evaluation of Prestressed Concrete Bridges through Load Testing*. PhD thesis, Missouri University of Science and Technology, Rolla, MO.

Hernandez, E. S. & Myers, J. J. (2015) In-situ field test and service response of Missouri Bridge A7957. *European Bridge Conference*. Edinburgh, UK.

Hernandez, E. S. & Myers, J. J. (2016) Initial in-service response and lateral load distribution of a prestressed self-consolidating concrete bridge using field load tests. In: Bakker, J., Frangopol, D. M. & Brangopol, K. V. (eds.) *The Fifth International Symposium on Life-Cycle Civil Engineering (IALCCE 2016)*. CRC Press, Delft, The Netherlands.

Hernandez, E. S. & Myers, J. J. (2018a) Diagnostic test for load rating of a prestressed SCC bridge. *ACI Special Publication*, 323.

Hernandez, E. S. & Myers, J. J. (2018b) Strength evaluation of prestressed concrete bridges by dynamic load testing. *Ninth International Conference on Bridge Maintenance, Safety and Management (IABMAS 2018)*. CRC Press, Melbourne, Australia.

Hodson, D. J., Barr, P. J. & Pockels, L. (2013) Live-load test comparison and load ratings of a post-tensioned box girder bridge. *Journal of Performance of Constructed Facilities*, 27, 585–593.

Jáuregui, D. V. & Barr, P. J. (2004) Nondestructive evaluation of the I-40 bridge over the Rio Grande river. *Journal of Performance of Constructed Facilities*, 18, 195–204.

Jauregui, D. V., Licon-Lozano, A. & Kulkarni, K. (2010) Higher level evaluation of a reinforced concrete slab bridge. *Journal of Bridge Engineering*, 15, 172–182.

Jeffrey, A., Breña, S. F. & Civjan, S. A. (2009) *Evaluation of Bridge Performance and Rating Through Nondestructive Load Testing*. University of Massachusetts Amherst, Amherst, MA.

Jones, B. P. (2011) *Reevaluation of the AASHTO Effective width Equation in Concrete Slab Bridges in Delaware*. MSc thesis, University of Delaware.

Kirkpatrick, J., Long, A. E. & Thompson, A. (1984a) Load distribution characteristics of spaced M-beam bridge decks. *The Structural Engineer*, 62B, 86–88.

Kirkpatrick, J., Rankin, G.I.B. & Long, A. E. (1984b) Strength evaluation of M-beam bridge deck slabs. *The Structural Engineer*, 62B, 60–68.

Konda, T. F., Klaiber, F. W., Wipf, T. J. & Schoellen, T. P. (2007) Precast modified beam-in-slab bridge system: An alternative replacement for low-volume roads. *Transportation Research Record*, 1989, 335–346.

Lantsoght, E. O. L., Van Der Veen, C., De Boer, A. & Alexander, S.D.B. (2017a) Extended strip model for slabs under concentrated loads. *ACI Structural Journal*, 114, 565–574.

Lantsoght, E. O. L., Van Der Veen, C., Hordijk, D. A. & De Boer, A. (2017b) State-of-the-art on load testing of concrete bridges. *Engineering Structures*, 150, 231–241.

LaViolette, M., Sanchez, T. A. & Fiallo, M. (2017) Construction of Ecuador's first launched steel girder bridge. *Proceedings, National ABC Conference*. Florida International University, Miami, FL.

Lawver, A., French, C. & Shield, C. K. (2000) Field performance of integral abutment bridge. *Transportation Research Record*, 1740, 108–117.

Lim, S., Akiyama, M. & Frangopol, D. M. (2016) Assessment of the structural performance of corrosion-affected RC members based on experimental study and probabilistic modeling. *Engineering Structures*, 127, 189–205.

Lin, W., Taniguchi, N., Yoda, T., Hansaka, M., Satake, S. & Sugino, Y. (2018) Renovation of existing steel railway bridges: Field test and numerical simulation. *Advances in Structural Engineering*, 21, 809–823.

McGrath, T. J., Selig, E. T. & Beach, T. J. (1995) Structural behavior of three-sided arch span bridge. *Transportation Research Record*, 1541.

Ministerio De Fomento: Direccion General De Carreteras (1999) *Recomendaciones para la realización de pruebas de carga de recepción en puentes de carretera*. Ministerio de Fomento, Madrid, Spain.

Mordak, A. G. & Manko, Z. (2008) Static load tests of posttensioned, prestressed concrete road bridge over reservoir water plant. *Transportation Research Record*, 2050, 90–97.

Myers, J. J., Holdener, D. & Merkle, W. (2012) Chapter 9: Load testing and load distribution of fiber reinforced, polymer strengthened bridges: Multi-year, post construction/post retrofit performance evaluation. In: Jain, R. & Lee, L. (ed.) *FRP Composites and Sustainability: Focusing on Innovation, Technology Implementation and Sustainability.* Springer, New York, NY.

NCHRP (1998) Manual for Bridge Rating through Load Testing. Washington, DC.

NCHRP (2014) *NCHRP Synthesis 453: State Bridge Load Posting: Processes and Practices. A Synthesis of Highway Practice.* Transportation Research Board, Washington, DC.

Nilimaa, J., Blanksvärd, T. & Taljsten, B. (2015) Assessment of concrete double-trough bridges. *Journal of Civil Structural Health Monitoring,* 2015, 29–36.

Ohanian, E., White, D. & Bell, E. S. (2017) Benefit analysis of in-place load testing for bridges. *Transportation Research Board Annual Compendium of Papers,* 14.

Orban, Z. & Gutermann, M. (2009) Assessment of masonry arch railway bridges using non-destructive in-situ testing methods. *Engineering Structures,* 31, 2287–2298.

Ponton, M. E., Robalino, A. F. & Sanchez, T. A. (2016) Stability considerations for the construction of steel I-girder bridges using the incremental launching method. *Annual Stability Conference Structural Stability Research Council.* Orlando, FL.

Rijkswaterstaat (2013) *Guidelines Assessment Bridges: Assessment of Structural Safety of an Existing Bridge at Reconstruction, Usage and Disapproval* (in Dutch), RTD 1006:2013 1.1. Utrecht, the Netherlands.

Robalino, A. F. & Sanchez, T. A. (2017) Global lateral-torsional buckling of I-girder systems in cantilever. *Proceedings, Annual Stability Conference, Structural Stability Conference, Structural Stability Research Council.* San Antonio, TX.

Russo, F. M., Wipf, T. J. & Klaiber, F. W. (2000) Diagnostic load tests of a prestressed concrete bridge damaged by overheight vehicle impact. *Transportation Research Record,* 1696, 103–110.

Sanayei, M., Reiff, A. J., Brenner, B. R. & Imbaro, G. R. (2016) Load rating of a fully instrumented bridge: Comparison of LRFR approaches. *Journal of Performance of Constructed Facilities,* 2.

Sanchez, T. A., LaViolette, M. & Fiallo, M. (2017) Construction of Ecuador's first launched steel girder bridge. *Proceedings, the International Bridge Conference, Engineers' Society of Western Pennsylvania.* National Harbor, MD.

Sanchez, T. A., Robalino, A. F. & Graciano, C. (2018) Interaction between patch loading, bending, and shear in steel girder bridges erected with the incremental launching method. *Proceedings, Annual Stability Conference, Structural Stability Conference,* Structural Stability Research Council, Baltimore, MD.

Saraf, V. K. (1998) Evaluation of existing RC slab bridges. *Journal of Performance of Constructed Facilities,* 12, 20–24.

Taylor, P., Hosteng, T., Wang, X. & Phares, B. (2016) Evaluation and testing of a lightweight fine aggregate concrete bridge deck in Buchanan County, IA.

Taylor, S. E., Rankin, B., Cleland, D. J. & Kirkpatrick, J. (2007) Serviceability of bridge deck slabs with arching action. *ACI Structural Journal,* 104, 39–48.

Velázquez, B. M., Yura, J. A., Frank, K. H., Kreger, M. E. & Wood, S. L. (2000) Diagnostic load tests of a reinforced concrete pan-girder bridge. The University of Texas at Austin, Austin, TX, USA.

Wipf, T. J., Ritter, M. A. & Wood, D. L. (2000) Evaluation and field load testing of timber railroad bridge. *Transportation Research Record,* 1696, 323–333.

Yang, Y. & Myers, J. J. (2003) Live-load test results of Missouri's first high-performance concrete superstructure bridge. *Transportation Research Record,* 1845, 96–103.

Appendix

Determination of Experimental Rating Factor According to Barker

In Barker's method, the experimental rating at the inventory level Exp_INV is based on the separate effects that cause differences between the actual structure and the analytical model, and is expressed as the ratio of the experimental rating factor at the inventory level Exp_INV to the analytical rating factor at the inventory level Ana_INV:

$$\frac{Exp\_INV}{Ana\_INV} = \left[\left(\frac{I_{mA}}{I_{mE}}\right) \left(\frac{M_E}{M_T}\right) \left(\frac{M_{LE}}{M_E}\right) \left(\frac{DF_A}{DF_E}\right) \left(\frac{M_{WL}}{\frac{M_{LE}}{DF_E}\frac{M_{RVW}}{M_{TRK}}}\right) \left(\frac{S_A^{ADIM}}{S_A}\right) \left(\frac{S_E}{S_A^{ADIM}}\right) \right] \tag{8.1}$$

The factor $\dfrac{I_{mA}}{I_{mE}}$ represents the contribution from the impact factor, with I_{mA} the analytical impact factor and I_{mE} the experimental impact factor. The value of the impact factor I_{mE} is determined in the field, whereas I_{mA} is calculated based on the AASHTO impact factor (American Association of State Highway and Transportation Officials, 2011). The effect of the measured impact factor can only be taken into account if the field test can cover the representative effect of different truck configurations and weights. The ratio $\dfrac{M_E}{M_T}$ represents the contribution from bearing restraint force effects, with M_E defined as the elastic measured moment with the bearing restraint moments removed and M_T the experimental total moment, which for steel-concrete composite bridges is the sum of the bending moment about the neutral axis of the steel girder M_L, the bending moment about the neutral axis of the concrete deck, M_U, and a force couple representing the inter-action composite action $N \times a$. The factor $\dfrac{M_{LE}}{M_E}$ represents the contribution from longitudinal distribution of moment, with M_{LE} the experimental elastic moment adjusted for longitudi-nal distribution and M_E the elastic measured moment as discussed previously. The factor $\dfrac{DF_A}{DF_E}$ gives the contribution from lateral load distribution, with DF_A the analytical distribu-tion factor and DF_E the distribution factor in the experiment. The factor $\dfrac{M_{WL}}{\left(\frac{M_{LE}}{DF_E}\right)\left(\frac{M_{RVW}}{M_{TRK}}\right)}$ takes into account the contribution from additional system stiffness, with M_{WL} the analytical wheel line moment for the RVW (rating vehicle weight) truck, M_{LE} the experimental elastic moment adjusted for longitudinal distribution, DF_E the experimental distribution factor, M_{RVW} the analytical wheel line RVW truck moment, and M_{TRK} the analytical wheel line test truck moment. The ratio M_{RVW}/M_{TRK} compares the actual test truck response to an equivalent rating response so that M_{RVW} can produce ratings for any analytical rating vehicle. The factor $\dfrac{S_A^{ADIM}}{S_A}$ gives the contribution from actual section dimensions for section modulus, with S_A^{ADIM} the analytical section modulus with actual measured dimensions and S_A the analytical section modulus with design dimensions. Similarly, the factor $\dfrac{S_E}{S_A^{ADIM}}$ accounts for the contribution from unintended or additional composite action, with S_E the experimental section modulus.

The experimental lateral load distribution factor is the percentage of total moment resisted by an individual girder, and can be determined as follows:

$$DF_E = \frac{2(\sigma_i S_{Ai})_{Critical Girder}}{\Sigma(\sigma_i S_{Ai})} \tag{8.2}$$

with σ_i the bottom flange stress for girder i, and S_{Ai} either the actual section modulus or the nominal design section modulus for girder i.

To determine M_E and M_{LE}, the resulting moments in the test need to be analyzed and different contributions need to be separated. The external moment at the support $M_{ext,sup}$ can be determined based on the measured strains and calculated stresses on the girders. When the stresses are known, the bearing force at an abutment $F_{bearing}$ is determined as:

$$F_{bearing} = A_{bf}\sigma_{bf} \tag{8.3}$$

with A_{bf} the area of the bottom flange at the bearing and σ_{bf} the measured stress on the bottom flange at the bearing. A bearing force $F_{bearing}$ at a pier requires taking into account the girders on both sides of the support:

$$F_{bearing} = \frac{(\sigma_{bf}^{pier2} - \sigma_{bf}^{pier1})A_{bf}}{2} \tag{8.4}$$

with σ_{bf}^{pier1} and σ_{bf}^{pier2} the measured stresses on the left and right side of the bearing. The external moment at the support, $M_{ext,sup}$, is then determined as:

$$M_{ext,sup} = F_{bearing} \times d_{NA} \tag{8.5}$$

with d_{NA} the depth of the neutral axis. When $M_{ext,sup}$ is known, the actual shape of the bending moment diagram is known to determine the factor $\frac{M_{LE}}{M_E}$. Assuming a constant moment of inertia in all spans, the moment distribution can be determined as follows:

$$DM = \frac{\left(\dfrac{1}{L_i}\right)_{Critical\_Span}}{\Sigma\left(\dfrac{1}{L_i}\right)} \tag{8.6}$$

with DM the moment distribution factor and L_i the span length on each side of the pier. Linear interpolation between the moments at the piers is used to find the bearing restraint moment M_{BR} at the critical section:

$$M_{BR}^{Critical\_Section} = \frac{M_{BR}^{Pier1} + M_{BR}^{Pier2}}{2} \tag{8.7}$$

178 Load Testing of Bridges

To find the total experimental moment M_T, the bending moment about the steel neutral axis M_L is necessary, for which the stresses in the girder need to be determined. When strain gages are placed over the height of a girder, the strain distribution can be derived based on linear interpolation. The stress profile in the girder, including the axial bearing restraint stress σ^{wa}, is then calculated as:

$$\sigma^{wa} = -\frac{1}{Sl}d + \frac{Int}{Sl} \tag{8.8}$$

with Int the neutral axis from the bottom flange, Sl the slope of the stress profile, and d the depth from the bottom flange. The axial stress from the bearing force σ_{axial} can be removed as follows:

$$\sigma_{axial} = \frac{BF}{A_{comp}} \tag{8.9}$$

with BF the bearing force and A_{comp} the equivalent composite area:

$$A_{comp} = A_{girder} + \frac{A_{Conc}}{n} \tag{8.10}$$

with A_{girder} the nominal or measured area of the steel girder section, A_{Conc} the nominal or measured area of the concrete deck, and n the modular ratio when A_{girder} is determined for a steel girder. Now, the effect of the bearing axial force can be removed from the stress profile as follows:

$$\sigma = -\frac{1}{Sl}d + \frac{Int}{Sl} - \sigma_{axial} \tag{8.11}$$

The experimental total moment M_T can be divided into three components (see Fig. 8.18):

1 Bending moment about the steel neutral axis, M_L
2 Bending moment about the concrete neutral axis, M_U
3 A couple representing the interaction composite action, $N \times a$

In other words, for steel girders with a concrete deck:

$$M_T = M_L + M_U + N \times a \tag{8.12}$$

$$M_L = (\sigma_0 - \sigma_{CG})S_{steel}^{ADIM} \tag{8.13}$$

$$M_U = \frac{(E_{conc}I_{conc})}{(E_{steel}I_{Steel}^{ADIM})}M_L \tag{8.14}$$

$$N \times a = \sigma_{CG}A_{steel}^{ADIM} \times a \tag{8.15}$$

with σ_0 the stress at bottom flange calculated with Equation (8.11), σ_{CG} the stress at the steel centroid calculated with Equation (8.11), d_{CG} the depth of the steel centroid from the bottom flange using the measured dimensions, E_{conc} the Young's modulus of the concrete,

Methodology for Diagnostic Load Testing   179

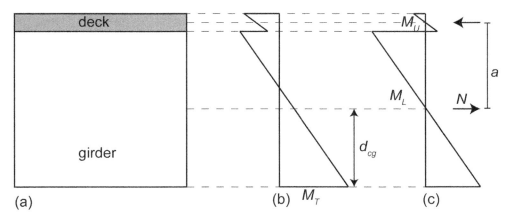

Figure 8.18 Total experimental moment: (a) composite section; (b) measured bending moments M_T; (c) elastic moments M_E

Source: Modified from Barker (2001).

E_{steel} the Young's modulus of the steel, I_{conc} the moment of inertia of the concrete slab, I_{steel}^{ADIM} the moment of inertia of the steel girder using measured dimensions, A_{steel}^{ADIM} the area of the steel girder using measured dimensions, and a the moment arm between the centroids of the steel girder and the concrete deck.

The elastic moments M_E with bearing restraint axial forces and bearing restraint moments M_{BR} removed are:

$$M_E = M_T - M_{BR} \tag{8.16}$$

The experimental section modulus S_E equals:

$$I_{Exp} = -(M_T)Sl \tag{8.17}$$

$$S_E = \frac{I_{Exp}}{Int} \tag{8.18}$$

The elastic longitudinal adjusted moment M_{LE} represents the elastic moment that should be at the section if the experimental longitudinal distribution would equal the moment from the analytical rating. The experimental moment diagram can be constructed with three measurement points and compared to the analytical moment diagram under the same truck load:

$$STAT_A = M_C^2 - (1-\alpha)M_C^1 - \alpha M_C^3 \tag{8.19}$$

$$STAT_E = M_E^2 - (1-\alpha)M_E^1 - \alpha M_E^3 \tag{8.20}$$

with superscripts 1, 2, and 3 to denote the point at the left support, at the maximum span moment, and at the right support, respectively; $STAT_A$ the analytical static moment for the load truck; $STAT_E$ the static moment from the experiment; M_c the analytical moment at left, α, and right sections; and α the percentage of length to the point of maximum moment (i.e. the distance from point 1 to point 2 divided by the distance from point 1 to point 3). The

180 Load Testing of Bridges

longitudinal adjustment moment at the critical section is then:

$$M_{LE} = \frac{STAT_E}{STAT_A} M_C^2 \tag{8.21}$$

with M_C^2 the analytical moment at the section with the largest span moment (point number 2).

The ratio from Equation (8.1) can be used for improving analytical ratings determined from hand calculations or spreadsheets by programming the procedures introduced in this section.

Chapter 9

Example Field Test to Load Rate a Prestressed Concrete Bridge

Eli S. Hernandez and John J. Myers

Abstract

Field tests have been widely adopted to monitor and validate the use of novel construction technologies and to perform an experimental evaluation of existing bridges. The AASHTO Manual for Bridge Evaluation defines two test methodologies: proof load tests and diagnostic load tests. Proof load tests are employed to estimate the maximum safe live load a bridge can withstand without undergoing inelastic deformation. Diagnostic load tests are used to better understand a bridge's in-service response. Diagnostic load tests have largely proven that existing bridge superstructures possess additional strength capacity than predicted by analytical methods. This difference can be explained by considering some in-situ parameters that are beneficial to the bridge's performance. This chapter introduces an example diagnostic load test conducted on the superstructure of Bridge A7957, built in Missouri, USA, to illustrate how experimental in-situ parameters can be included in the estimation of a bridge load rating. The experimental load rating resulted to be less conservative than the analytical load rating.

9.1 Introduction

Field load tests represent an alternative to conduct and effective evaluation of the in-service performance of a bridge structure and its live load-carrying capacity. In general, the American Association of State Highway and Transportation Officials (AASHTO) *Manual for Bridge Evaluation* (MBE) defines two different types of test methodologies: proof load tests and diagnostic load tests (AASHTO, 2010). Proof load tests are employed to estimate the maximum safe live load a bridge can withstand without undergoing inelastic deformation. Diagnostic tests are used to validate design assumptions and to verify the performance of a structure – most times improved – by implicitly considering in-situ parameters which are beneficial for the bridge's response (Cai and Shahawy, 2003). For instance, diagnostic load tests permit the verification of design and analysis assumptions such as the actual lateral load distribution and dynamic load allowance (impact factor) of a bridge.

Load rating is the strength evaluation procedure employed to estimate the in-service load that a bridge structure can withstand without suffering inelastic deformation. Traditionally, bridge evaluation standards (AASHTO, 2010) provide two approaches to load rating: analytical calculations and field testing. Analytical ratings are based on simplifying assumptions and may result in a conservative rating evaluation of a bridge. Conversely, experimental load ratings present a more realistic visualization of the live-load capacity of a bridge because it provides an in-service, as-built characterization of its performance.

DOI: https://doi.org/10.1201/9780429265426

The diagnostic load test of Bridge A7957, presented herein, describes (1) the bridge instrumentation plan; (2) the diagnostic load test program; (3) comparisons between the experimental and analytical lateral load distribution factors; and (4) comparisons of the experimental and analytical dynamic load allowance proposed by the AASHTO *LRFD Bridge Design Specifications* (AASHTO LRFD) (AASHTO, 2012). Finally, a comparison between the analytical and experimental load rating is provided to illustrate how incorporating site-specific data, measured in a diagnostic load test, results in a less conservative rating evaluation compared to the analytical strength evaluation procedure.

9.2 Sample bridge description

During the summer and fall of 2013, the two-lane expansion project of US Route 50 was conducted by the Missouri Department of Transportation (MoDOT). The project included the construction of Bridge A7957, which spans the Maries River in Osage County, located 10 miles west of Linn, Missouri, USA. Bridge A7957 is a three-span, precast, prestressed concrete (PC) bridge made continuous via a cast-in-place (CIP) reinforced concrete (RC) slab deck that has a 30-degree skew angle and a smooth surface condition (Figs. 9.1–9.3). The PC Nebraska University girders (NU53) are 1.35 m (53 in.) high at each span and were designed with different concrete mixtures. The first span has girders that are 30.48 m (100 ft) long and were fabricated using conventional concrete (CC) with a specified design strength of 55.2 MPa (8000 psi). The second span measures 36.58 m (120 ft) long and used HS-SCC with a target compressive strength of 68.9 MPa (10,000 psi). The girders of the third span are 30.48 m (100 ft) long and were cast using a normal-strength self-consolidating concrete (NS-SCC) mixture with a design compressive strength of 55.2 MPa (8000 psi). Prestressed concrete panels with a specified compressive strength of 41.4 MPa (6000 psi) extend between the top flanges of the girders in the transverse direction

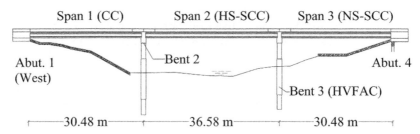

Figure 9.1 Bridge A7957 elevation. Conversion: 1 ft = 0.3048 m

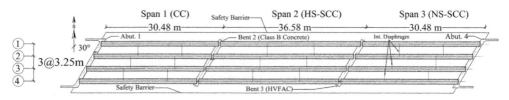

Figure 9.2 Bridge A7957 plan view. Conversion: 1 ft = 0.3048 m

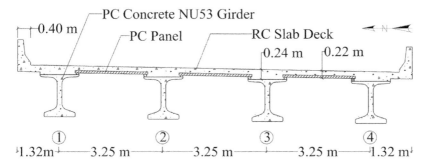

Figure 9.3 Bridge A7957 cross section. Conversion: 1 ft = 0.3048 m

(Fig. 9.3). The CIP deck was cast with a conventional concrete mix with a target design strength of 41.4 MPa (6000 psi). Two end abutments and two intermediate bents support the superstructure (Fig. 9.1).

The two end abutments and intermediate bent 2 were designed with conventional concrete using a 20% fly ash replacement of portland cement with design compressive strength of 20.7 MPa (3000 psi) (Fig. 9.2).

Intermediate bent 3 was constructed with a high-volume fly ash concrete (HVFAC) using a 50% fly ash replacement of portland cement with a specified compressive strength of 20.7 MPa (3000 psi).

9.3 Bridge instrumentation plan

An instrumentation plan was designed and implemented during the preconstruction of Bridge A7957. The structural elements that were selected to be instrumented included two PC girders per span and two PC panels. The PC panels were chosen at the second span midspan, between girder lines 2 and 3 and girder lines 3 and 4. The type of sensors employed and details about their location and installation are provided in the following subsections.

9.3.1 Installation of embedded sensors

A total of 86 vibrating wire strain gages (VWSG) with built-in thermistors (type EM-5) were utilized to record the strain and stress variations in the PC girders and RC deck slab from fabrication through service life. Before casting, a total of 62 VWSG were installed in all spans within the PC girders of lines 3 and 4. The PC girder cluster locations at which the VWSG were embedded are illustrated in Figure 9.4.

The instrumentation clusters were selected at two cross-sections within each girder of span 1 and span 3. One section was located at midspan, and the other was placed at approximately 0.60 m (2 ft) from the support centerline of bents 2 and 3. The instrumentation clusters for the center span (span 2) were arranged at three different cross-sections: one at midspan and two at approximately 0.60 m (2 ft.) from each support centerline. The cluster sections in the second span were arranged at three different cross-sections: one at the midspan and the other sections approximately 0.60 m (2 ft) from each support centerline. Details on VWSGs installed at the girders' near-support and midspan sections before

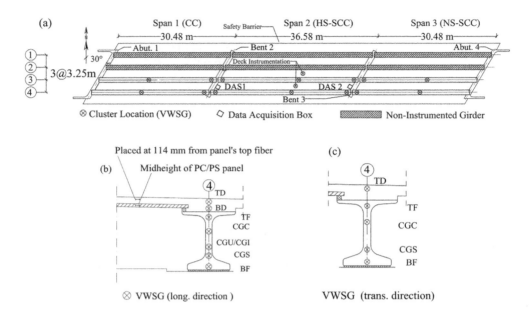

Figure 9.4 VWSG installation details: (a) cluster locations layout; (b) midspan section sensors; (c) near-end section sensors (Hernandez and Myers, 2016a). Conversion: 1 ft = 0.3048 m

concrete was cast are illustrated in Figures 9.4b and 9.4c. The following notation was employed to identify the layer where the VWSG sensors were installed across the girders' cross section:

- TD: 150 mm from the deck's bottom fiber
- BD: 50 mm from the deck's bottom fiber
- TF: 50 mm below the girder's top fiber
- CGC: center of gravity of composite section
- CGU/CGI: center of gravity of non-composite section (only at midspan clusters)
- CGS: center of gravity of prestressed tendons
- BF: 50 mm from girder's bottom fiber.

Twenty VWSGs were placed within the CIP RC slab deck (Fig. 9.4) in the longitudinal direction (sensors TD and BD). A VWSG was transversely deployed at the mid-height of two selected PC/PS panels (Figs. 9.4a and 9.4b). The last two VWSGs were located along the bridge's transverse direction, between girder lines 2 and 3 and girder lines 3 and 4. These two sensors were placed directly above the panel sensors, separated 114 mm from the panels' top fiber (Fig. 9.4c).

9.3.2 Data acquisition by non-contact and remote equipment

A non-contact, remote equipment was employed to collect (a) the static vertical deflection at midspan of girders 1–4 (for the three spans) as shown in Figure 9.5; (b) the static vertical deflection at several sections spaced 1/6 of the span length along girder 3 (for all the spans); and (c) the vertical dynamic deflection at girder 3's midspan (only spans 1 and 3). The next

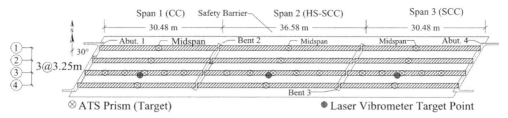

Figure 9.5 Bridge A7957 ATS Prism locations. Conversion: 1 ft = 0.3048 m

Figure 9.6 Non-contact remote data acquisition systems: (a) automated total station; (b) remote sensing vibrometer (RSV-150)

sections present details about the non-contact laser equipment employed to collect the experimental data described in (a)–(c).

9.3.2.1 Automated total station (ATS)

The automated total station (ATS), Leica TCA2003 (Fig. 9.6) was employed to record the vertical deflection along girder line 3 (Fig. 9.5) during the diagnostic load test (Hernandez and Myers, 2018). The ATS has an accuracy of 1 mm (0.039 in.) ± 1 ppm (parts per million) for distance measurements and a precision of 0.5 arc-seconds for angular measurements. The ATS collected three-dimensional coordinates of the prisms (targets) shown in Figure 9.5 by recording the horizontal and vertical angle as well as the distance between the ATS and target. The instrument was configured to collect three readings per target. This task is performed by four internal diodes installed to optically read a fine bar code set on a glass ring inside the Leica TCA2003. During monitoring, the equipment continuously read the bar codes on the horizontal and vertical planes by sending a laser ray that reflects on the targets mounted on the structure. The accuracy of the vertical deflections estimated using the ATS has been reported to be ±0.1 mm (0.005 in.) by Merkle and Myers (2006). Twenty-four sections were selected to monitor the vertical deflection of the girders. Fifteen ATS prisms were deployed along the third girder at $1/6L$, $1/3L$, $1/2L$, $2/3L$, and $5/6L$ of each span. Three additional prisms were mounted at the other girders' midspan (at $L/2$) in each span (Fig. 9.5). The ATS prisms have an internal magnet that kept them fixed to steel plates that were attached to the girders' bottom flange with an epoxy adhesive.

9.3.2.2 Remote sensing vibrometer (RSV-150)

The remote sensing vibrometer (RSV-150), shown in Figure 9.6b, was used to record the bridge's dynamic vertical deflection of the exterior spans' girder 3 (midspan). The RSV-150 has a bandwidth up to 2 MHz for nondestructive test (NDT) measurements and can detect the vibration and displacement of distant points of structures with limited access. The precision of the RSV-150 is ±0.025 mm (0.001 in.) when it records the dynamic response of a member.

9.4 Diagnostic load test program

A diagnostic load test program was designed to obtain the maximum static and dynamic response of Bridge A7957. Six H20 dump trucks were employed to load the bridge during the static load test. These tests were conducted in three parts (see test days 1–3 in Table 9.1).

The trucks were loaded with sand and gravel before the tests were performed. The dynamic tests were performed on test day 3 (Table 9.1). The test procedures and load configurations used during the diagnostic test program have been reported by the authors elsewhere (Hernandez and Myers, 2016a, 2016b, 2017, 2018). The average dimensions of the trucks employed during the diagnostic tests are shown in Figure 9.7.

Table 9.1 Truck weights

Test Day	Truck	Rear (kN)	Front (kN)	Total (kN)
1, 2	1	158.2	74.0	232.2
1, 2	2	161.6	57.2	218.8
1, 2	3	150.3	56.0	206.3
1, 2	4	178.0	75.3	253.3
1, 2	5	170.2	77.9	248.1
1, 2	6	166.4	71.6	238.0
3	1	164.6	61.1	225.7
3	2	180.3	70.8	251.1
3	3	169.1	70.4	239.5

Note: During days 1 and 2, the trucks were loaded with the same weight (Hernandez and Myers, 2018). Conversion factor: 1 kN = 0.2248 kip.

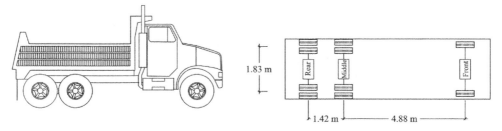

Figure 9.7 H20 dump truck average dimensions. Conversion factor: 1 m = 3.28 ft

9.4.1 Static load test

Figure 9.8 shows the 13 static load configurations employed to obtain the maximum static vertical deflection of the bridge when a single lane or two lanes were loaded. For the first three load configurations (stops 1–3), two lanes of trucks were driven from east to west and parked separately at the centers of spans 3, 2 and 1 (Figs. 9.8a–9.8c).

For load configurations 4–6 (stops 4–6), the trucks were driven from west to east, and parked separately at the centers of spans 1, 2, and 3 (Figs. 9.8d–9.8f). For these first six load configurations, the center of the trucks' exterior wheels was separated 3.25 m (10.67 ft) from the safety barrier's edge (Fig. 9.9a). For stops 7–9, the trucks were driven along the west-east direction as illustrated in Figures 9.8g–9.8i. The trucks' exterior axles were placed at 0.60 m (2 ft) from the safety barrier's edge (Fig. 9.9b). These first nine stops simulated two-lane load cases.

For the next three load configurations (stops 10–12), shown in Figures 9.8j–9.8l, a lane of three trucks was driven along the west-east direction on the south side of the bridge. The trucks were parked at 0.60 m (2 ft) from the barrier's edge. For the last load configuration (stop 13), the three trucks were driven from east to west along the north side of the bridge. The lane of trucks, loading the central region of the second span (Fig. 9.8m), was separated 1.63 m (5.33 ft) from the barrier's edge (Fig. 9.9c).

9.4.2 Dynamic load test

Dynamic tests were conducted by driving a truck at different speeds that ranged from 16 km/h (10 mph) to 97 km/h (60 mph). For each test the truck speed was constant, starting with 16 km/h (10 mph). Then the speed was increased at a rate of 16 km/h (10 mph) until the maximum speed of 96 km/h (60 mph) was attained for the last test. The maximum dynamic and static responses were compared to estimate the experimental dynamic load allowance (impact factor).

Experimental data was recorded with the RSV-150 (Fig. 9.6b) at a sampling rate of 120 Hz. The truck was driven over the south side of the bridge (along the west-east and east-west directions), separated 0.60 m (2 ft) from the safety barrier's edge as illustrated in Figure 9.9c.

9.5 Test results

The bridge's static and dynamic responses, recorded during the diagnostic test, are presented in the following subsections.

9.5.1 Static load tests

The girders' vertical deflections and longitudinal strains recorded at midspan and used to compute the load distribution factors (LDFs) are reported next.

9.5.1.1 Vertical deflection

Table 9.2 reports the vertical deflections estimated with data recorded by the ATS. The maximum deflections values were obtained at midspan 2, the longest span. In general, larger deflections were obtained for the exterior and interior girders located near the applied loads. It was observed that for the stop configurations loading the superstructure

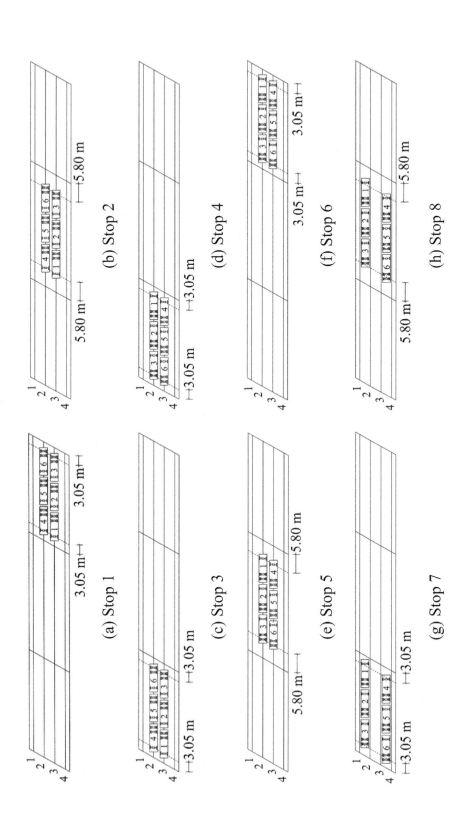

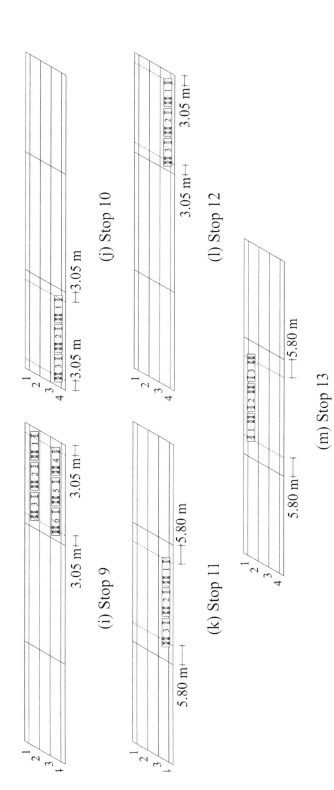

Figure 9.8 Static test configurations (stops). Conversion factor: 1 m = 3.28 ft

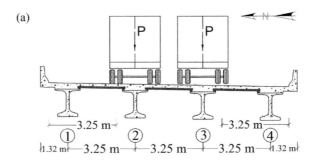
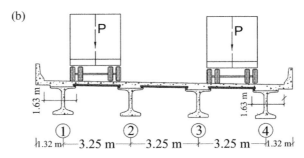
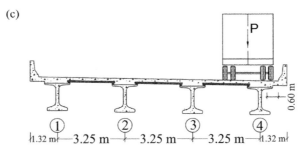

Figure 9.9 Distance from safety barrier to trucks' exterior axle: (a) stops 1–6; (b) stops 7–9; (c) stops 10–13. Conversion factor: 1 m = 3.28 ft

near the safety barriers (stops 7–9), the load was distributed more uniformly and the midspan deflections were comparable for the girders of the same span. For the load configurations acting directly over the interior girders 2 and 3 (stops 1–6), a larger value of deflection was obtained for the interior girders than for the exterior girders. A symmetric response was observed in span 1 when the loads applied corresponded to the configurations of stops 3 and 4 (Table 9.2). Span 3's responses were also compared for the load configurations of stops 1 and 6, and no symmetry was observed in the vertical deflection values. This difference might be related to a variation in the line of application of the loads when the trucks were placed in the central region of the first and third spans. That is, the trucks might have been placed at a different location that did not correspond to the load lines indicated in Figures 9.8a and 9.8f (3.05 m [10 ft] from the center lines of bents 3 and 4).

Example Field Test 191

Table 9.2 Midspan vertical deflections

Stop	Span	δ_{G1} (mm)	δ_{G2} (mm)	δ_{G3} (mm)	δ_{G4} (mm)
Two lanes loaded					
1	3	4.2	7.1	6.9	4.6
2	2	6.3	9.7	9.5	6.2
3	1	5.1	6.9	6.7	4.9
4	1	4.2	6.7	6.9	4.4
5	2	6.4	9.8	10.1	6.4
6	3	4.9	8.4	7.8	5.2
7	1	4.9	5.1	5.5	5.7
8	2	7.3	7.8	8.1	7.6
9	3	4.4	5.5	5.9	5.9
One lane loaded					
10	1	0.1	1.3	3.5	5.0
11	2	0.8	2.0	4.9	7.7
12	3	1.2	2.1	3.5	5.4
13	2	8.6	5.4	2.6	1.0

Note: Experimental values were truncated to the ATS's accuracy for vertical deflection measurements (Hernandez and Myers, 2015, 2018). Conversion factor: 1 in. = 25.4 mm.

9.5.1.2 *Lateral distribution factor (deflection measurements)*

The LDFs were estimated from field deflection measurements. Distribution factors obtained from the experimental results are herein referred to as load distribution factors (LDF) (Cai and Shahawy, 2003). Distribution factors computed using the AASHTO LRFD equations are herein referred to as girder distribution factors (GDF). The load distribution factors were computed using Equation (9.1).

$$LDF_{\delta Gi} = n \frac{\delta_{Gi}}{\sum_{i=1}^{k} \delta_{Gi}} \tag{9.1}$$

Where $LDF_{\delta Gi}$ = load distribution factor of the ith girder obtained from deflection measurements; δ_{Gi} = midspan deflection of the ith girder; k = number of girders; and n = number of lanes loaded. The LDFs are presented in Table 9.3 for the cases of one and two lanes loaded, respectively.

9.5.1.3 *Girders' longitudinal strain*

The longitudinal strains collected at the girders' bottom flange during the 13 load configurations are listed in Table 9.4. These values were collected at the midspan cluster locations (Fig. 9.5a) for 25–30 minutes to allow the bridge superstructure to undergo the total deformation response. These values correspond to the two-lane and one-lane load cases depicted in the previous section. Larger strain values were caused at the exterior and interior girders' midspan located near the applied load. The recorded longitudinal strain values, obtained for two-lane load stop configurations acting on spans 1 and 3 (i.e. stops 1 and 3, stops 4 and 6, and stops 7 and 9), were comparable. No significant difference was noted in the exterior and interior girders' response of spans 1 and 3. For load stops 7 and 9 (two-lane load

192 Load Testing of Bridges

Table 9.3 Load distribution factors estimated from deflection measurements

Stop	Span	$LDF_{\delta G1}$	$LDF_{\delta G2}$	$LDF_{\delta G3}$	$LDF_{\delta G4}$
Two lanes loaded					
1	3	0.369	0.623	0.605	0.403
2	2	0.398	0.612	0.597	0.393
3	1	0.430	0.584	0.571	0.415
4	1	0.380	0.602	0.620	0.398
5	2	0.391	0.598	0.618	0.393
6	3	0.372	0.639	0.593	0.397
7	1	0.463	0.482	0.516	0.540
8	2	0.473	0.506	0.524	0.496
9	3	0.407	0.507	0.544	0.542
One lane loaded					
10	1	0.010	0.131	0.354	0.505
11	2	0.052	0.130	0.318	0.500
12	3	0.098	0.172	0.287	0.443
13	2	0.489	0.307	0.148	0.057

Table 9.4 Girders' longitudinal strain

Stop	Span	$\varepsilon_{G1}(\mu\varepsilon e)$	$\varepsilon_{G2}(\mu\varepsilon)$	$\varepsilon_{G3}(\mu\varepsilon)$	$\varepsilon_{G4}(\mu\varepsilon)$
Two lanes loaded					
1	3	45	83	89	48
2	2	55	95	92	54
3	1	46	84	87	49
4	1	49	87	84	46
5	2	54	92	95	55
6	3	48	89	83	45
7	1	–	–	73	65
8	2	–	–	80	75
9	3	–	–	67	58
One lane loaded					
10	1	–	–	44	64
11	2	4	17	51	78
12	3	–	–	43	65
13	2	78	51	17	4

cases), the difference in the reported strain values, for the interior and exterior girders, was about 10%.

This difference may be related to two possible reasons. The first, as in the case of deflection measurement data collection, might be attributed to errors committed during the placement of the trucks in the central region of the spans 1 and 3 for the stops shown in Figures 9.8g and 9.8i. Second, the load stops might not have lasted long enough to allow the bridge to undergo its total flexural response. Both sources of errors will be investigated and controlled in future series of load tests. However, the data collected for the two-lane stop configurations loading spans 1 (CC girders) and span 3 (SCC girders) were very close. These results also suggest that the initial flexural response of spans 1 and 3 is independent of the materials employed to fabricate the PC girders. This is in agreement with what Hernandez and Myers (2018) found when they conducted a comparison of the deflection response of these spans.

9.5.1.4 Lateral distribution factor (strain measurements)

The LDF, for the interior and exterior girders, was also estimated using strain experimental data and Equation (9.2).

$$LDF_{\varepsilon Gi} = n \frac{\varepsilon_{Gi}}{\sum_{i=k}^{k} \varepsilon_{Gi}} \tag{9.2}$$

Where LDF_{ei} = load distribution factor of the ith girder obtained from strain measurements; ε_{Gi} = strain of ith girder at midspan; k = number of girders; and n = number of lanes loaded. The longitudinal strain, measured by the girders' bottom sensors installed at the midspan cluster (Table 9.4), were necessary to compute the LDFs reported in Table 9.5. For the LDF values obtained with the longitudinal strain measurements, no significant difference was distinguished when the interior and exterior girders' LDF of spans 1 and 3 were compared. In addition, the LDF values reported in Table 9.5 are comparable to the LDF results reported in Table 9.3 which gives validity to the deflections collected with the ATS. It is important to recall that the ATS data acquisition method results in a more practical fashion to record the bridge's deflection response and to estimate experimental load distribution factors. This is particularly useful in the case of bridges with a limited access that hinders the installation of strain gage or LVDT sensors underneath the superstructure.

9.5.2 Dynamic load tests

The dynamic load allowance (DLA) or impact factor has been defined as the ratio of the maximum dynamic and static responses, regardless of whether the two maximum responses occur simultaneously (Bakht and Pinjarkar, 1989; Deng et al., 2015; Paultre et al., 1992). The experimental DLA for the interior girder 3 of Bridge A7957 was estimated using

Table 9.5 Load distribution factors estimated from strain measurements

Stop	Span	LDFε_{G1}	DFε_{G2}	LDFε_{G3}	LDFε_{G4}
Two lanes loaded					
1	3	0.340	0.626	0.672	0.362
2	2	0.372	0.642	0.622	0.365
3	1	0.346	0.632	0.654	0.368
4	1	0.368	0.654	0.632	0.346
5	2	0.365	0.622	0.642	0.372
6	3	0.362	0.672	0.626	0.340
One lane loaded					
2	11	0.027	0.113	0.340	0.520
2	13	0.520	0.340	0.113	0.027

Equation (9.3).

$$DLA^{exp} = \frac{D_{dyn}^{max} - D_{sta}^{max}}{D_{sta}^{max}} \qquad (9.3)$$

Where DLA^{exp} = experimental dynamic load allowance; D_{dyn}^{max} = maximum dynamic (measured) vertical deflection (mm); and D_{sta}^{max} = maximum static deflection (mm).

Paultre et al. (1995) reported that the maximum static response of a bridge can be obtained by (1) conducting a quasi-static test where vehicles move across the bridge at a low speed of 5–16 km/h (3–10 mph); (2) filtering the measured dynamic response with a low-pass filter to eliminate the dynamic components of signal; or (3) using finite element models (FEM) to calculate the static response when the vehicle weight and loading position are known. For the diagnostic load test presented herein, the first option was adopted to estimate the bridge's DLA (i.e. the values of the D_{dyn}^{max} and D_{sta}^{max} were recorded with the RSV-150 and used to estimate the DLA^{exp}). Dynamic and quasi-static deflection values presented by Hernandez and Myers (2017) were used to estimate the DLA. Figure 9.10 shows the maximum static and dynamic vertical deflection recorded with the RSV-150 when the truck passed over the bridge at a crawling speed of 16 km/h (10 mph) and at a speed of 96 km/h (60 mph), respectively. The bridge's static and dynamic maximum deflections recorded for the different speeds are listed in Table 9.6.

Equation (9.4) was employed to experimentally estimate the dynamic amplification factor, DAF^{exp}, reported in Table 9.6.

$$DAF^{exp} = (1 + DLA^{exp}) \qquad (9.4)$$

The maximum estimated experimental dynamic load allowance was found for the maximum truck speed of 96 km/h (60 mph) and corresponded to a value of 0.175.

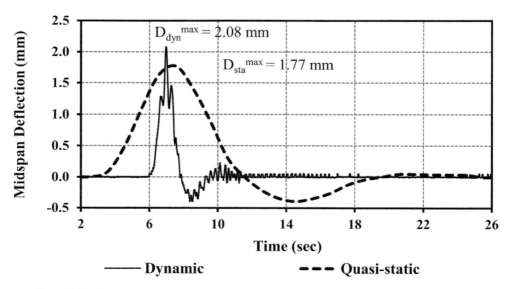

Figure 9.10 Dynamic and quasi-static vertical deflection. Conversion factor: 1 in. = 25.4 mm

Table 9.6 Dynamic load allowance

Truck speed (km/h)						
	16	32	48	64	80	96
D_{sta}^{max} (mm)	1.77	1.77	1.77	1.77	1.77	1.77
D_{dyn}^{max} (mm)	1.77	1.79	1.79	1.77	2.03	2.08
DLA^{exp} (mm)	0.000	0.010	0.010	0.000	0.150	0.175
DAF^{exp} (mm)	1.000	1.010	1.010	1.000	1.150	1.175

Conversion factors: 1 in. = 25.4 mm; 10 mph = 16 km/h.

Comparing this experimental *DLA* to the current design and evaluation value proposed by AASHTO LRFD (AASHTO, 2012), it was observed that the dynamic load allowance value recommended by AASHTO LRFD Design Specifications (*DLA* = 0.33) is conservative, at least for this initial stage of the service life of Bridge A7957. Differences between the experimental and analytical *DLA* values may have repercussions in the load rating of a bridge structure.

9.6 Girder distribution factors

The AASHTO LRFD (AASHTO, 2012) equations were used to estimate the interior and exterior girder distribution factors (GDF) for single and multiple loaded lanes. The GDF for an interior girder with two or more (multiple) design lanes loaded can be estimated using Equation (9.5).

$$GDF_{int}^m = 0.075 + \left(\frac{S}{2900}\right)^{0.4} \left(\frac{S}{L}\right)^{0.2} \left(\frac{K_g}{Lt_s^3}\right)^{0.1} \tag{9.5}$$

where S = girder spacing (mm); L = span length (mm); t_s = deck thickness; K_g = stiffness parameter (mm^4); $K_g = n(I_g + e_g^2 A_g)$; e_g = girder eccentricity (vertical distance from the girder's centroid to the slab's centroid); n = modular ratio (E_{girder}/E_{slab}); E = modulus of elasticity of the concrete, estimated as $0.043 w^{1.5}(f_c')^{0.5}$ (MPa); w = unit weight of concrete (kg/m^3); I_g = girder's moment of inertia (mm^4); and A_g = girder's cross-sectional area (mm^2). The GDF of an interior girder with a single lane loaded was computed with Equation (9.6).

$$GDF_{int}^s = 0.06 + \left(\frac{S}{4300}\right)^{0.4} \left(\frac{S}{L}\right)^{0.3} \left(\frac{K_g}{Lt_s^3}\right)^{0.1} \tag{9.6}$$

The GDF of an exterior girder with two or more design lanes loaded was estimated with the following equations:

$$GDF_{ext}^m = e(GDF_{int}^m) \tag{9.7}$$

$$e = 0.77 + \frac{d_e}{2800} \geq 1 \tag{9.8}$$

where d_e = horizontal distance from the barrier's edge to the exterior girder's centroid (mm). The simple static distribution method (lever rule) was employed to estimate the GDF of an

196   Load Testing of Bridges

exterior girder subjected to a single-lane load as Equation (9.9) presents. This equation was developed by assuming an internal hinge at an interior support (girder 3) and by summing moments, produced by the acting forces and girder's reactions, about girder 3 (support selected to estimate the GDF). For example, the forces acting to the left side of girder 3 are girder 4's reaction and the load P producing moment about girder 3 (Fig. 9.9c).

$$GDF_{ext}^s = m_p \left(\frac{S + d_e - 1524}{S} \right) \tag{9.9}$$

where m_p = multiple presence factor (equal to 1.2 for a single lane loaded). A skew modifying factor was estimated using Equations (9.10) and (9.11). Equation 9.10 is used to modify the girder distribution factors when the skew angle, θ, has values in the range $30° \leq \theta \leq 60°$.

$$SF = 1 - C_1 (\tan\theta)^{1.5} \tag{9.10}$$

$$C_1 = 0.25 \left(\frac{K_g}{Lt_s^3} \right)^{0.25} \left(\frac{S}{L} \right)^{0.5} \tag{9.11}$$

where SF = skew correction factor and θ = skew angle. Table 9.7 summarizes the bridge's design parameters employed to compute the GDF of the exterior and interior girders of each span. Table 9.8 presents the GDF values estimated according to the AASHTO LRFD approach (AASHTO, 2012). The load distribution factor of an exterior or interior girder corresponds to the maximum value obtained for the two-lane and one-lane load cases. It is

Table 9.7 Bridge A7957's design parameters

Variable	Spans 1, 3	Span 2
A_g (mm^2)	479.9×10^3	479.9×10^3
I_g (mm^4)	1.2383×10^{11}	1.2383×10^{11}
K_g (mm^2)	702.207×10^9	785.936×10^9
d_e (mm)	914	914
e_g (mm)	880	880

Table 9.8 Summary of AASHTO LRFD girder distribution factors

Span	GDF_{int}	GDF_{int}†	GDF_{ext}	GDF_{ext}†
Two or more lanes loaded				
1, 3	0.819	0.783	0.901	0.861
2	0.788	0.756	0.866	0.832
One lane loaded				
1, 3	0.558	0.533	0.975	0.932
2	0.528	0.507	0.975	0.936

Note:
†GDF values were modified by the skew correction factor.

observed that GDFs found with the AASHTO LRFD equations are larger than the experimental LDF values. The difference may be related to different causes.

The AASHTO LRFD method used to compute the lateral load distribution was proposed to be applied to a wide range of bridges with different (1) span lengths, (2) girder spacing, (3) stiffness, and (4) construction material. The experimental LDFs implicitly consider in-situ conditions that are typical of the bridge under study. These conditions may include (1) unintended support restraints, (2) skew angle, (3) contribution of secondary members, and (4) multiple presence factors. It is fundamental to conduct further research to isolate the contribution of each of these factors and to determine to what extent the load rating of a bridge can be affected and relied upon on each of these in-situ bridge conditions.

9.7 Load rating of Bridge A7957 by field load testing

The AASHTO *LRFD Bridge Design Specifications* (AASHTO, 2012) were presented as the primary design method for highway bridges in 1998. These specifications represent the first AASTHO effort to integrate modern principles of structural reliability and probabilistic models of loads and resistance into the design of highway bridges. In addition, the AASHTO LRFD specifications introduced reliability-based limit states concepts into the design philosophy using calibrated load and resistance factors that satisfy uniform safety levels corresponding to each limit state. The approach was extended to the evaluation of bridges with the completion of the *Manual for Condition Evaluation and Load and Resistance Factor Rating (LRFR) of Highway Bridges* (MCE), published in 2003 (AASHTO, 2003). The MCE proposed the first bridge strength evaluation approach in the United States presenting a structural reliability format (LRFR). A more recent update of the LRFR procedure is found in the AASHTO *Manual for Bridge Evaluation* (MBE), updated in 2010 (AASHTO, 2010). The rating factor of a bridge component according to the AASHTO LRFR evaluation approach (AASHTO, 2003, 2010) is given by the following equation:

$$RF = \frac{C - \gamma_{DC}DC - \gamma_{DW}DW \pm \gamma_P P}{\gamma_{LL}LL(1 + IM)} \tag{9.12}$$

where RF = rating factor; C = capacity = $\varphi\varphi_C\varphi_S R_n \geq 0.85\varphi R_n$ (strength limit states); R_n = nominal member resistance; $C = f_R$ (service limit states); f_R = allowable stress specified in LRFD specifications (AASHTO, 2012; Minervino et al., 2004); φ = LRFD resistance factor; φ_C = condition factor; φ_S = system factor; DC = dead-load effect due to structural components and attachments; DW = dead load effect due to wearing surface and utilities; P = permanent loads other than dead loads (typically, post-tensioning loads); LL = live-load effect; IM = dynamic load allowance (impact factor); γ_{DC} = LRFD load factor for structural component and attachments; γ_{DW} = LRFD load factor for wearing surfaces and utilities; γ_P = LRFD load factor for permanent loads other than dead loads; and γ_{LL} = evaluation live-load factor. An analytical load rating may be enhanced if experimental results obtained from a diagnostic load test are used in Equation (9.12).

To illustrate how these improvements can be achieved in the analytical load rating of Bridge A7957, the experimental LDFs (reported in Table 9.5) and the experimental dynamic load allowance (DLA^{exp} reported in Table 9.6) were employed to obtain the experimental load rating of Bridge A7957 using Equation (9.12). Some calculations are omitted for the sake

198   Load Testing of Bridges

Table 9.9 Bridge A7957 load rating data

Parameter	Span 1	Span 2	Span 3
Mn (kN-m)	11,116.0	13,606.8	11,122.0
M_{DC} (kN-m)	3299.7	4654.0	3299.7
M_{DW} (kN-m)	309.4	280.7	309.4
M_{LL} (kN-m)	2839.7	2935.0	2839.7
Theoretical parameters (estimated using AASHTO LRFD approach)			
GDF_{int}	0.783	0.756	0.783
GDF_{ext}	0.932	0.936	0.932
IM	0.33	0.33	0.33
Field parameters (estimated using diagnostic test results)			
LDF_{int}	0.672	0.654	0.672
LDF_{ext}	0.520	0.520	0.520
DLA^{exp}	0.175	0.175	0.175

Note: Conversion factor: 1 m = 3.28 ft. 1 kN = 0.2248 kip.

Table 9.10 Bridge A7957's analytical and experimental evaluation results

LR	Span 1	Span 2	Span 3
Theoretical load rating			
LR_{int}	1.26	1.43	1.26
LR_{ext}	1.06	1.15	1.06
Experimental load rating			
LR_{int}	1.66	1.87	1.66
LR_{ext}	2.15	2.35	2.15

of brevity; however, the information used in the subsequent computations is summarized in Table 9.9. The values of the live-load moment (M_{LL}) were obtained using the AASHTO LRFD design truck (HL-93) and correspond to the strength I limit state (at inventory level). The analytical and experimental load rating results are presented in Table 9.10.

Rows 3 and 4 list the load rating of the interior and exterior girders of each span. The controlling analytical load rating corresponded to the exterior girder (LR = 1.06) of the end spans 1 and 3. By using the experimental load distribution factors (LDF) and experimental dynamic load allowance (DLA^{exp}), the exterior girder's load rating of spans 1 and 3 was increased to 2.15, which represents an approximate 100% increment. Similarly, the experimental load rating of the exterior girder of span 2 increased approximately 100% with respect to the analytical load rating (2.35 vs. 1.15). In the case of the interior girder of the three spans, the experimental load rating was enhanced approximately 31% in comparison to the analytical load rating. The experimental LDFs had a beneficial effect on the load rating capacity of Bridge A7957. In the case of the exterior girders, the average improvement was close to 44%. In the case of the interior girder, the average enhancement due to the LDF effect was close to 14%. The difference between the experimental and analytical amplification factor, estimated as $1 + IM$, was close to 12%. These findings demonstrate how the analytical load rating of Bridge A7957 could be improved by conducting a field load test.

9.8 Recommendations

The following recommendations were drawn from this example field test:

- Special supervision must be followed when the test loads are applied to a bridge. A misplacement of the truck axles over a span will increase the chance of committing errors when predicting the response of the bridge or interpreting the test results. Similarly, it is important to avoid errors when reporting the test trucks' weight, as reported by Merkle and Myers (2006), that can result in poor correlations between the experimental and predicted response of the different spans. Other researchers (Cai and Shahawy, 2004; Lantsoght et al., 2016) attained successful test results by employing standardized test trucks in diagnostic and proof load tests.
- It is fundamental to allow sufficient duration of the load application to act on the bridge main carrying members. This not only allows the bridge to undergo the total expected deformation caused by the load, but it also permits the measurement equipment to record more precisely the bridge's total response.
- Finally, it is very important that the loads applied to the structure during the test be close to the service load level. This ensures that recorded response values are larger than the accuracy of the instruments and sensors employed in the test. Consequently, the error committed by the data acquisition equipment is kept within an acceptable level and the test results can be relied on to estimate parameters such as the load distribution factors and the dynamic load allowance (impact factor) of the bridge.

9.9 Summary

The first diagnostic load test was successfully performed on the superstructure of Bridge A7957 to corroborate some design and analysis assumptions and to perform an experimental evaluation of the bridge's superstructure. By incorporating site-specific data measured during the diagnostic load test, a less conservative load rating of Bridge A7957 was obtained in comparison to the analytical load rating values obtained following the AASHTO LRFD procedure. The controlling analytical load rating corresponded to the exterior girder (LR = 1.06). By conducting a diagnostic load test, the exterior girder's load rating was improved by approximately 100% compared to the analytical load rating. In the case of the interior girders, the experimental load rating was enhanced approximately 31% compared to the ana-lytical load rating. The contribution of the experimental LDF on the enhancement of the load rating capacity of Bridge A7957 was close to 44% for the exterior girders. In the case of the interior girder, this benefit was close to 14%. The difference between the experimental and analytical amplification factor, estimated as $(1+IM)$, was close to 12%. The diagnostic load test presented herein proved that Bridge A7957's superstructure possesses a larger strength capacity than the predicted by the theoretical approach. This difference can be explained by the fact that a diagnostic load test incorporates in-situ parameters that are beneficial to the bridge's service response. These parameters are not considered by the current AASHTO LRFD evaluation specifications. Additional research must be conducted to quantify and estimate the influence of these parameters in the load rating evaluation and to isolate those beneficial parameters that are always present, independent of the level of load applied to the bridge.

References

AASHTO (2003) *Manual for Condition Evaluation and Load and Resistance Factor Rating (LRFR) of Highway Bridges*. American Association of State Highway and Transportation Officials, Washington, DC.

AASHTO (2010) *The Manual for Bridge Evaluation*, 2nd edition with 2011, 2013, 2014 and 2015 Interim Revisions. American Association of State Highway and Transportation Officials, Washington, DC.

AASHTO (2012) *LRFD Bridge Design Specifications*, 6th edition. American Association of State Highway and Transportation Officials, Washington, DC.

Bakht, B. & Pinjarkar, S. G. (1989) Dynamic testing of highway bridges: A review. *Transportation Research Record 1223, TRB, Washington, DC*, 93–100.

Cai, C. S. & Shahawy, M. (2003) Understanding capacity rating of bridges from load tests. *Practice Periodical on Structural Design and Construction*, 209–216.

Cai, C. S. & Shahawy, M. (2004) Predicted and measured performance of prestressed concrete bridges. *Journal of Bridge Engineering*, 9(1), 4–13.

Deng, L., Yu, Y., Zou, Q. & Cai, C. S. (2015) State-of-the-art review of dynamic impact factors of highway bridges. *Journal of Bridge Engineering*, 20(5), 04014080.

Hernandez, E. S. & Myers, J. J. (2015) In-situ field test and service response of Missouri Bridge A7957. *16th European Bridge Conference (EBC16)*, 23–25 June, Edinburgh, Scotland, UK.

Hernandez, E. S. & Myers, J. J. (2016a) Initial in-service response and lateral load distribution of a prestressed self-consolidating concrete bridge using field load tests. In: Bakker, J., Frangopol, D. M. & Van Breugel, K. (eds.) *The Fifth International Symposium on Life-Cycle Civil Engineering (IALCCE 2016), Oct. 2016*. CRC Press, Delft, The Netherlands. pp. 1072–1079.

Hernandez, E. S. & Myers, J. J. (2016b) Monitoring the initial structural performance of a prestressed self-consolidating concrete bridge. In: Khayat, K. H. (ed.) *8th International RILEM Symposium on Self-Compacting Concrete (SCC2016), May 2016*. RILEM Publications SARL, Washington, DC. pp. 401–411.

Hernandez, E. S. & Myers, J. J. (2017) Dynamic load allowance of a prestressed concrete bridge through field load tests. *SMAR 2017 Fourth Conference on Smart Monitoring, Assessment and Rehabilitation of Civil Structures*, September, Zurich, Switzerland.

Hernandez, E. S. & Myers, J. J. (2018) Diagnostic test for load rating of a prestressed SCC bridge. *Evaluation of Concrete Bridge Behavior Through Load Testing: International Perspectives*, ACI Technical Publication, SP-323 11.1–11.16.

Lantsoght, E. O. L., Yang, Y., van der Veen, C., de Boer, A. & Hordijk, D. A. (2016) Ruytenschildt Bridge: Field and laboratory testing. *Enginerring Structures*, 128, 111–123.

Merkle, W. J. & Myers, J. J. (2006) Load testing and load distribution response of Missouri bridges retrofitted with various FRP systems using a non-contact optical measurement system. In: TRB (eds.) *Transportation Research Board 85th Annual Meeting*. Washington, DC. pp. 1679.1–22.

Minervino, C., Sivakumar, B., Moses, F., Mertz, D. & Edberg, W. (2004) New AASHTO guide manual for load and resistance factor rating of highway bridges. *Journal of Bridge Engineering*, 43–54.

Paultre, P., Chaallal, O. & Proulx, J. (1992) Bridge dynamics and dynamic amplification factors: A review of analytical and experimental findings. *Canadian Journal of Civil Engineering*, 19(2), 260–278.

Paultre, P., Proulx, J. & Talbot, M. (1995) Dynamic testing procedures for highway bridges using traffic loads. *Journal of Structural Engineering*, 362–376.

Chapter 10

Example Load Test
Diagnostic Testing of a Concrete Bridge with a Large Skew Angle

Mauricio Diaz Arancibia and Pinar Okumus

Abstract

Bridge load testing is widely used for assessing bridge structural behavior and may be preferred over other means, since it is capable of capturing the actual response of structures. However, load testing is complex and requires careful consideration of several activities that precede its execution. This chapter describes the planning, coordination, scheduling, execution, and data analysis of a load test using an example of a highly skewed prestressed concrete girder/reinforced concrete deck bridge under service loads. The load test allowed the evaluation of the effects of high skew angles and mixed pier support fixity arrangements on girder load distribution behavior and deck performance.

10.1 Summary

This chapter provides details on planning, instrumentation, execution, and results of a load test on a prestressed concrete girder bridge. The unique feature of the bridge that necessitated the load test was its very high (64°) skew angle. The goal of the diagnostic test was to understand the impact of the large skew angle on load distribution and bridge performance. An instrumentation plan was created to address these project goals and was selected considering site conditions, site accessibility, target measurement magnitudes, sensor type, and sensor availability. A preliminary finite element model guided the instrumentation plan. Selected girders and deck were instrumented with strain gages. Bearing pads and girders were instrumented with displacement sensors. A preliminary review of data showed that displacement measurements were not reliable due to the sensitivity of small magnitudes of displacements to measurement errors. Execution of the load test required coordination between the research team, the bridge owner (the department of transportation), the local government (county), the contractor, and the subcontractors. The bridge was tested under live load created by two loaded dump trucks in four configurations during a night road closure. Duplicate gages and load configurations were used to ensure repeatability of results and redundancy against data interruption. Data collected were used to validate analytical models, which provided information on the entire bridge domain. The results showed that moment and shear distribution predictions of AASHTO *LRFD Bridge Design Specifications* (BDS) (2014) were conservative despite the large skew angle, with the exception of moment for exterior girders. Deck strains caused by long-term loading are more significant than the ones by live load. Lack of deck cracks may be due to the bearing fixity selected to limit superstructure displacements.

DOI: https://doi.org/10.1201/9780429265426

10.2 Characteristics of the bridge tested

The feature of the bridge which necessitated a load test was its large skew angle (64°). The bridge was built in 2016. Construction of structural components (including parapets) were completed shortly before load testing. Nonstructural components such as lane marking and lighting were completed after the load test. It had three spans (33.3 m [109.2 ft], 38.2 m [125.4 ft], and 26.2 m [85.8 ft]); was 28.9 m (94.9 ft) wide; carried five lanes of interstate highway traffic, and crossed over a bicycle path. The superstructure is composed of 15 pretensioned concrete girders, each 1.4 m (54 in.) deep, composite with a 203 mm (8 in.) thick, reinforced concrete cast-in-place deck. The bridge is catalogued as B-40-870 in the Highway Structures Information System of the Wisconsin Department of Transportation (DOT). More information on the bridge can be found elsewhere (Wisconsin Department of Transportation, n.d.). A unique feature of the bridge was the mixed use of girder expansion and fixed bearings over the same pier. A photograph of the bridge under construction is shown in Figure 10.1.

10.3 Goals of load testing

Early identification of the desired outcomes of the load test informed the instrumentation and testing plan. This load test was necessitated due to the unique features of the bridge (i.e., 64° skew and the mixed girder bearing types over the same support). Large skew angles affect live-load paths to supports, altering distribution of live load to each girder. For example, higher shear will be transmitted to exterior girders near obtuse corners than in a bridge with no skew. Near the acute corners, reduced shear at exterior girders may cause uplift. Large skew also may invalidate simple, one-dimensional bridge analyses, which utilize girder live-load distribution factors given by AASHTO LRFD BDS (2014). Therefore, goal 1 of load testing was to validate analysis methods and understand live-load distribution to each girder. Instrumentation was selected to capture response of girders and deck that can be used to validate analysis methods.

In addition to changes in load distribution, large skew may lead to performance problems. The two prevalently reported performance problems are deck cracks near the acute corners and horizontal movements of the bridge. These likely occur over time under long-term loading such as temperature and shrinkage, and understanding the response requires long-term monitoring of deck strains or bridge displacements. Acute deck

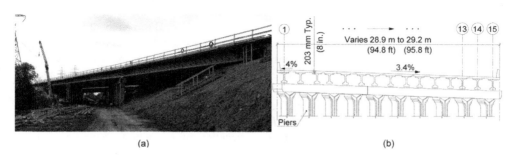

Figure 10.1 (a) Photograph of the bridge under construction, and (b) cross section of the bridge looking east

corner cracks may further alter live load distribution to girders as they may change deck stiffness. Goal 2 of this investigation was to understand the factors behind performance problems, the effect of the unique bearing fixity configuration on them, and the impact of acute deck corner cracks on load distribution. In this chapter, the focus will be on short-term (one month after deck pour) performance of the bridge. Long-term measurements are given in detail elsewhere (Diaz Arancibia & Okumus, 2018).

10.4 Preliminary analytical model

A preliminary analytical model was built before load testing. This model was used to predict magnitudes of response (i.e. strain and displacement) and identify target areas for instrumentation. The locations of sensors, types of sensors, location of load, and magnitude of load were determined partially based on the key information extracted from the model. The preliminary model was a finite element model where girders and deck were simulated as frame and shell elements, respectively. Refinements to the model through inclusion of secondary bridge elements (diaphragms and parapets) and expansion bearing stiffnesses were not needed at this stage since approximate magnitudes of strain and displacement were adequate. Figure 10.2a shows an extruded view of the preliminary model. Figure 10.2b shows positive flexure and negative vertical shear strain envelopes under two trucks arranged in series, with a distance of 3.0 m (9.8 ft) between front and rear axles and 0.6 m (2.0 ft) away from the interior face of the south parapet. Additional load configurations, with varying number and location of trucks, were also considered but are not

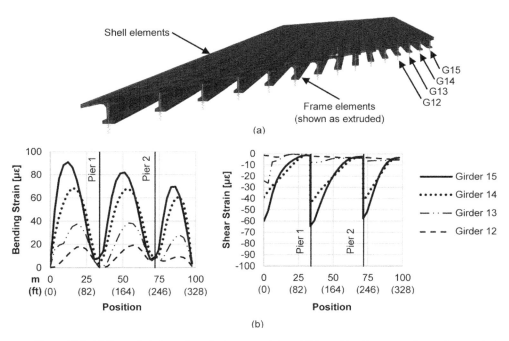

Figure 10.2 (a) Extruded view of preliminary model, and (b) positive flexure and negative vertical shear strain envelopes under two trucks arranged in series

presented here. Note that at this stage the magnitude of loading was not known and a reasonable load type was selected (three-axle, 5.5 m [18.0 ft] long by 1.8 m [5.9 ft] wide truck, with 89 kN [20 kip] rear and middle and 67 kN [15 kip] front axle weights, with 1.5 m [4.9 ft] spacing between rear axles).

The preliminary analyses showed that the maximum magnitudes of shear and moment in girders were 91 and 67 microstrains, respectively. Maximum deck strains were expected to be 160 microstrains. Girder midspan displacements were anticipated to be up to 10 mm (0.39 in.). Bearing pads were not included in the preliminary model, and therefore expected vertical displacements of bearing pads were not known prior to testing.

Creating refined models (including features such as secondary members and material properties obtained from testing, etc.) prior to load testing is recommended, particularly for bridges with unusual features. Inclusion of these features is generally expected to yield a stiffer response and lower measurement magnitudes. For this test, a more detailed model (not described here) was constructed before testing. This model allowed predicting the stiffer response and finalizing the instrumentation plan.

10.5 Coordination of the load test

Planning was critical for a successful load test. Several parties had to be involved: the research team, the bridge owner, the local government, the contractor, and the subcontractors. The contractor and subcontractors were involved because the bridge was under construction during instrumentation and shortly before the load test. The role of the parties involved in load testing is discussed in this section.

The research team was responsible for proposing an instrumentation plan, proposing load configurations, coordinating the load test among parties, executing the instrumentation and the load test, analyzing data, and making recommendations to the bridge owner regarding bridge design, analysis, and performance.

The bridge owner was the Wisconsin DOT. Within the Wisconsin DOT, a project oversight panel was formed as part of the Wisconsin Highway Research Program. The oversight panel reviewed the instrumentation and load test plan proposed by the research team, and they provided input based on their expected outcomes of the load test. For example, an initial version of the proposed plan focused mainly on short-term load testing. Input from the panel led the research team to include additional instrumentation to measure the long-term response.

In Wisconsin, local government (counties) typically coordinates traffic control with relevant parties such as the state patrol and local department of transportation. These include announcements of road or lane closures, signage, directing traffic, and determining alternate routes. The research team would inform the county on the duration, time, and extent (number of lanes, rolling closures vs. full closures) of traffic control they need. However, for this research the bridge was already scheduled to be closed to traffic due to construction within the same project on the same route. Therefore, the research team coordinated the road closure with the contractor and the department of transportation, instead of the local government. Milwaukee County Department of Transportation was the local government and assisted the research team by providing trucks to be used as live load during load testing. They also provided drivers and measured the weight of the trucks prior to testing.

The contractor provided safety training to the research team and informed the research team on the construction schedule updates. They also influenced the instrumentation plan by providing input on site accessibility and availability of equipment (power, generators, and equipment such as forklifts). The contractor assumed these responsibilities because the bridge was instrumented while it was being constructed and load tested right after it was constructed.

10.6 Instrumentation plan

The instrumentation plan included decisions on sensor type and location. The following factors determined the instrumentation plan: desired outcomes and goals (goals 1 and 2, described in Section 10.3), accessibility of bridge site and bridge elements, the number of sensors available, capabilities of the data acquisition system, the expected response of the bridge to loads, load magnitude, load type, load rate, and load duration.

10.6.1 Sensor types and application methods

Sensors selected to fulfill goal 1 (described in Section 10.3) were electrical resistance strain gages, vibrating wire embedment strain gages, displacement transducers and optical displacement measurement targets.

Electrical resistance strain gages were used to measure flexure and shear strains. The precision of these gages was ±0.02 microstrains based on the excitation voltage and strain range setup in the data acquisition and was verified by identifying the smallest strain change measured during testing. Expected strains, determined from the preliminary analyses, were small (less than 100 microstrains for flexure and 70 microstrains for shear), which is typical for prestressed concrete bridges. Gage precision was small enough to measure these small strains. Eighteen and nine gages were placed on the bottom and side surfaces for flexure and shear, respectively, of three selected prestressed girders at four locations along each girder length. These gages require surface preparation since the measurements depend on the bond between gage and girders. Prestressed girder surfaces were prepared by clearing the location of strain gages, applying a thin layer of quick-dry epoxy, sanding and smoothing the hardened epoxy, and gluing strain gages using cyanoacrylic glue. Once the gages were installed, they were protected from the environment by a thin coat of rubber, rubber mastic tape, and duct tape. Gage wires were secured by attaching them to girders at multiple locations. The load test took place approximately one month after the installation of gages. During this period, gage coatings were able to protect gages from mechanical impact and water. Figure 10.3 shows the electrical resistance strain gages after coatings were applied.

Vibrating wire strain gages were embedded in deck concrete, which was possible since instrumentation took place during bridge construction. The precision of these gages is ±0.1 microstrains. Expected strains in the deck were 160 microstrains, which were larger than the precision available. The gages were placed right below or above deck reinforcement at 11 locations near an acute corner (southeast corner) of the bridge and bonded directly to concrete. These gages were either directly attached to reinforcing bars or through an apparatus prepared for each gage. No coating was necessary as these gages are waterproof. Each vibrating wire gage also measures temperature. In addition, a single gage was placed in a load-free, concrete-filled cylinder to measure changes in strain due to temperature and shrinkage. This gage is hereinafter referred to as a "stress-free" gage.

The deck was poured within five days of vibrating wire gage installation. Vibrating wire gage locations were marked with color to alert construction workers so that they can avoid

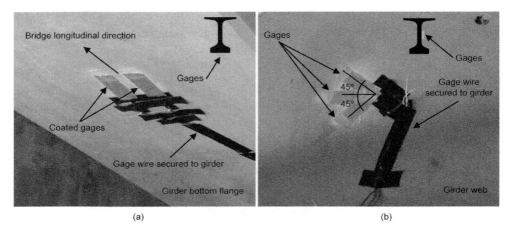

Figure 10.3 Electrical resistance surface strain gages on prestressed girders after coating to measure (a) bending and (b) shear strains

Figure 10.4 Vibrating wire gages installed (a) in the deck before deck pour and (b) in a no-stress cylinder enclosure

placing concrete vibrators directly over gages. All vibrating wire gages were protected from damage and were able to provide readings during the load test that took place approximately a month after their installation. Figure 10.4 shows the vibrating wire gages before the deck pour and the cylinder in which the stress-free gage was placed. In addition to providing deck strains during load testing, these gages also provide long-term (over a one-year period) strains from the deck to understand long-term response of the deck (goal 2). In this example, measured strains are reported between the time gages were installed and time of load testing. For long-term measurements on this bridge (Diaz Arancibia & Okumus, 2018) and others (Fu et al., 2007), the reader can see other references.

Displacement transducers were placed at the bearing pads of four selected girders to measure the vertical displacement of elastomeric bearing pads. Even though these displacements were expected to be small, the intent was to understand the proportion of the load (assumed proportional to bearing pad displacement) that goes to each of the instrumented girders. Each bearing pad had four displacement transducers (one at each bearing pad corner). The transducers were attached to metallic plates on top of the elastomeric bearing pad using magnets. A better choice for measuring girder end reaction would be load cells. However, load cells were deemed cost prohibitive since large reaction forces were expected (251 kN [56 kip]) and since load cells installed under bearing pads cannot be removed or reused. The displacement transducers were attached one day before the load test. Data collected by displacement transducers were inconsistent due to the sensitivity of very small displacements (expected to be less than 0.30 mm [0.012 in.]) to unintended movements in displacement transducers and the limited precision of the transducers (0.10 mm [0.004 in.]). Higher precision displacement transducers such as linear variable differential transducers (LVDTs) or string potentiometers may lead to satisfactory readings. However, regardless of which transducer type is used, proper connection between them and bearing plates must be guaranteed.

Finally, displacement targets were placed at midspan of three selected girders to measure girder displacements using an optical instrument (a total station). Displacement targets were glued on the bottom surface of prestressed concrete girders approximately one month prior to load testing. Displacements measurements were intended to help understand load distribution between instrumented girders. However, due to the very small displacements (less than 7 mm [0.28 in.] over a 38.2 m [125.4 ft] span) and the inadequate precision of the measurement method, the displacement results were evaluated to be unreliable. Figure 10.5 shows the displacement transducers and displacement targets. Surveying systems with higher precision or, when there is access under the bridge, use of LVDTs or string potentiometers are recommended for better measurements. A typical setup of LVDTs for deflection monitoring requires a fixed reference frame, where the deflection transducer is located, and a wire that connects the transducer with the girder bottom face.

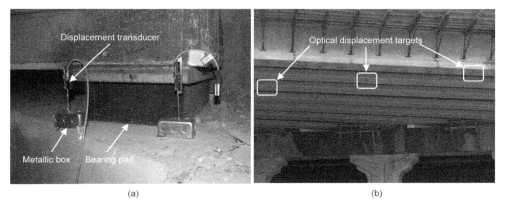

Figure 10.5 (a) Displacement transducers on bearing and (b) optical displacement target on girders

10.6.2 Sensor locations

Due to the large size of the bridge and use of data acquisition systems that required wiring from sensors, locations of sensors had to be selected strategically. Sensors were concentrated in areas that were expected to be affected by the high skew angle. For example, the gages that measure shear strains were placed on girders near the obtuse corner of the bridge so that any increase in shear due to skew could be captured. Gages that measure flexure were also placed in this region because shear and bending gages were wired to the same data acquisition system. Both shear and bending gages were near the planned locations of loading so that measurable strains could be obtained and precision errors could be minimized. Gages were placed on three adjacent girders at the same locations along their spans to obtain load distribution between girders. In addition, flexure gages were placed at three locations along the span length to capture influence lines under moving loads. Figure 10.6 shows the locations of shear and bending strain gages. Resources were available to only instrument one span of the bridge with shear and bending gages.

Vibrating wire gages in the deck were placed near the acute corner of the deck since deck cracks have been reported near acute corners for bridges with high skew (Diaz Arancibia and Okumus, 2018; Fu et al., 2007; Diaz Arancibia et al., 2017). Gages were located along two lines (lines 1 and 2), and oriented along or perpendicular to girder directions so that trends of strains can be captured in these directions. Gages measured transverse or longitudinal (direction of the bridge) strains in the deck. Gage locations and directions are shown in Figure 10.7.

The locations of displacement transducers and optical displacement targets are shown in Figure 10.8. These gages were placed to measure the maximum response: near the obtuse

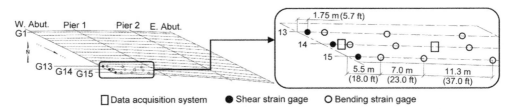

Figure 10.6 Locations of shear and bending strain gages and their data acquisition systems

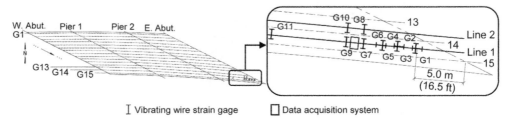

Figure 10.7 Vibrating wire gage locations and directions and the data acquisition system location

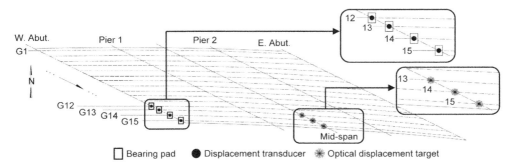

Figure 10.8 Displacement transducer and data acquisition system locations and optical displacement target locations

corner for displacement transducers and midspan for optical displacement targets under live loads planned to be placed on the bridge during the load test.

10.7 Data acquisition

The data acquisition available to the research team included 36-channel and 16-channel systems from Pacific Instruments for electrical resistance strain gages and displacement transducers. These two data acquisition systems were located near the abutment (see Fig. 10.6) at a location that minimized the required lead wire length for electrical resistance strain gages and displacement transducers. The longest wire length was 14 m (46 ft). Gage wires and channels were carefully numbered prior to installation of gages. These data acquisition systems were powered using a generator on site. Data acquisition rates were 10 Hz and 30 Hz for the 36-channel and 16-channel systems, respectively.

A Campbell Scientific data logger with 16 channels was used for the vibrating wire gages. The longest lead wire for the vibrating wire gages was 15 m (49 ft). This data acquisition system was powered by a 12 V battery and was equipped to remotely transmit data over a mobile phone network. The system was also powered by a solar panel for long-term data acquisition. Data was averaged and recorded every 30 seconds during the load test and every hour during long-term monitoring.

10.8 Loading

10.8.1 Load type and magnitude

The loading was composed of two identical dump trucks provided by Milwaukee County. These trucks were filled to capacity to generate the highest possible service level response in the bridge, in order to minimize the percentage of error in strain and displacement measurements. The trucks were weighed by the county before they arrived at the bridge site. The dimensions (distance between axles and wheels) were measured carefully on site by the research team. The dimensions and axle weights of trucks are given in Figure 10.9. The middle and rear axle weights were measured together. In Figure 10.9, their individual weights are assumed to be identical.

210 Load Testing of Bridges

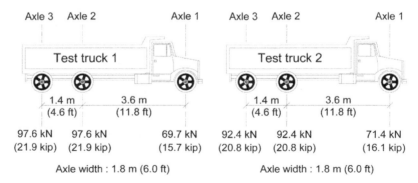

Figure 10.9 Trucks used for load testing

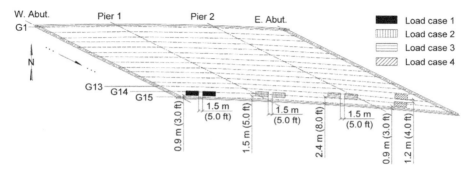

Figure 10.10 Locations and number of trucks used in each load case (LC): LC1–LC4

Table 10.1 Purposes of load cases 1–4

Load Case	Purpose
1	To maximize exterior girder response under one-lane loading.
2	To maximize deck response under one-lane loading.
3	To maximize interior girder response under one-lane loading.
4	To maximize exterior girder response under two-lane loading.

10.8.2 Load configurations and locations

Four load configurations were used to load test the bridge. The locations and number of trucks involved in each load case are shown in Figure 10.10. All load configurations were located near the south end of the bridge, where sensors were placed.

Each load case was selected to maximize response at a selected bridge member or to understand live-load distribution when one or more lanes were loaded. The goal of each load case is summarized in Table 10.1.

Trucks were driven on the bridge from west to east at crawl speed (less than approximately 10 km/h [6.2 mph]). This speed was assumed to create static loading on the bridge. The trucks were also brought to a full stop for at least five seconds at the midspan of span 1 (west exterior span) and 2 (central span) to obtain readings under true static loading. Strains and bearing displacements were measured continuously during the test. When trucks were stopped at the midspan of span 2, girder vertical displacements of midspan were measured using displacement targets.

The intended paths of trucks were marked on the bridge prior to the arrival of the trucks on site. During the load test, one researcher visually tracked the front wheel location and directed truck drivers so that the trucks remained on the load path as they crossed the bridge. Slight horizontal curvature of the bridge created a challenge for trucks to remain on the exact path. However, the trucks were able to remain on path with ±15 cm (5.9 in.) of error.

Midspan points of span 1 and 2, where the trucks came to a complete stop, were marked to guide trucks to a stop. In addition, the longitudinal location of trucks with time was tracked by marking every 3 m (9.8 ft) of the bridge and attaching a camera to trucks that captured 3 m (9.8 ft) markings with respect to time. The longitudinal truck location can be used to create approximate influence lines. Video recordings allowed the research team to track truck locations with ±1.5 m (4.9 ft) accuracy.

10.9 Planning and scheduling

The first coordination meeting between the researchers, bridge owner, project oversight committee, contractor, and subcontractors took place in June 2016. The bridge was instrumented in mid-October 2016 and load tested on 11 November 2016. The instrumentation plan evolved according to feedback from the parties involved and analyses over six months leading to the load test.

Three researchers were present on site to perform instrumentation. Instrumentation was completed within approximately 72 hours. The road was completely closed to traffic to perform the load test. Due to the heavy traffic on this route, road closure had to take place at night between 10:00 p.m. and 5:00 a.m. Marking the road for truck locations took approximately an hour. Load testing was completed within approximately four hours. Four researchers participated in the load testing. Based on this project, one additional researcher is recommended to be on site for documentation of work through time stamps and photographs for future load tests.

10.10 Redundancy and repeatability

To account for unexpected circumstances on site, redundant gages and load cases were used. Fragile electrical resistance strain gages had to be installed approximately a month before the load test date. To account for potential malfunctioning of gages due to environmental or mechanical damage, two gages were installed at the location of each flexure gage. The redundant gages were also used to check the repeatability of measurements. Only one gage was lost. Redundant gages measured the same strains as the original gages, indicating that the gages were sufficiently precise to measure bending strains.

Similarly, to ensure the repeatability of results and to account for potential malfunctioning of data acquisition, each load case was run twice. All measurements, except for girder displacements measured by a total station, were the same between the first and the second runs. This indicated that the displacement target measurements were not reliable. These results were discarded.

10.11 Results

10.11.1 Preliminary evaluation of results

All measurements, including shear strains, bending strains, deck strains, bearing pad displacements, and superstructure midspan displacements, were reviewed for errors before data were analyzed further. The patterns and magnitudes of shear and bending strains were as expected. In addition, girder and deck strains measured through duplicate gages or load cases were the same as the ones originally measured. These evaluations showed that strain measurements were reliable.

On the other hand, bearing pad displacement measurements were deemed to be erroneous and discarded because duplicate measurements were not the same as the originals, and magnitudes of measurements were not consistent with loading. Errors in measurements were attributed to very small displacements at bearing pads caused by loading and accidental movements of displacement sensors during loading. Similarly, duplicated load cases did not result in the same girder midspan displacements. Low precision of total station and displacement targets compared to the very small measurements likely caused error in these measurements. Midspan displacement measurements were also discarded and will not be presented here.

10.11.2 Shear strain influence lines and shear distribution

After shear strain data were reviewed for errors and deemed accurate, shear strain influence lines were created using the approximate locations of trucks along the span. Vertical shear strains were calculated using two (i.e. the diagonally oriented gages) of the three gages arranged in a 45° rosette configuration. The horizontal gage served as a redundant gage. Figure 10.11 shows the influence line for shear strains on the three instrumented girders for the four load cases. Girders 13, 14 and 15 were adjacent. Girder 15 is the exterior girder as shown in Figure 10.6.

Since only three girders were instrumented, shear load distribution cannot directly be calculated from load testing. Therefore, to understand the response of the remaining girders, refined finite element models were built. Refined models had girders and deck modeled

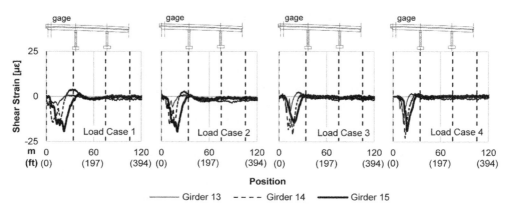

Figure 10.11 Shear strain influence lines for girders 13, 14, and 15 for load cases 1–4

as frame and shell elements, respectively, and included secondary elements. Models were validated using the available test data and were used to obtain the shear girder distribution factors for the exterior and interior girders. A comparison of model and test results can be found elsewhere (Diaz Arancibia & Okumus, 2018). Models indicate that the governing shear distribution factors for an exterior and interior girder are 0.818 and 0.662, respectively, under the loads exerted by AASHTO trucks. These are 99.8% and 69.0% of the exterior and interior girder distribution factors predicted using AASHTO LRFD BDS (2014).

10.11.3 Bending strain influence lines and moment distribution

Bending strain influence lines for the instrumented girders are shown in Figure 10.12 for each load case and for each gage location. In Figure 10.12 and throughout the chapter, positive and negative magnitudes of bending strains indicate tension and compression, respectively. These influence lines are an indicator for how bending moments were distributed among the three instrumented girders that were located closest to loading. The moment distribution in other girders was obtained from the refined finite element models. The maximum bending moment distribution factors under AASHTO trucks were 0.596 and 0.478 for exterior and interior girders, respectively. These factors are 114.3% and 98.0%

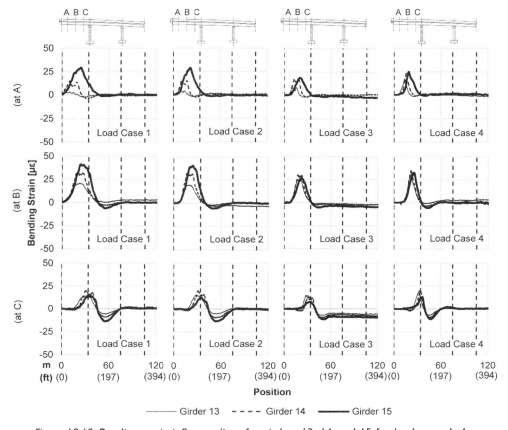

Figure 10.12 Bending strain influence lines for girders 13, 14, and 15 for load cases 1–4

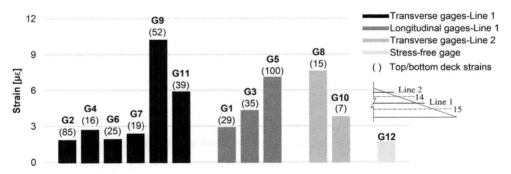

Figure 10.13 Maximum deck strains in absolute value from all load cases at all gages and at deck extreme fibers for all gage locations

of the ones predicted by AASHTO LRFD BDS (2014) for exterior and interior girders, respectively.

10.11.4 Deck strains under short-term loading

Changes in deck strains were small during load testing under all load cases. Figure 10.13 shows the maximum strain in absolute value measured at each gage during load testing. The figure also shows the maximum strain in absolute value measured at the stress-free gage. A comparison of strains shows that strains created by truck loading in the deck were small and, in various cases, comparable to the ones created by small temperature changes that took place during the load test (in approximately four hours). The short span of deck between wide flange prestressed girders may have resulted in the small strains. In addition, vibrating wire gages were embedded in concrete and therefore did not measure the maximum tensile and compression stresses at extreme fibers of the deck. The average depth of gages in deck, measured from the top surface of the deck, was approximately 112 mm (4.4 in.), close to the mid-height of the deck. The maximum extreme fiber strains in absolute value were calculated from the measured strains, assuming a linear strain profile along the deck height, and are shown in parentheses for each gage location in Figure 10.13.

At the time of load testing, which took place 28 days after the placement of deck concrete, there were no visible cracks in the deck. The maximum change in deck strains between the time at the installment of gages and the load test was −262 microstrains. Measured strains were caused by creep, shrinkage, and temperature effects. During the same period, the stress-free gage, having a different volume-to-surface ratio, reinforcement, and restraint conditions than the deck, measured a change in strain of −134 microstrains. The observation for lack of cracks and small values of deck strains may indicate that the unique arrangement of bearings over piers helped prevent high tension stresses in deck, and potentially the acute corner deck cracking.

10.12 Conclusions and recommendations

In this chapter, the load testing of a prestressed concrete girder/reinforced concrete deck bridge was discussed. The following are the conclusions regarding planning, scheduling, and execution of load test and data analysis.

- Communication between bridge owners, transportation agencies, local transportation agencies, local government, and contractors (if there is construction) regarding site access, instrumentation plan, traffic control, construction schedule, and scheduling of the load test is crucial. For this project, the communication was initiated as early as six months before load testing.
- Any field test is susceptible to unexpected weather, field conditions, malfunctioning of equipment or sensors, and time delays. Redundancy in measurements can help alleviate the impact of all or some of these factors on the results. For the bridge discussed in this chapter, redundancy was provided by adding duplicate gages for flexure strain measurements and by running every load case twice. Planning for a research team larger than needed will also help navigate unexpected circumstances.
- Electrical resistance surface strain gages and vibrating wire gages were durable and provided reliable data, even after being exposed to the environment for approximately one month before load testing. Reliability of measurements were confirmed by checking results using duplicate gages and duplicate load cases. On the other hand, vertical displacements at the bearings were too small to measure using selected displacement transducers. Similarly, midspan deflections could not be reliably measured using a total station and reflective targets. Careful selection of sensors for the required precision is of utmost importance for testing under service loads where responses are expected to be small in magnitude. Vertical bearing displacements and midspan deflection measurements were discarded as they did not provide repeatable data. Preliminary review of data allows detection of errors.
- Although sensors with wires were used in this study in accordance with the budget and available data acquisition systems, wireless sensors should be considered when available. These sensors have the potential to reduce instrumentation time and allow better distribution of sensors over bridge area.
- Preliminary finite element models are very beneficial in estimating the expected strains, determining areas that need instrumentation, and identifying the most suitable sensor types. In addition, most load tests will be limited by number of sensors or number of load cases. Therefore, it is recommended that finite element models be updated using the test data to obtain a holistic view of bridge response. For the bridge tested, finite element models were validated using load test data.
- Load testing was requested by the bridge owner for the bridge chosen because of the very large skew angle. The first goal of load testing was to understand the distribution of shear and moment between girders, which is known to change due to high skew angles. The load test results, supported by finite element modeling results, showed that even though the skew angle on this bridge was larger than the limit over which distribution factors given by AASHTO LRFD BDS (including skew correction factors) are not valid, AASHTO LRFD BDS predicted both shear and moment distribution factors conservatively, with the exception of moment distribution factors for exterior girders. Distribution factors for exterior shear, interior shear, exterior moment and interior moment were 99.8%, 69.0%, 114.3%, 98.0% of the ones predicted by specifications, respectively. Designers may consider excluding skew correction factors for moments while calculating girder live-load distribution.
- The second goal of load testing was to understand the impact of the unique girder bearing arrangement over piers on the bridge response, particularly on deck strains. To address this, deck strains were measured after deck pour for approximately one year and

during the load test that took place in that period (i.e. one month after deck pour). Changes in strains due to transient loading were very small (less than 10 microstrains), and were smaller than or similar to the ones caused by temperature alone. The short span of deck (close girder spacing and wide girder top flanges) and the effective location of gages (i.e. close to deck mid-height) may have caused the small strains from live load. Average deck strains measured for the first month after the deck pour were −323 microstrains in the deck and −174 microstrains in the stress-free cylinder gage. There were no visible cracks in the deck approximately one month after pouring. These observations and measurements indicate that the unique bearing fixity arrangement may be beneficial for deck performance.

Ackowledgements

This research project was funded through WisDOT and the USDOT. Neither WisDOT nor the USDOT assume any liability for the contents or the use thereof nor does this paper reflect official views, policies, standard specifications or regulations of either department.

References

AASHTO (2014) *AASHTO LRFD Bridge Design Specifications, Customary U.S. Units*, 7th edition. American Association of State Highway and Transportation Officials, Washington, DC.

Diaz Arancibia, M. & Okumus, P. (2018) Load testing of highly skewed concrete bridges. *ACI Special Publication, SP-323: Evaluation of Concrete Bridge Behavior through Load Testing-International Perspectives*, 2.1–2.18.

Diaz Arancibia, M., Okumus, P. & Oliva, G. M. (2017) Review of skew effects on prestressed concrete girder bridges. *Proceedings of the PCI Convention and National Bridge Conference*, 28 February– 4 March, Cleveland, United States. Precast/Prestressed Concrete Institute. p. 18.

Fu, G., Feng, J., Dimaria, J. & Zhuang, Y. (2007) *Bridge Deck Corner Cracking on Skewed Structures*. Construction and Technology Division, Michigan Department of Transportation. Report number: RC-1490. Detroit, MI.

Wisconsin Department of Transportation (n.d.) *Highway Structures Information System (HSI)*. [Online] Available from: http://wisconsindot.gov/Pages/doing-bus/eng-consultants/cnslt-rsrces/strct/hsi.aspx [accessed June 2016].

Chapter 11

Diagnostic Load Testing of Bridges – Background and Examples of Application

Piotr Olaszek and Joan R. Casas

Abstract

This chapter presents principles and justification of diagnostic load tests of bridges as normally performed. Normally, diagnostic tests serve to verify and adjust the predictions of an analytical model. However, as presented in the chapter, the results of a diagnostic load test in a bridge can also serve other objectives. Several examples of application are presented with the main objective not only to show how the tests are carried out, but chiefly to introduce which are the main issues to take into account to obtain accurate and reliable results that can be used in the assessment of the actual capacity of the bridge. In the case of static tests, conclusions regarding the measurement stabilization time are presented. For dynamic tests in railway bridges, the extrapolation to lower and higher speeds is also discussed.

11.1 Background

11.1.1 *Definition*

A diagnostic load test is aimed to supplement and check the assumptions and simplifications made in the theoretical assessment. This type of test is used for a better structural idealization and appraisal of material properties and structural behavior. Diagnostic tests serve to verify and adjust the predictions of an analytical model. At the end, it must be considered that the origin of diagnostic test loading comes from the necessity of investigating the ability of the bridge structure to carry the design loads before entering into service. Load testing has been performed in many countries for a long time and is connected with the ancient tradition that obliged the designer to stand underneath the bridge while the test was being carried out. Now the main objective of diagnostic load testing is to use its results as a research tool, providing better understanding of the way in which loads are carried by and distributed through the bridge structure (COST 345, 2004).

Based on the load level and traffic restriction requirements, the diagnostic load testing can be divided in two types: soft load testing and diagnostic load testing.

The soft load test uses the actual traffic in the bridge as the loading source. Using a weigh-in-motion (WIM) system, not only are the main characteristics of exciting traffic obtained but also information about the structural behavior of the bridge, through the calculation of experimental influence lines, dynamic amplification factors, and load distribution to different structural members. In this way, the test is aimed to supplement and check the assumptions and simplifications made in the theoretical assessment. Therefore, the main

DOI: https://doi.org/10.1201/9780429265426

218 Load Testing of Bridges

objective is to optimize the structural model used for design or safety assessment. The execution of the test does not require the closure of the bridge to normal traffic. It is also shown (Casas et al., 2009) that the soft load test can be used to evaluate the characteristic total load effect of traffic action in a quite simple way, provided sufficient time data of traffic records are available.

Similar to the soft load test, diagnostic tests serve to verify and adjust the predictions of an analytical model. However, in this case the level of load in the bridge is higher and introduced by different devices (trucks, water tanks, ballast, jacks) with accurately measured weight. Normally the bridge is closed to traffic during the execution of the test to better control the relationship between the load level and the bridge response. The loading source may be static or dynamic. The following information can be obtained: experimental influence lines, dynamic amplification factors, load distribution, and dynamic parameters (natural frequencies, mode shapes, damping).

11.1.2 Objectives

The main objectives when carrying out a diagnostic test are the following:

1   In new bridges, to check the assumptions in the design and the correct performance before opening to traffic. In old bridges, without drawings and information about the design and construction details and behavior under loading, to derive longitudinal and transverse influence lines (distribution factors) under normal traffic load.
2   In posted bridges, to check if the posting (limiting of the traffic loading) is justifiable or if it can be released or removed.
3   To provide input data for the efficient management of heavy vehicles with special permits.
4   To obtain experimental dynamic amplification factors to be used for the advanced assessment of existing bridges under normal traffic.
5   To assess site-specific bridge characteristic total load (including dynamic effects) caused by traffic.
6   To update an analytical model in bridges where there are doubts on the analytical load rating model and where the model can or should be further developed to be accurate enough in the process of bridge load-carrying capacity assessment.
7   In bridges with sufficient data and information on as-built bridge details, dimensions, and materials or, alternatively, sufficient data obtained through inspections and material testing, the diagnostic test can confirm the obtained data.
8   To investigate the real behavior in bridges with some doubts about unintended composite action of the bridge main or even secondary members, which could appear during testing with low load level but can disappear at higher load ratios.
9   To assess bridges that should be rated versus dynamic forces (earthquake, wind, vehicle collision, mining damage, etc.).

11.1.3 Planning and execution

The first step in planning the test is to visit the site and have a close look at the structure. The load level is one of the most important parameters when planning the diagnostic load testing. The bridge structure is loaded with static and (or) dynamic loads during the load testing. Highway bridges are usually statically loaded with vehicles (lorries). The main parameter of the load test is the magnitude of the load, the way it could be placed on the bridge, and the tools and equipment needed. As a rule, the load quantity is chosen in relation to some percentage of the characteristic or design traffic loads. The load level achieved in the test must be representative of serviceability conditions. It is recommended to achieve a load level corresponding to a five-year return period. In practice, this means to raise the test load up to approximately 60% of the characteristic live load present in the design code. It is recommended never to go beyond the 70% of the design load or 100% of the unfactored traffic load (Casas et al., 2009). Another approach is to determine the load level based on the normal day-to-day load experienced by the structure. Static load quantities used in diagnostic load tests in different European countries are presented in COST 345 WG 2 and WG 3 Report (COST 345, 2004).

The scope of investigation and measurements during load testing is nearly the same in different countries and contains (COST 345, 2004):

Investigation range

- Visual examination before load testing
- Measurements under the load testing:

  - Applied load
  - Deflection
  - Support displacement
  - Strain/stress
  - Crack width
  - Secondary effects like temperature, sun exposure, wind

- Visual examination during the load testing
- Visual examination after the load testing.

Measurement methods

- Applied load:

  - Load cells

- Deflection:

  - Inductive mechanic gages
  - Dial gages
  - Inclinometer (by calculating deflection curve on the base of measured angles in several sections)
  - Accelerometer (dynamic loading)
  - Laser technique

- Strain:

  - Vibrating wire strain gages,
  - Resistance gages
  - Optical fibers

220 Load Testing of Bridges

- Temperature:

  - Thermocouples
  - Optical fibers.

As discussed later in this chapter, one of the main parameters to fix in the execution of a static test is the sampling time between measurements in order to keep a good estimate of the stabilization time after application of a load increment. This point is of importance to gather reliable and accurate measurements, which are really representative of the bridge response.

During deflection and strain/stress measurements and future comparison with analytical calculation it is necessary to take into account other factors, which should be measured or estimated:

- Applied load (real weight and its position)
- Support displacement
- Bearing displacement
- Measurement uncertainty
- Secondary effects:

  - Temperature
  - Sun exposure.

A very important part of the static test is the zero load case measurement, preferably completed sometime before the test start to identify possible trends in temperature response. During load increase and maximum load phases, it is recommended to continuously observe the load-displacement plot to control an early indication of any nonlinear response. After the load phase, it is necessary to continue the measurements to detect any permanent quantities.

11.1.4 Results and safety assessment

Load-carrying capacity based on diagnostic load testing results should optimize bridge assessment by using load tests together with computational analysis to find reserves in load-carrying capacity.

The ARCHES Project deliverable *Internet database of load test results and analytical calculations* (Olaszek and Karkowski, 2009) contains information on the correlation (comparison factor) between the real structure behavior (load testing results) with corresponding results of analytical calculation of different types of bridges obtained from different countries (available data from national resources and from other projects, including tests of bridges before putting them into service, assessment of load-carrying capacity of existing bridges and load tests done for research purposes). The analytical modeling of a structure is characterized by a lot of simplifications and inaccuracies in relation to real structure behavior. This is clearly shown in the examples presented in the deliverable (more than 100 bridges with different type and materials), where the comparison factor review pointed out that average comparison factors of nearly all bridges have an unacceptable match. This means that nearly all analytical models presented in the database required calibration. The hypothetical assessment of bridge load capacity with the use of those models without

Diagnostic Load Testing of Bridges 221

previous updating and calibration by a diagnostic test would have an unacceptable error. Only 3%–36% (depending on the bridge structural material) of presented comparisons in the database contain loaded bridge members with an acceptable match.

Therefore, in the case of diagnostic loading tests, the integration of the results from the test in the safety assessment process is achieved through the updating and improvement of the structural model of the bridge. The methodology is different depending on the nature of the test (static or dynamic).

11.1.4.1 Static tests

In the case of a static test, as long as the bridge exhibits linear behavior, the test can be used to validate and update the analytical model and bridge load capacity. There are two ways to incorporate the results of the static tests in the assessment process:

1   By updating the structural model and calculation of the new bridge capacity (reliability index, load factor) based on the new model. The idea is to change the bridge properties (area, inertia, modulus of elasticity, etc.) so that the theoretical model matches as well as possible the results of the load test. In many tests, the deflections in different sections of the bridge are measured. An acceptable match is considered to have been reached when the differences between the site-measured maximum deflections and the analytical values are within the following limits:

    ±10% for prestressed concrete and metallic bridges
    ±15% for reinforced concrete and composite bridges

    Once the model is updated, the assessment calculations are carried out using the revised model and it can be used in the recalculation of the bridge safety (reliability index, load factor, etc.).

2   Direct calculation of the load capacity from the test results. In this case, it is assumed that the bridge assessment is carried out using the partial safety factor format and the load capacity is the value for which the rating live load should be multiplied to reach the failure limit state. This is the method followed by AASHTO (AASHTO, 2003), which proposes the following equation:

$$LC_T = (1 + K_a K_b) \times LC_C \tag{11.1}$$

    where LC_T is the load capacity based on the result of the load test and LC_C is the load capacity based on calculations and before incorporating the results of the load test. K_a can be positive or negative depending on the results of the load tests and is calculated as:

$$K_a = \frac{\varepsilon_C}{\varepsilon_T} - 1 \tag{11.2}$$

    where ε_T is the maximum member strain measured during the load test and ε_C is the calculated strain due to the test vehicle at its position on the bridge which produced ε_T. It should be calculated using a section factor (area, inertia, etc.) which most closely approximates the member's actual resistance during the test.

222 Load Testing of Bridges

K_b is a factor that takes into account the possibility that the bridge has adequate reserve capacity beyond the rating load level and also the load level (compared to the rating load) that the bridge has faced during the test. If the relationship between the unfactored test vehicle effect (T) and the unfactored gross rating load effect (W) is less than 0.4, it is recommended to take $K_b = 0$. If this relationship is higher than 0.7, then a value of 1.0 is recommended if the behavior of the member during the load test can be extrapolated for a load level of $1.33W$; if not, the value is 0.5.

11.1.4.2 Dynamic tests

In the case of a diagnostic dynamic test, the results are usually a measure of stiffness rather than strength. The results may be used to validate the prediction of calculations. The structural model updating and calculation of the new bridge dynamic properties should be done together with the updating of the structural model in the static tests range. The other application of dynamic testing is a comparison of results over time that may be used to monitor any deterioration or serious damage to the structure. However, it should be noted that the modification of dynamic parameters may be not sufficient to detect damages. Variations of these parameters may be caused by changes of environmental conditions such as temperature or humidity. This effect has to be deeply considered when evaluating the results of a dynamic test. The importance of dynamic tests is caused by several additional reasons. For instance, earthquake response is dependent on bridge frequency and damping. Dynamic behavior with repeated stress oscillations may have a large influence on the fatigue assessment. The assessment of the bridge safety under these dynamic effects requires in many cases the execution of a dynamic test.

An important issue of diagnostic load testing is the analysis of the uncertainty of the tests results and its influence on the final assessment of the bridge capacity. The notion of measurement uncertainty is relatively well-known and often taken into consideration (JCGM, 2010). This is not, however, the only reason for the uncertainty of the whole load test. Each element of the measurement process, design of the structure's calculation model, and comparative analysis of the measurement results and calculations brings an unavoidable uncertainty. In the case of dynamic load testing there is a standard describing possible reasons of uncertainty (ISO, 2004). In the case of static load testing, one of the uncertain elements is the correct estimation of the elastic and permanent values of the measured displacement. The possibilities of the estimation will be presented in the next section with the example of several bridge tests, when different speeds of stabilization of the measured displacements were observed. This is an important issue, as the incorrect sampling frequency during measurements could distort the final result of the measured deflections and, as a consequence, the bridge model updating and the final decision regarding the correct performance.

The dynamics of railway bridges is one of the most complex areas of theoretical analyses and experimental studies (Frýba, 1996, Yang et al., 2004). The results of dynamic analysis of railway structures are considerably uncertain as compared with static analyses and as a rule require experimental verification. The basic element of these analyses is the determination of the dependency of the maximum amplitudes of displacements or accelerations on the train's speed. It is important to analyze the structure's behavior during the train passages at the speed increasing from the quasi-static one to the maximum one permissible for the tested section.

The load applied during the dynamic testing of railway bridges is similar or identical to the load applied during the static tests. Each train ride takes place in almost identical conditions, and the only difference is the travel speed. In the case of railway bridge tests and due to budget restrictions it is normal to limit the number of train passages, and few of them are at the quasi-static speed as well as at speeds close to the maximum ones permissible for the tested sections. This chapter presents examples showing how significant mistakes can be connected with the estimation of the maximum displacement amplitudes by extrapolation of train passages at speeds higher and lower than the registered ones.

11.2 Examples of diagnostic load testing

The presented examples of tests are a summary of the analysis of possible mistakes in analyzing and evaluating the results of static and dynamic diagnostic tests carried out on bridge structures. These works are a continuation of those presented in (Olaszek et al., 2016; Olaszek, 2016).

11.2.1 Static load testing

While selecting measurement methods for the diagnostic load testing, it is important to consider not only the measurements' uncertainty, but also the possibilities of making an analysis of the structure's displacements in on-line mode. The load is applied in accordance to a loading scheme and held for a certain time period. The duration of the test and the accuracy of the results will depend on the time the load should be held until stabilization of the outputs. The early removal of the load before the permanent or stationary value is reached can lead to important errors. Therefore, it is important to have some criteria to decide when the stabilized value of the displacements is obtained. This issue is discussed next.

11.2.1.1 The estimation of the elastic and permanent values

For over 20 years, the Road and Bridge Research Institute in Poland has been trying to apply practically an ongoing observation of the vertical displacements of girders. The displacements were registered with high sampling frequency of the measured signal (5 Hz) in order to make a detailed analysis of the quasi-static variation during the static load testing of bridge structures. The following load phases were taken into account (Fig. 11.1):

1 Phase **a** of "zero load" preceding the phases of applying loads to the structure
2 Phase **b** of applying loads to the structure
3 Phase **c** of a constant load applied to the structure
4 Phase **d** of removing the load from the structure
5 Phase **e** of "zero load" following the removal of the load from the bridge structure.

The elastic values of the displacement d_{ev} are determined on the basis of the difference between displacement at phases **c** d_c and **e** d_e:

$$d_{ev} = d_c - d_e \qquad (11.3)$$

224 Load Testing of Bridges

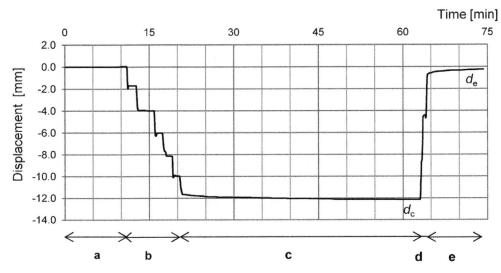

Figure 11.1 An example of displacements observed in the time function recorded during the diagnostic static load test with marked phases of the diagnostic test. Conversion: 1 mm = 0.0394 in.

Source: From Olaszek et al. (2016).

while the permanent values of the displacement d_{pv} – on the basis of phase **e** d_e:

$$d_{pv} = d_e \quad (11.4)$$

Phases **b** and **d** include elements connected with quasi-dynamic loading of the structure, such as entry and exit of vehicles or locomotives to and from the bridge, pouring water in or out, and adding or removing concrete slabs or steel blocks. Phases **a, c,** and **e** are typical phases of a bridge structure static load test where the load remains constant.

11.2.1.2 Examples of application to different types of bridges

The time history presented in Figure 11.1 is characterized by a fast stabilization of displacements both after applying a load and after removing it. Different speeds of stabilization can be observed during the tests in different bridge types, thus the optimal duration of each individual tests is also different. This chapter presents examples of tests during which a relatively slow stabilization of vertical displacements was registered for different types of concrete and steel bridge structures:

1 Steel bridge with a composite-reinforced concrete bridge deck
2 Reinforced concrete slab bridge
3 Reinforced concrete frame bridge
4 Cable-stayed bridge: steel girders with reinforced concrete deck
5 Steel arch bridge with a reinforced concrete bridge deck
6 Prestressed concrete box girder bridge
7 Steel girders with reinforced concrete deck
8 Prestressed concrete girders with reinforced concrete deck.

Figures 11.2 and 11.3 present photographs of the tested structures.

In order to evaluate the speed of stabilization, the absolute increments $\Delta d_{\Delta t}(t)$ of vertical dislocations were determined in the time function with different time intervals Δt:

$$\Delta d_{\Delta t}(t) = d(t) - d(t - \Delta t) \tag{11.5}$$

also, the relative increments $\dfrac{\Delta d_{\Delta t}}{d(t - \Delta t)}(t)$ were obtained:

$$\frac{\Delta d_{\Delta t}}{d(t - \Delta t)}(t) = \frac{d(t) - d(t - \Delta t)}{d(t - \Delta t)} \tag{11.6}$$

Figure 11.2 Lateral view of tested bridges, from above. (1) Steel bridge with a composite-reinforced concrete bridge deck; (2) reinforced concrete slab bridge.

Source: Photos 1 and 2 from Olaszek (2015).

Figure 11.2 (3) reinforced concrete frame bridge; (4) cable-stayed bridge – steel girders with reinforced concrete deck.

Source: Photos 4 from Olaszek (2015), 3 from © Google.

the calculations were made at intervals between the readouts equal to $\Delta t = 1, 5, 10,$ and 15 minutes.

In order to compare the relative increments, they were normalized with respect to the time increment:

$$\frac{\frac{\Delta d_{\Delta t}}{d(t-\Delta t)}(t)}{\Delta t} = \frac{d(t) - d(t - \Delta t)}{d(t - \Delta t)\Delta t} \tag{11.7}$$

Figure 11.3 Lateral view of tested bridges, from above. (5) Steel arch bridge with a reinforced concrete bridge deck; (6) prestressed concrete box girder bridge; (7) steel girders with reinforced concrete deck.

Source: Photos 5 and 6 from © Google, 7 from Olaszek (2015).

Table 11.1 shows the results of the analysis of the speed of vertical displacements stabilization in phase **c** of the load test, obtained for the tested structures, while Figures 11.4, 11.5, and 11.6 show examples of displacements together with displacement increments in phase **c** of the load and phase **e** after removing the load, obtained at intervals $\Delta t = 1, 5, 10, 15$ minutes. The table shows the results of an analysis of each bridge, starting from the

Figure 11.3 8) prestressed concrete girders with reinforced concrete deck.
Source: Photo 8 from Olaszek (2015).

lowest speeds of stabilization. The results show extremely different speeds of stabilization depending on the bridge type.

In the first case (bridge 1), we can see an exceptional behavior where significant deflection increments and no tendency of displacement stabilization after the application of the load were observed (Fig. 11.4). The tests were carried out after renovation of the old structure, which consisted of strengthening its steel structure and completing a composite-reinforced concrete bridge deck. Considerable gradients of deflection increments visible on the presented time history prove the inappropriate performance of the structure. The bigger the load, the higher the gradients. The gradients did not bear features connected with plasticity of the carrying elements material (steel) and, what is more important, the calculated stress values were fairly low. The analysis of the measured deflections suggested that during loading, uncontrollable structure clearances and tolerances were eliminated. Following these observations, a hypothesis that the source of these clearances was an inappropriate functioning of frictional joints could be formulated. Fortunately, the test was stopped and the spans were unloaded. If the process of loading had been continued, clearances between the bolts of friction joints and the holes in structure elements could have disappeared (Olaszek et al., 2014).

In case of the other bridges, as we can see from the table, the relative increments normalized with respect to time and obtained 30 minutes after the application of the load have similar values independently of the considered time interval. For instance, in the continuous bridge composed by a reinforced concrete slab (bridge 2), these increments are close to 0.2%/minute and do not significantly depend on the time between the readouts Δt.

This confirms the fact that in most presented tests the speed of the increments is almost identical in the analyzed intervals Δt after 30 minutes of the application of the load. There is, however, a considerable variation of the increments during the first 15 minutes after the application of the maximum test load.

In case of the evaluation of displacements stabilization in phase **e**, the increments recorded after removing the load can be considerably burdened with measurement errors and noise. Thus relative increment parameters can have considerable values due to measurement errors but not a lack of stabilization. This is also presented in Figures 11.5 and 11.6. To evaluate the process of displacements stabilization after removing of the load,

Diagnostic Load Testing of Bridges 229

Table 11.1 Results of the analysis of the speed of vertical displacements stabilization in phase c of the load test of concrete bridges

Types of bridge structures Spans length (m) (marked tested span)	Readouts time	Intervals between the readouts (min)							
		15	10	5	1	15	10	5	1
		Relative increments $\frac{\Delta d}{d(t-\Delta t)}(t)$ (%)				Relative increments time normalized $\frac{\Delta d}{d(t-\Delta t)}(t)/\Delta t$ (%/min)			
1. Steel bridge with a composite-reinforced concrete bridge deck	$t_{cp} + \Delta t$	19.96	15.06	9.41	2.35	1.33	1.51	1.88	2.35
Simply supported spans: **29.0** + 21.2 + 29.0	$t_{cp} + 30$	17.42	11.87	4.29	0.75	1.16	1.19	0.86	0.75
2. Reinforced concrete slab bridge	$t_{cp} + \Delta t$	18.59	16.15	11.39	4.34	1.24	1.61	2.28	4.34
Continuous spans: **18.0** + 21.0 + 21.0 + 18.0	$t_{cp} + 30$	4.32	2.57	1.17	0.21	0.29	0.26	0.23	0.21
3. Reinforced concrete frame bridge	$t_{cp} + \Delta t$	8.91	4.38	2.97	0.16	0.59	0.44	0.59	0.16
Single span: **22.0**	$t_{cp} + 30$	2.30	1.57	0.56	0.14	0.15	0.16	0.11	0.14
4. Cable-stayed bridge: steel girders with reinforced concrete deck	$t_{cp} + \Delta t$	7.98	7.81	7.71	5.85	0.53	0.78	1.54	5.85
Continuous spans: 48.0 + **77.0** + 250.0 + 77.0 + 48.0	$t_{cp} + 30$	0.42	0.33	0.18	0.08	0.03	0.03	0.04	0.08
5. Steel arch bridge with a reinforced concrete bridge deck	$t_{cp} + \Delta t$	2.90	2.87	2.61	1.97	0.19	0.29	0.52	1.97
Single span: **45.0**	$t_{cp} + 30$	0.04	0.04	0.00	0.00	0.00	0.00	0.00	0.00
6. Prestressed concrete box bridge	$t_{cp} + \Delta t$	2.16	1.64	1.24	0.47	0.14	0.16	0.25	0.47
Continuous spans: 42.0 + 55.0 + **80.0** + 136.0 + 80.0 + 3 × 55.0 + 42.0	$t_{cp} + 30$	1.15	0.84	0.48	0.26	0.08	0.08	0.10	0.26
7. Steel girders with reinforced concrete deck	$t_{cp} + \Delta t$	1.68	1.23	0.82	0.19	0.11	0.12	0.16	0.19
Continuous spans: 23.7 + 35.3 + **38.0** + 35.0 + 33.6	$t_{cp} + 30$	0.75	0.51	0.19	0.01	0.05	0.05	0.04	0.01
8. Prestressed concrete girders with reinforced concrete deck	$t_{cp} + \Delta t$	0.56	0.45	0.23	0.00	0.04	0.05	0.05	0.00
Continuous spans: **20.2** + 20.2	$t_{cp} + 30$	0.78	0.33	0.11	0.00	0.05	0.03	0.02	0.00

Note: t_{cp} – time of the phases c beginning
Conversion factor: [1 m = 39.37 in].

we cannot use relative increment parameters but we have to analyze the history of absolute displacement. Comparing Figures 11.5 and 11.6, we can also see that the displacement stabilizes much faster in the prestressed concrete girder bridge than in the reinforced concrete slab bridge.

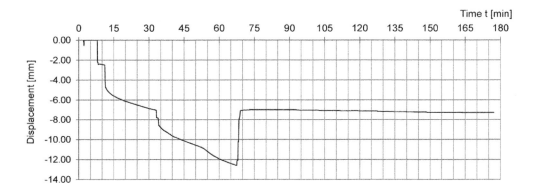
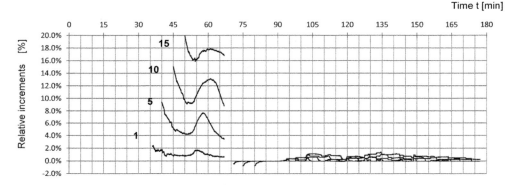

Figure 11.4 Analysis of the vertical displacements stabilization during static load test in case of the steel bridge with a composite-reinforced concrete bridge deck (bridge 1). Upper figure: time history of the displacements. Lower figure: time history of the displacement increments in phases **c** and **e** of the load tests, determined at intervals Δt = 1, 5, 10, 15 minutes. Conversion: 1 mm = 0.0394 in.

Summing up the analysis of the measured time histories of displacements, we can conclude that to guarantee a reliable and accurate process of bridge assessment through diagnostic load testing, it is necessary to use measuring techniques which not only ensure appropriate accuracy of the measurements themselves but also enable to monitor the tested structure practically on an ongoing or on-line basis.

Thanks to the use of quasi-continuous displacement monitoring (measurements at least once a minute) as compared with measurements every 15 minutes, we have much more possibilities to control the correctness of the final measurements as well as much better control over the displacement stabilization process during and after the load test of the bridge.

It has been also shown that for bridges of any type, after 30 minutes of the application of the load, the decay of the displacement curve is almost independent of the time interval considered. In relation to the above, and also due to the necessity of carrying out a series of readouts minimizing the impact of noise, the minimum duration of the load

Diagnostic Load Testing of Bridges 231

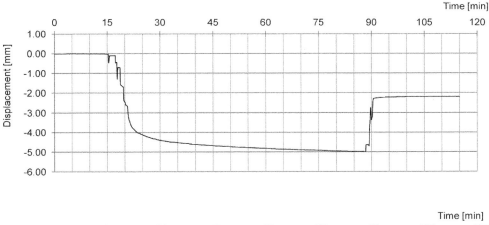

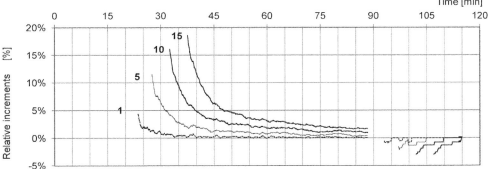

Figure 11.5 Analysis of the vertical displacements stabilization during static load test in case of the reinforced concrete slab bridge (bridge 2). Upper figure: time history of the displacements. Lower figure: time history of the displacement increments in phases **c** and **e** of the load tests, determined at intervals Δt = 1, 5, 10, 15 minutes. Conversion: 1 mm = 0.0394 in.

Source: Adapted from Olaszek et al. (2016).

ΔTc, in case of intervals between the readouts equal to $\Delta t = 1$ min, should be $\Delta T_c = 30$ min, and for $\Delta t = 5$, 10, and 15 min, $\Delta T_c = 35$, 40 and 45 min, respectively. In case of short intervals between the readouts ($\Delta t \leq 5$ min), we can consider the moment when the relative increments will be lower than 0.2% as the minimum condition for stabilization.

In case of static load tests, the readouts carried out with a proper sample frequency allow to obtain a better evaluation of the stabilization process. Extrapolation of the measured signals is a useful solution connected with the use of digital techniques, aiming to forecast the stabilization time and values determined in steady state. This is presented in Figure 11.7, where on the time history of the displacements from the static load tests an extrapolation on a time history section was performed and then its accuracy was checked by comparing it with the measured values. The section of the actual time history of displacements was selected from the end of loading to the moment of recording an increase of 2.6%. On the basis of a section of 30 minutes, an equation was determined describing the curve and

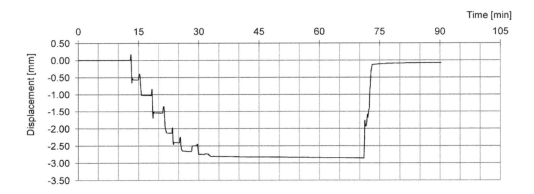

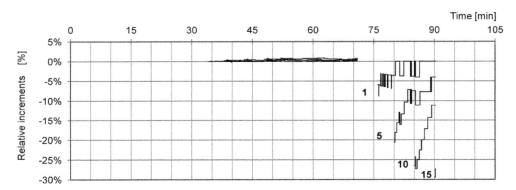

Figure 11.6 Analysis of the vertical displacements stabilization during static load test in case of the prestressed concrete girders with reinforced concrete deck (bridge 8). Upper figure: time history of the displacements. Lower figure: time history of the displacement increments in phases **c** and **e** of the load tests, determined at intervals Δt = 1, 5, 10, 15 minutes. Conversion: 1 mm = 0.0394 in.

Source: Adapted from Olaszek et al. (2016).

then the curve was extrapolated up to 95 minutes of the real-time history. After the extrapolated curve obtained an increase in the displacements below 2% an average value from 1 minute was determined. The relative error of the extrapolated average value in relation to the value determined on the basis of the actual measurement does not exceed 0.7% (0.03 mm [0.001 in]). Obtaining this high accuracy in the extrapolation results from making it at the moment close to stabilization. If the sampling interval had been 15 minutes, the resulting stabilization time would be around 82 minutes, which means that the loading time of the structure could have been shortened by about 15 minutes.

It must be noted that due to the possible occurrence of significant nonlinearities of the changes in the time history of displacement during the test load (or even step changes), extrapolation can be only of supporting character to evaluate the stabilization speed and cannot be used to determine the values of the displacement in the steady state.

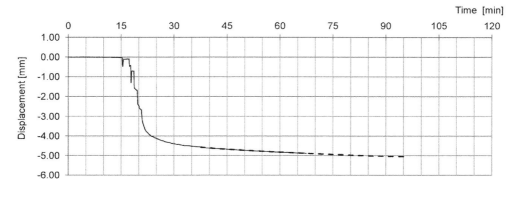

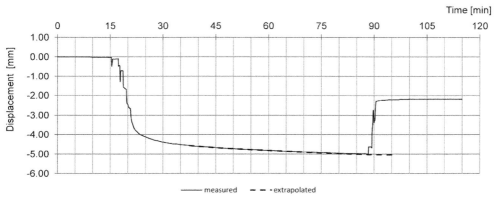

	Measured	**Extrapolated**
Time [min]	Relative increment of displacements [%/15 min]	
37.5	6.22	
67.5	2.57	
Time [min]	Average value of displacements [mm]	
87.0÷88.0	-4.98	-5.01

Figure 11.7 Bridge 2: example of using extrapolation to forecast excess displacements. From top to bottom: time history of the displacements with interrupted registration together with an interpolated curve on the basis of a fragment from the last 30 minutes and extrapolated curve; actual time history of the displacements together with an extrapolated curve from the previous figure; summary of results obtained on the basis of an analysis of the actual and extrapolated time history. Conversion: 1 mm = 0.0394 in.

234 Load Testing of Bridges

11.2.2 *Dynamic load testing*

Analyses of forced and free vibration are carried out in the case of dynamic load tests. As a rule, a free vibration analysis is carried out in the time domain, and mode shapes and frequencies of free vibrations as well as damping are determined. Analysis of forced vibration is a more complex issue. It is especially important for railway bridges. The results of theoretical dynamic analyses of bridge structures are determined with a considerable uncertainty in relation to static analyses and as a rule require experimental verification. The dynamic coefficient or dynamic amplification factor, in the case of railway bridges, is of significant importance depending on the speed of ride, in contrast to road structures. The purposefulness of determining these coefficients is justified mainly by the fact that during dynamic load testing of railway structures the applied load is close to or identical to the load applied during static load tests.

11.2.2.1 *Extrapolation of values for quasi-static speed*

The extrapolation of values under lower speed is presented with an example of quasi-static displacements time history $d(v_{sta},t)$ made on the basis of displacements time history $d(v_{max},t)$ registered during the train ride at the maximum permissible speed v_{max}. Quasi-static displacements history $d(v_{sta},t)$ is obtained by filtering the dynamic displacements time history $d(v_{max},t)$:

$$d(v_{max},t) \xrightarrow{\text{filtering}} d(v_{sta},t) \tag{11.8}$$

The value d_{sta} was determined as the extreme value of filtered dynamic displacements time history:

$$d_{sta} = MAX(|d(v_{sta},t)|) \tag{11.9}$$

In case of road bridges the method of obtaining quasi-static displacements history by means of filtering was presented in (Paultre et al., 1992). According to this publication, a low-pass digital filter, applied to the recorded data, is used to smooth out the dynamic frequencies in the signal. The filtering can be done with a moving average filter or finite-impulse response filter. The applied filter must have a passband of f_{pb} frequency:

$$f_{pb} = \frac{v}{L} \tag{11.10}$$

where v is the vehicle speed and L is the span length. The stopband with a cutoff frequency f_{co} must be below the bridge's first fundamental frequency f_{F1}:

$$f_{pb} < f_{co} < f_{F1} \tag{11.11}$$

In order to analyze the effectiveness of the filtering method in the case of railway bridges, three types of low-pass filters, significantly different in frequency characteristics, were tested:

- Bessel filter
- Finite impulse response (FIR) filter
- Moving average filter.

These are typical low-pass digital filters used for smoothing signals in the time domain (Smith, 2003; Lyons, 2011). The first two are relatively difficult to use: they require specialist software. The Bessel filter is derived from the analog filters and has a characteristic optimized for maximally linear phase response, that is, minimal distortion of the wave shape of the filtered signals in the passband. This filter is characterized by a relatively wide transition region from passband to stopband. On the other hand, the finite impulse response (FIR) filter has characteristics optimized for a narrow transition region from passband to stopband and for maximally linear phase response. The moving average filter is the easiest digital filter to use. This filter is optimal for reducing random noise while retaining a sharp step response. However, the moving average is the worst filter for frequency domain encoded signals, with little ability to separate one band of frequencies from another.

The results in case of the Bessel filter and FIR filter were analyzed by means of three cutoff frequencies:

- f_{c1}, equal to the average of the so called static frequency f_{pb} (Equation 11.10), and first fundamental frequency of free vibration f_{F1}:

$$f_{c1} = \frac{f_{pb} + f_{F1}}{2} \tag{11.12}$$

- f_{c2}, using the method of successive approximations (filtering using variable f_{c2}) in order to obtain no free vibration in the filtered signal:

$$d(v_{max}, t) \xrightarrow{\text{filtering with } f_{c2}} d(v_{sta}, t) \cap no\ free\ vibration \tag{11.13}$$

- f_{c3}, using the method of successive approximations (filtering using variable f_{c3}) in order to obtain value d_{sta}, which was consistent with the extreme value d_{v10} registered during the train ride at the quasi-static speed of 10 km/h (6.21 mph):

$$d(v_{max}, t) \xrightarrow{\text{filtering with } f_{c3}} d(v_{sta}, t) \cap d_{sta} = d_{v10} \tag{11.14}$$

where

$$d_{v10} = MAX(|d(v_{10}, t)|) \tag{11.15}$$

In case of moving average filtering method, the results were analyzed by using three averaging periods:

- T_{ave1}, equal to the inverse of average from the so-called static frequency f_{pb} (Equation 11.10) and first fundamental frequency of free vibration f_{F1}:

$$T_{ave1} = \frac{2}{f_{pb} + f_{F1}} \tag{11.16}$$

- T_{ave2}, using the method of successive approximations (filtering using variable T_{ave2}) in order to obtain no free vibration in the filtered signal:

$$d(v_{max}, t) \xrightarrow{\text{filtering with } T_{ave2}} d(v_{sta}, t) \cap no\ free\ vibration \tag{11.17}$$

- T_{ave3}, using the method of successive approximations (filtering using variable T_{ave3}) in order to obtain value d_{sta}, which was consistent with the extreme value dv_{10} (Equation 11.15) registered during the train ride at the quasi-static speed of 10 km/h (6.21 mph):

$$d(v_{max}, t) \xrightarrow{\text{filtering with } T_{ave3}} d(v_{sta}, t) \cap d_{sta} = d_{v10} \tag{11.18}$$

236 Load Testing of Bridges

To sum up, the frequency f_{c1} and the averaging period T_{ave1} were determined on the basis of one-time calculations, while the frequencies f_{c2} and f_{c3} and the averaging periods T_{ave2} and T_{ave3} were obtained using the method of successive approximations in order to yield no free vibration or the compatibility of values d_{sta} and d_{v10}.

The frequencies f_{c1} and f_{c2} and the averaging periods T_{ave1} and T_{ave2} were determined to analyze the effectiveness of the filtering method to obtaining quasi-static displacements. The frequency f_{c3} and the averaging period T_{ave3} were determined to find the optimal parameters of the filtrations.

11.2.2.2 Extrapolation of values under higher speed

The extrapolation concerned the extreme values (minimal and maximal) obtained from the registered time history of displacement during the train passage with the speeds v_i:

$$d(v_i, t) \xrightarrow{maximum} d_{max}(v_i) \tag{11.19}$$

$$d(v_i, t) \xrightarrow{minimum} d_{min}(v_i) \tag{11.20}$$

The extrapolation is applied for values of the train passage with a speed reaching m of measured values:

$$\{d_{max}(v_{sta}), ..d_{max}(v_i), ..d_{max}(v_k)\} \xrightarrow{Extrapolation} d_{max}(v_{max}) \tag{11.21}$$

$$\{d_{min}(v_{sta}), ..d_{min}(v_i), ..d_{min}(v_k)\} \xrightarrow{Extrapolation} d_{min}(v_{max}) \tag{11.22}$$

where $v_{max} = mv_k$.

The extrapolation is realized with the use of polynomial extrapolation.

11.2.2.3 Examples of dynamic testing

This subchapter presents possible applications of digital signal processing techniques for extrapolation of measurement results during dynamic testing of high-speed railway bridges. Extrapolation concerns calculation of measurement results of high- and low-speed train rides, which either exceed or do not reach the speed values registered during the test carried out. The extrapolation of values under lower speed is presented with the example of quasi-static value calculation made on the basis of displacements registered during train rides at the maximum permissible speed. This activity is applied mainly to the calculation of dynamic amplification factor. The extrapolation of values under higher speed is presented with the example of calculation of displacements at a speed reaching 125% of the measured values. The applied extrapolation methods are presented with the examples of three bridges:

1   Steel truss bridge with reinforced concrete deck
2   Steel plate girder bridge with orthotropic deck
3   Steel arch bridge with reinforced concrete bridge deck.

Figure 11.8 presents photographs of the tested structures.

The tests on bridges 1 and 2 were carried out as diagnostic tests of old bridges which had been in use for a long time in order to evaluate their load-carrying capacity and to

Diagnostic Load Testing of Bridges 237

Figure 11.8 View of tested bridges, from above. (1) Steel truss bridge with reinforced concrete deck and curved railway track.

Source: Photo 1 from Olaszek (2015).

determine the maximum permissible speed of trains. The tests on bridge 3 were carried out as acceptance tests of a newly constructed bridge.

Bridge 1 is located on a curved alignment. The railway track is curved over the whole length of the span with a radius of $R = 2600$ m (8530 ft). Because of this, while testing this bridge a nonlinear increase in the vertical and horizontal displacements was recorded as the train speed increased.

During the tests on bridges 2 and 3, high values of the dynamic amplification factor were observed, which resulted from the bridge's dynamic susceptibility. Bridge 2 shows an exceptionally high level of forced and free vibrations. The damping was relatively low. Bridge 3 shows a high level of vibration in the hangers both for forced and free vibration cases.

The dynamic tests were conducted using a special train consisting of two locomotives and four railway cars placed between them. In the case of bridge 2, the train rides were at speeds from $v_{sta} = 10$ km/h (6.21 mph) to $v_{max} = 150$ km/h (93.21 mph), with intermediate speeds of $v_i = 80, 120,$ and 130 km/h (49.71, 74.56, and 80.78 mph). In the case of bridges 1 and 3, the train rides were at speeds from $v_{sta} = 10$ km/h (6.21 mph) to $v_{max} = 200$ km/h (124.27 mph), with intermediate speeds of $v_i = 80, 120, 160,$ and 180 km/h (49.71, 74.56, 99.42, and 111.85 mph).

During these tests, the measurements of the vertical and horizontal (for bridge 1 only) displacements as well as the accelerations of span and the speed of crossing test train were carried out. The examples of measured time histories of the vertical displacements during the train passage with speeds v_{10} and v_{max} are presented in Figure 11.9 (bridge 1), Figure 11.10 (bridge 2), and Figure 11.11 (bridge 3).

2)

3)

Figure 11.8 (2) steel plate girder with orthotropic deck; (3) steel arch bridge with reinforced concrete bridge deck.

Source: Photo 2 from Olaszek (2015).

11.2.2.3.1 EXTRAPOLATION OF VALUES FOR QUASI-STATIC LOAD

In all bridges, quasi-static displacements time history $d(v_{sta},t)$ was made on the basis of the displacements time history $d(v_{max},t)$ registered during the train ride at the maximum permissible speed $-v_{max}$: for bridge 2 $-d(v_{150},t)$, and for bridges 1 and 3 $-d(v_{200},t)$.

The filtering methods described in Section 11.2.2.1 (low-pass Bessel filter, FIR filter, and moving average filter) were used. The obtained values of cutoff frequency f_{1c} and averaging period T_{ave1} were as follows:

- Bridge 1: $f_{c1} = 1.35$ Hz $T_{ave1} = 0.74$ s
- Bridge 2: $f_{c1} = 3.20$ Hz $T_{ave1} = 0.31$ s
- Bridge 3: $f_{c1} = 0.83$ Hz $T_{ave1} = 1.24$ s

Diagnostic Load Testing of Bridges 239

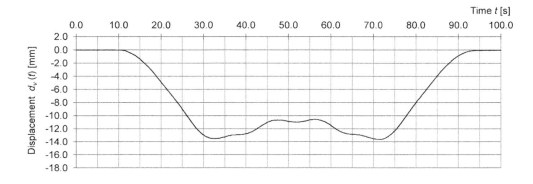

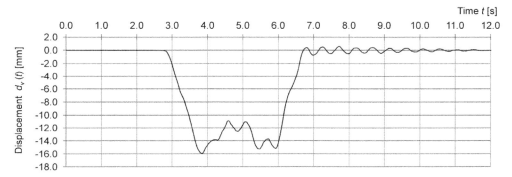

Figure 11.9 Bridge 1: measured time histories of the vertical displacements for train passages at the speed v = 10 km/h (6.21 mph) (upper) and v = 200 km/h (124.27 mph) (lower). Conversion: 1 mm = 0.0394 in.

Source: From Olaszek (2016).

Figures 11.12–11.14 present examples of train passage results obtained after using FIR filters with the cutoff frequencies f_{c1}, f_{c2}, and f_{c3} from the time histories of displacements registered during the train rides at 150 km/h (93.21 mph) (bridge 2) and 200 km/h (124.27 mph) (bridges 1 and 3). The extreme levels of displacements registered during the train ride at 10 km/h (6.21 mph) are also shown.

Table 11.2 presents the results of quasi-static values estimation obtained from the filtering mentioned above. The estimated values are compared with the values obtained directly from the train rides at the quasi-static speed of 10 km/h (6.21 mph). The measured and estimated quasi-static values and the relative deviation are also presented. The table also displays the dynamic amplification factors obtained with all methods.

In the case of bridge 1, the best result (1% relative deviation) was obtained by using the Bessel filtration method with the filter parameters corresponding with the decay of free vibrations (practically no vibration). The worst result, on the other hand (−42% relative deviation), was obtained after using the moving average with the filter parameters corresponding to the decay of free vibration. In the case of bridge 2, the best result (−5% relative deviation) was obtained by using the moving average with the decay of free

240 Load Testing of Bridges

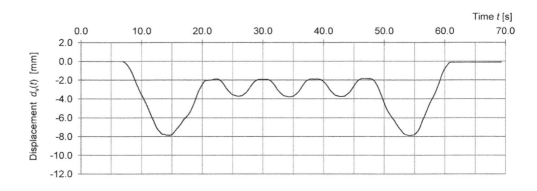

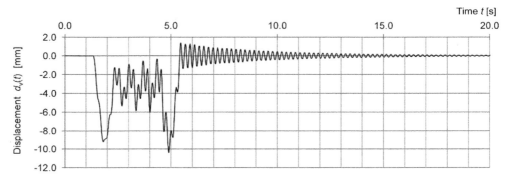

Figure 11.10 Bridge 2: measured time histories of the vertical displacements for train passages at the speed v = 10 km/h (6.21 mph) (upper) and v = 150 km/h (93.21 mph) (lower). Conversion: 1 mm = 0.0394 in.

Source: From Olaszek (2016).

vibrations. The worst result (21% relative deviation) was obtained after using the Bessel filtering method with the cutoff frequency between the static frequency and the first fundamental frequency of the structure's free vibrations. In the case of bridge 3, the best result ($\approx$0% relative deviation) was obtained by using the FIR filtration with the parameters corresponding with the decay of free vibrations. The worst result (−83% relative deviation) was obtained after using the moving average with the filtration parameters corresponding with the decay of free vibrations.

In the majority of the analyzed cases of bridges 1 and 2, the value of dynamic amplification factor was considerably underestimated. In the case of bridge 1, the dynamic amplification factor determined on the basis of train rides at 10 km/h (6.21 mph) and 200 km/h (124.27 mph) was 1.17. After the filtration used to estimate the quasi-static value, four (out of six) analyzed filters gave values of the dynamic amplification factor equal to 1.03 ÷ 1.10, and one was close to the real value. In the remaining case, the obtained value was 1.55. In the case of bridge 2, the dynamic amplification factor obtained on the basis of train rides at 10 km/h (6.21 mph) and 150 km/h (93.21 mph) was 1.32. After the filtration used to estimate the quasi-static value, five (out of six) analyzed filters gave values of the dynamic amplification factor equal to 1.09 ÷ 1.22. In the remaining case, the value was

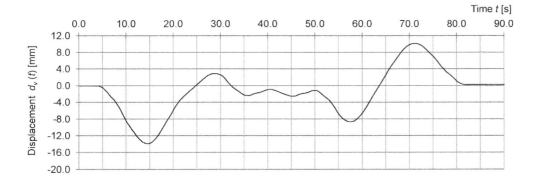

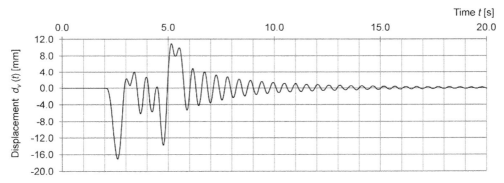

Figure 11.11 Bridge 3: measured time histories of the vertical displacements for train passages at the speed $v = 10$ km/h (6.21 mph) (upper) and $v = 200$ km/h (124.27 mph) (lower). Conversion: 1 mm = 0.0394 in.

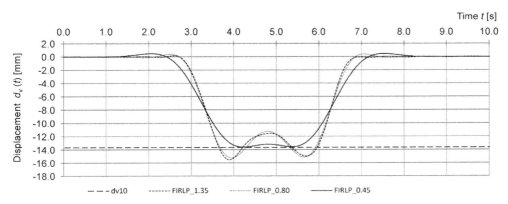

Figure 11.12 Bridge 1: an analysis of the quasi-static value estimation using the low-pass FIR filters (cutoff frequency $f_{c1} = 1.35$ Hz, $f_{c2} = 0.80$ Hz, and $f_{c3} = 0.45$ Hz) for the time histories of the displacements registered during the train ride at 200 km/h (124.27 mph). Conversion: 1 mm = 0.0394 in.

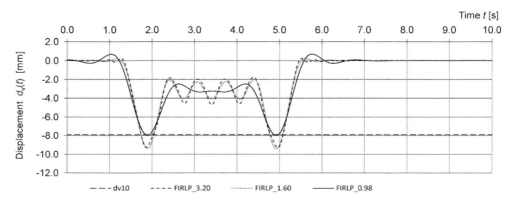

Figure 11.13 Bridge 2: an analysis of the quasi-static value estimation using the low-pass FIR filters (cutoff frequency f_{c1} = 3.20 Hz, f_{c2} = 1.60 Hz, and f_{c3} = 0.98 Hz) for the time histories of the displacements registered during the train ride at 150 km/h (93.21 mph). Conversion: 1 mm = 0.0394 in.

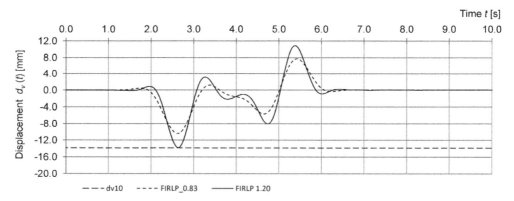

Figure 11.14 Bridge 3: an analysis of the quasi-static value estimation using the low-pass FIR filters (cutoff frequency f_{c1} = 0.83 Hz, f_{c2} = f_{c3} = 1.20 Hz) for the time histories of the displacements registered during the train ride at 200 km/h (124.27 mph). Conversion: 1 mm = 0.0394 in.

1.38. In the case of bridge 3, we can see a reverse trend. The dynamic amplification factor determined on the basis of train rides at 10 km/h (6.21 mph) and 200 km/h (124.27 mph) was 1.23. After the filtration used to estimate the quasi-static value, five (out of six) analyzed filters gave overvalued values of the dynamic amplification factor equal to 1.64 ÷ 7.19. Only one estimated value is close to the real value.

Table 11.3 shows the cutoff frequencies used as filter parameters compared to the first frequency of free vibration and the static frequency, as defined in Section 11.2.2.1.

As we can see in Table 11.3, there is not a trend in the relation of cutoff frequencies to the static frequency to estimate the quasi-static values closer to the real ones. We can state, then, that the estimation of the quasi-static value on the basis of the displacements registered during the train rides at the speeds close to the maximum ones instead of the

Diagnostic Load Testing of Bridges   243

Table 11.2 The results of quasi-static values measured at 10 km/h (6.21 mph) and estimated with the use of the different filtering methods

Types of bridge structures Span length (m)	Quasi-static vertical displacements d_{v10}			Value (mm)	Relative deviation[1] (%)	Dynamic amplification factor[2] d_{vmax}/d_{v10} (−)
	Designation method					
	Measured at 10 km/h			−13.63	−	1.17
1. Steel truss bridge with reinforced concrete deck and curved railway track Single span: 93.0 m	Estimates based on driving speed 200 km/h	Bessel filter	f_{c1}	−15.30	12%	1.04
			f_{c2}	−13.72	1%	1.16
		FIR filter	f_{c1}	−15.52	24%	1.03
			f_{c2}	−15.22	20%	1.05
		moving average gilter	T_{1ave1}	−14.57	12%	1.10
			T_{1ave2}	−10.29	−42%	1.55
	Measured at 10 km/h			−7.90	−	1.32
2. Steel plate girder bridge with orthotropic deck Single span: 24.0 m	Estimates based on driving speed 150 km/h	Bessel filter	f_{c1}	−9.57	21%	1.09
			f_{c2}	−8.54	8%	1.22
		FIR filter	f_{c1}	−9.29	18%	1.12
			f_{c2}	−9.42	19%	1.10
		moving average filter	T_{1ave1}	−8.85	12%	1.18
			T_{1ave2}	−7.52	−5%	1.38
	Measured at 10 km/h			−13.87	−	1.23
3. Steel arch bridge with reinforced concrete deck Single span: 75.0 m	Estimates based on driving speed 200 km/h	Bessel filter	f_{c1}	−8.65	−38%	1.97
			f_{c2}	−9.25	−33%	1.85
		FIR filter	f_{c1}	−10.42	−25%	1.64
			f_{c2}	−13.81	≈0%	1.24
		moving average filter	T_{1ave1}	−6.03	−57%	2.83
			T_{1ave2}	−2.37	−83%	7.19

[1]Relative deviation from measured value d_{v10}.
[2]Dynamic amplification factor was calculated with d_{vmax}
  Bridge 1: $d_{200} = -15.97$ mm
  Bridge 2: $d_{150} = -10.41$ mm
  Bridge 3: $d_{200} = -17.06$ mm
Note: 1 mm = 0.0394 in.; 1 m = 39.37 in.; 150 km/h = 93.21 mph; 200 km/h = 124.27 mph.

speed of approximately 10 km/h (6.21 mph) can result in significant errors. It seems impossible to define the filtering principles in order to obtain reliable results in different types of bridges. The biggest errors are in bridge 3, where the static frequency is close to the natural frequency of the first vibration mode. This result is compatible with Paultre et al. (1992), where it was found that when the static frequency becomes equal to (and even higher than) the one corresponding to the first vibration mode of the bridge, dynamic components of the signal can no longer be removed without impact on the static component. However, in this case the application of the FIR filter led to the best result: the relative deviation was close to 0%. Significant errors appeared also in the case of bridge 2, where there was a larger difference between static frequency and first frequency of free vibration. In this case, all methods gave errors higher than or equal to 5%.

Table 11.3 Cutoff frequencies used as filtration parameters compared to first frequency of free vibration and static frequency

Types of bridge structures	First frequency of free vibration	Static frequency	Filter parameter						
			No free vibration was observed			Optimal parameter $d_{sta} = d_{10}$		
			Bessel filter	FIR filter	moving average filter	Bessel filter	FIR filter	moving average filter
	f_1 (Hz)	f_s (Hz)	f_{c2} (Hz)	f_{c2} (Hz)	$1/T_{ave2}$ (Hz)	f_{c3} (Hz)	f_{c3} (Hz)	$1/T_{ave3}$ (Hz)
1. Truss bridge	2.10	0.60	0.50	0.80	0.25	0.50	0.45	0.91
2. Plate girder bridge	4.69	1.74	1.40	1.60	1.56	0.80	0.98	1.72
3. Arch bridge	0.92	0.74	0.90	1.20	0.20	1.75	1.20	2.33

Note: The marked values of cutoff frequencies give results close to the real value.

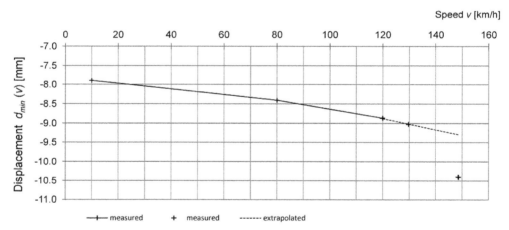

Figure 11.15 Bridge 2: an analysis of the accuracy of the extreme displacement amplitude extrapolation for the train ride speed of 150 km/h (93.21 mph) on the basis of the results of the measurements taken for the train ride speeds up to 120 km/h (74.56 mph) (points of two rides of the speed 150 km/h [93.21 mph], practically overlapped). Conversion: 1 mm = 0.0394 in.

Source: From Olaszek (2016).

11.2.2.3.2 EXTRAPOLATION OF VALUES UNDER HIGHER SPEED

In the case of bridge 2, an analysis of extrapolation accuracy was carried out by using the measurements for the speed equal to 80% of the maximum speed, that is, up to 120 km/h (74.56 mph). The extrapolation was made for speeds up to 150 km/h (93.21 mph) (Fig. 11.15).

Next, the extrapolated values were compared with the values obtained during the train rides at speeds close to 150 km/h (93.21 mph) (two such rides took place). The value of the relative deviation from the measured values was between 10.6% and 10.7%.

Diagnostic Load Testing of Bridges 245

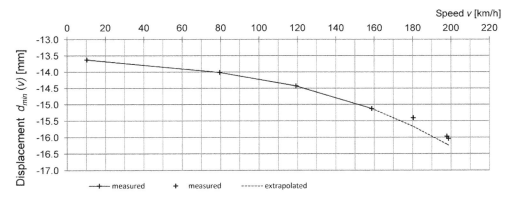

Figure 11.16 Bridge 1: analysis of the accuracy of the extreme displacement amplitude extrapolation for the train ride speed of 200 km/h (124.27 mph) on the basis of the results for the speeds up to 160 km/h (99.42 mph). Conversion: 1 mm = 0.0394 in.

Source: From Olaszek (2016).

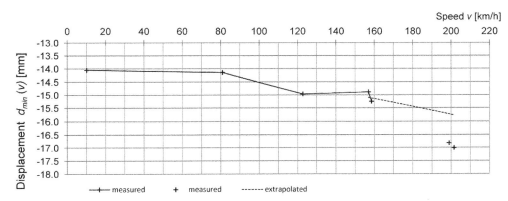

Figure 11.17 Bridge 3: analysis of the accuracy of the extreme displacement amplitude extrapolation for the train ride speed of 200 km/h (124.27 mph) on the basis of the results obtained for speeds up to 160 km/h (99.42 mph). Conversion: 1 mm = 0.0394 in.

For bridges 1 and 3, a similar extrapolation was done, but in this case for the speed values of 160 and 200 km/h (99.42 and 124.27 mph). Two passages at speed close to 200 km/h (124.27 mph) were carried out to compare with the extrapolated value (Figs. 11.16 and 11.17).

In case of bridge 1, the value of the differences in relation to the measured values was between 1.3% and 1.5%, and in case of bridge 3, between 6.5% and 7.3%.

On the basis of the three tests, we can state that the extrapolation of values under higher speed obtained good results only in the case of bridge 1. In this case, the extrapolation was concerned with the nonlinear deflection increase due to the horizontal curvature. The nonlinear increase of deflection with train speed was caused by the influence of centrifugal

246 Load Testing of Bridges

force. In the case of bridge 2, the extrapolation was not correct because of the deflection increase due to the vibration of the orthotropic deck with the increase of train speed. In the case of bridge 3, the extrapolation was not correct due to the increase of vibration of the hangers and bridge deck with increasing train speed.

11.3 Conclusions and recommendations for practice

From the examples presented in this chapter, relating static and dynamic diagnostic load tests, the following conclusions can be made:

The eight examples showing the speed of displacements stabilization during static tests obtained in different types of bridges do not provide a common guide or general conclusion about the speeds of stabilization that may be obtained depending on the bridge type. They show, instead, how different speeds of stabilization can be observed during the tests in different bridge types.

An example is presented of an exceptional behavior where significant deflection increments and no tendency of displacements stabilization after the application of load were observed. In case of the other bridges, the relative increments normalized with respect to time and obtained 30 minutes after the application of the load have similar values independent of the considered time interval. This confirms the fact that in most presented tests the speed of the increments is almost identical in the analyzed intervals Δt after 30 min of the application of the load. There is, however, a considerable variation of the increments during the first 15 min after the application of the maximum test load.

The evaluation of displacement stabilization after removal of the load can be considerably burdened by measurement errors and noise. Thus relative increment parameters can have considerable values due to measurement errors but not a lack of stabilization. To evaluate the process of displacement stabilization after removal of the load, we cannot use relative increment parameters but we have to analyze the history of absolute displacement. Therefore, if speeds of stabilization are not correctly monitored during the test, the danger is to finish the test when the final value is not yet obtained, thus leading to inaccurate conclusions regarding the bridge performance.

It should be recommended that to guarantee a reliable and accurate process of bridge assessment through diagnostic load testing, it is necessary to use measuring techniques which not only ensure appropriate accuracy of the measurements themselves but also enable to monitor the tested structure practically on an ongoing or on-line basis. Thanks to the use of quasi-continuous displacement monitoring (measurements at least once a minute) as compared with measurements every 15 min, we get much higher possibilities to control the correctness of the final measurements as well as much better control over the displacements stabilization process during and after the load test of the bridge.

Thanks to the continuous monitoring of the displacements, it is possible to control the process of stabilization of the results during and after applying a load increment. Thus, in this way it is possible to minimize the error in the elastic and permanent values of the measured quantities. Based on the results presented here, it is possible then to shorten the time of applying a load to a structure and the time of waiting for stabilization after the removal of the load.

In the case of railway bridges, the estimation of the quasi-static value based on the displacements recorded during the train rides at speeds close to the maximum should not be used, as they may lead to significant underestimation of the dynamic amplification factor. In the majority of the analyzed cases of bridges 1 (truss bridge with and curved railway track) and 2 (plate girder bridge with orthotropic deck), the value of the dynamic amplification factor was considerably underestimated. After the filtration used to estimate the quasi-static value, nine (out of 12) analyzed filters gave values of the dynamic amplification less than the actual value, one was close to the actual value, and two were larger. In the case of bridge 3 (arch bridge), we can see a reverse trend. After the filtration used to estimate the quasi-static value, five (out of six) analyzed filters gave too large values of the dynamic amplification factor, and only one estimated value was close to the actual value.

It is recommended that the calculation of the dynamic amplification factor should be always based on the results obtained from a quasi-static test, and not from digital filtration.

The extrapolation of the results of the maximum measured displacements at the maximum speed during the test to higher (than obtained) speeds can also lead to some important errors. They are, however, much smaller than the errors connected with the estimation of the quasi-static values.

For bridge 1, the value of the relative differences between extrapolated values and measured values was between 1.3% and 1.5%; for bridge 2, between 10.6% and 10.7%; and for bridge 3, between 6.5% and 7.3%. We can state that the extrapolation of values under higher speed obtained good results only for bridge 1. In this case, the extrapolation was concerned with the nonlinear deflection increase due to the horizontal curvature. The nonlinear increase of deflection with train speed was caused by the influence of the centrifugal force. For bridge 2, the extrapolation was not correct because of the deflection increase due to the vibration of the orthotropic deck with the increase of train speed. For bridge 3, the extrapolation was not correct due to the increase of vibration of the hangers and bridge deck with increasing train speed. It cannot be recommended to use extrapolation of the results of the dynamic test to higher speeds than obtained during the bridge load test.

References

AASHTO (2003) *Manual for Condition Evaluation and Load and Resistance Factor Rating (LRFR) of Highway Bridges*. American Association of State Highway and Transportation Officials, Washington, DC.

Casas, J. R., Znidaric, A. & Olaszek, P. (2009) Recommendation on the use of soft, diagnostic and proof load testing. Deliverable D16. EU project ARCHES. Brussels. Available from: http://arches.fehrl.org.

COST 345 (2004) Joint report of Working Group 2 and 3: Methods used in the European States to inspect and assess the condition of highway structures. *Procedures Required for Assessing Highway Structures*, European Commission Directorate General Transport. Available from: http://cost345.zag.si/final_re-ports.htm.

Frýba, L. (1996) *Dynamics of Railway Bridges*. Thomas Telford Ltd, London.

International Organization for Standardization (2004) ISO18649:2004. *Mechanical Vibration: Evaluation of Measurement Results from Dynamic Tests and Investigations on Bridges*. ISO, Geneva.

Joint Committee for Guides in Metrology (2010) *Evaluation of Measurement Data: Guide to the Expression of Uncertainty in Measurement*. JCGM: Joint Committee for Guides in Metrology, International Organization of Legal Metrology. Edition 2008 (E), Corrected version 2010, Paris.

Lyons, R. G. (2011) *Understanding Digital Signal Processing*, 3rd edition. Pearson, Prentice Hall, Upper Saddle River.

Olaszek, P. (2015) *Application of Digital Measurement Methods to Bridge Research* (in Polish). Polska Akademia Nauk Komitet Inżynierii Lądowej i Wodnej, Warszawa.

Olaszek, P. (2016) Extrapolation of measurement results during dynamic testing of high speed railway bridges. In: Bittencourt, T. N., Frangopol, D. M. & Beck, A. (eds.) *Maintenance, Monitoring, Safety, Risk and Resilience of Bridges and Bridge Networks: Proceedings of the Eighth International Conference on Bridge Maintenance, Safety and Management (IABMAS 2016), 26–30 June, 2016, Foz do Iguaçu, Brazil.* Taylor & Francis Group, London. pp. 1664–1671.

Olaszek, P. & Karkowski, M. (2009) Internet database of load test results and analytical calculations. Deliverable D07. EU project ARCHES. Brussels. Available from: http://arches.fehrl.org/? m=7&id_ directory=1614.

Olaszek, P., Łagoda, M. & Casas, J. R. (2014) Diagnostic load testing and assessment of existing bridges: Examples of application. *Structure and Infrastructure Engineering*, 10(6), 834–842.

Olaszek, P., Casas, J. R. & Świt, G. (2016) Some relevant experiences from proof load testing of concrete bridges. In: Bakker, J., Frangopol, D.M. & van Breugel, K. (eds.) *Life-Cycle of Engineering Systems: Emphasis on Sustainable Civil Infrastructure: Proceedings of the Fifth International Symposium on Life: Cycle Civil Engineering (IALCCE), 16–20 October 2016, Delft, The Netherlands.* Taylor & Francis Group, London. pp. 1–8.

Paultre, P., Chaallal, O. & Proulx, J. (1992) Bridge dynamics and dynamic amplification factors: A review of analytical and experimental findings. *Canadian Journal of Civil Engineering*, 19(2), 260–278.

Smith, S. W. (2003) *Digital Signal Processing: A Practical Guide for Engineers and Scientists.* Newnes, Elsevier Inc., Amsterdam.

Yang, Y. B., Yau, J. D. & Wu, Y. S. (2004) *Vehicle-Bridge Interaction Dynamics: With Applications to High-Speed Railways.* World Scientific Publishing Co., New Jersey.

Chapter 12

Field Testing of Pedestrian Bridges

*Darius Bačinskas, Ronaldas Jakubovskis
and Arturas Kilikevičius*

Abstract

The chapter contains basic information related to field static and dynamic testing of pedestrian bridges. A brief overview of test objectives and test classification is presented. The chapter also covers the stages of preparation for testing, aspects of test program creating, methods of static and dynamic loading of footbridges. A sequence of test organization and execution is also presented. The second part of the chapter discusses the aspects of processing and evaluation of static and dynamic test results as well as aspects of comparing them with results obtained by the theoretical modeling of the bridge. The chapter presents requirements for pedestrian bridges specified in the design codes and recommendations of various countries. These need to be taken into account in assessing the results of the tested bridges. The chapter concludes with a discussion on the presentation of the test results and aspects of assessment of the bridge condition according to the test data. The presented material may be useful to researchers and experts involved in the design, construction, and maintenance of pedestrian bridges.

12.1   Introduction

Transport infrastructure plays a vital role in the economy and society of any country. The development and expansion of various industries is accompanied by the intensity and growth of transport flows. Studies conducted in this area have shown that in Europe, between 2010 and 2050, transport flows will increase by 42% in passenger and by 60% in freight transport (European Commission, 2017). The latter tendencies considerably complicate the sustainable development of transport infrastructure and make it even more challenging (European Commission, 2017).

Increasing transport flows are closely linked to the safety of pedestrians and cyclists on the roads. Although significant progress has been achieved in recent decades in making roads and railways more safe, the number of deaths and serious injuries and the associated economic costs are still high (European Commission, 2017). According to available statistics, in 2016, 25,500 people lost their lives and 135,000 people were seriously injured on European roads (European Commission, 2017), where pedestrians accounted for a significant proportion of the deaths. In 2013, pedestrians and cyclists accounted for nearly 30% of all road accident deaths in the European Union (EU). These tendencies vary significantly between countries, from under 10% in Iceland and Norway to nearly 40% in Latvia and Romania (Eurostat, 2018a). Most of the dead and injured on the railways (68%) are

DOI: https://doi.org/10.1201/9780429265426

unauthorized persons illegally present in the areas of the operating railways. Together with the level crossing of a railway track in unauthorized and non-designated areas (31%), these two types of accidents account for 99% of all deaths occurring on railways in the EU (Eurostat, 2018b). The latter tendencies resulting in loss of human life or determining their injury revealed the further need for the development of transport infrastructure (bridges, underground passages, etc.) with the purpose to ensure greater safety for pedestrians and cyclists.

The cyclist and pedestrian bridges form an important part of the transport infrastructure in cities (Fig. 12.1a), on motorways (Fig. 12.1b), and on railways (Fig. 12.1c). They are often built for the development of recreational and entertainment areas (Fig. 12.1d). Various static schemes are used for pedestrian structures, whereas architecturally, these structures are usually distinguished from other bridge types (e.g. roadway or railway bridges). An overview of the different pedestrian bridges in the world and the requirements for them can be found in a number of textbooks (Schlaich et al., 2005; Idelberger, 2011; Keil, 2013; Humar, 2014).

From the structural point of view, pedestrian bridges are exceptional structures. Nowadays innovative footbridges most often are designed as slender structural systems that are particularly sensitive to human-induced effects such as vibrations, kinematic displacements, and so forth (van Nimmen et al., 2014). Effects caused by crowds of people are replaced by simplified equivalent models of static and dynamic loads in the design codes or recommendations governing the design of these structures. A brief overview of the standards and recommendations used in practical design is presented in Roos (2009) and Eriksson (2013). The above design documents also contain the essential requirements to ensure the conformity of structures for ultimate and serviceability limit states (SLSs).

Figure 12.1 Different types of footbridges

Bridges constructed in the past can be attributed to the other group. After many years of operation, defects and damages occur in bridge structures that cause changes in their basic parameters (stiffness, load-bearing capacity, etc.), loss of structural integrity, and subsequent decrease in reliability and safety of structures. Physical deterioration, aging, maintenance, and structural interventions of transport structures are the most significant issues in bridge engineering during recent decades (Sánchez-Silva et al., 2016). It is necessary to properly assess the load-carrying capacity and serviceability of the bridge to ensure bridge safety, taking into account the current condition of structures as well as the code requirements. As an example, the first all-composite Aberfeldy Footbridge built in Scotland in 1992 can be mentioned (Burgoyne and Head, 1993). Despite some durability problems with parapets and a glass fiber-reinforced polymer surface protection layer, the primary structural components after 20 years in service were in good condition with no obvious degradation in performance (Stratford, 2012). However, the natural frequencies of the bridge have decreased to 9%. This may be caused by the changes of mechanical properties as well as local structural damages. Although recent changes are not significant in terms of load-bearing capacity; however, they may be related to noncompliance with the requirements for comfort parameters (e.g. vibrations). The latter tendency is particularly important for bridges whose comfort indicators at the beginning of operation were close to the threshold values.

The behavior of both new and existing bridges can be analyzed using modern computer software programs based on the finite element technique. Despite advanced modern design technologies, numerical models of pedestrian bridges in most cases have a fair number of uncertainties (Živanović et al., 2006). The latter is determined by the interaction between structural and nonstructural elements, differences in the approved material design parameters, errors in approved models and methods, physical and geometrical nonlinearities, composite actions, load distribution effects, and other aspects. Field testing of bridges can be performed for identification of the above uncertainties. Bridge tests are a reliable way to assess the technical condition of the bridge since the results of the test show the real structural behavior under static or dynamic loads. Factors that cannot be taken into account when designing or building a bridge are evaluated by tests. With systematic accumulation of data and experience in testing various bridges, the methods of calculation and analysis of bridges, design standards, concepts, and technologies of new bridges are being improved (Živanović et al., 2006; Kamaitis, 1995).

The present chapter contains basic information related to field static and dynamic testing of pedestrian bridges. A brief overview of test objectives and test classification is presented. The chapter also covers the stages of preparation for testing, aspects of test program creation, and methods of static and dynamic loading of footbridges. A sequence of test organization and execution is also presented. The second part of the chapter discusses the aspects of processing and evaluation of static and dynamic test results as well as aspects of comparing these with results obtained by the theoretical modeling of the bridge. The chapter presents requirements for pedestrian bridges specified in the design codes and recommendations of various countries. These need to be taken into account when assessing the results of the tested bridges. The chapter concludes with a discussion on the presentation of the test results and aspects of assessment of the bridge condition according to the test data. The presented material may be useful to researchers and experts involved in the design, construction, and maintenance of pedestrian bridges.

252 Load Testing of Bridges

12.1.1 *Types of the tests*

The field test of a bridge is considered as a reliable tool for the assessment of the behavior and technical condition of the newly constructed, strengthened, or repaired bridge or the bridge operated without additional intervention measures when its structures are subjected to static and/or dynamic loading (Lithuanian Road Administration, 2005). Normally, such studies are aimed at assessing the compliance of bridge parameters and overall behavior with the requirements specified in design codes and bridge project, and to ensure the safety and reliability of structures. Tests of existing bridges are usually carried out to assess structural response without causing significant damages. Therefore, sometimes field load tests are referred to as nondestructive load testing (Cai et al., 2012).

Bridge testing can be of two types: acceptance testing and proof load testing.[1] Acceptance testing is the verification of the bridge and structural response after the construction of a new bridge or after its repair or strengthening. The latter testing determines whether the behavior of the bridge meets the assumptions or hypotheses adopted at the design stage. It should be emphasized that information obtained during this testing is of particular importance. First, it allows validation and updating of the theoretical numerical model according to the testing results. Second, testing results are the starting point for the bridge owner to assess adequately the impact of potential damage on the safety and reliability of the bridge during degradation in the bridge structure over time. These studies are particularly important for pedestrian bridges, as defects due to flexibility of structures or loss of structural integrity can lead to significant changes in the load-bearing capacity, displacements, vibrations, or other parameters of the bridge. In the absence of initial information, at the beginning of operation of the bridge, or after the introduction of intervention measures, the subsequent adequate assessment of the structural degradation by experimental methods is not possible.

Proof load testing is carried out on existing bridges in the case of possible unsafe operation ensuring to reduce uncertainties in bridge resistance or calculated internal effects caused by external mechanical actions. These tests are usually carried out in order to increase bridge safety and reliability (Faber et al., 2000) or to provide for possible intervention measures to ensure a higher level of these parameters. Proof load testing carried out together with acceptance testing during the service life of the bridge complements the bridge information system and can serve in developing and improving prediction models of bridge degradation of the corresponding structural system.

The level of the test load in acceptance testing is usually higher than that of the proof load testing, as the design resistance of the structures is known (Lithuanian Road Administration, 2005). Proof load testing is carried out by applying a more cautious strategy for determining the resistance of complicated structures due to the occurrence of damage and other aspects.

Acceptance and proof load testing may be classified depending on the size, the direction and the type of the test load, the importance of testing, and the location of testing (Lithuanian Road Administration, 2005). According to these criteria, tests can be classified into routine, special, and destructive tests subjected to vertical and/or horizontal external loads. Special tests are performed when it is intended to increase the operational load of the bridge due to possibly higher load-bearing capacity or redundancy of the bridge. For example, in the case of a pedestrian bridge, these tests may be carried out when a vehicle that causes higher internal forces than a pedestrian load intends to move over a bridge. This type of testing is not widely used for footbridges. Failure load tests in most case are carried out on old bridges before they are demolished in order to determine the critical resistance limit and to assess the adequacy of the corresponding theoretical models (Lantsoght et al., 2017a). Scientifically valuable information that is obtained during these tests can be used for the future prevention of failure of bridges (Dymond et al., 2013). Vertical static and/or dynamic loads are usually applied during the testing of bridge structures. In exceptional cases, pedestrian bridge decks and bridge supports can be tested by horizontal loading. Since footbridges are often slender structures, vibrations of footbridges become a decisive criterion in their design. This in turn determines the exceptional importance of dynamic testing for footbridges (Živanović et al., 2005).

According to the importance of tests, the tests are classified into initial proof load tests and basic tests. The initial proof load static test is carried out by placing small loads on the bridge structures (e.g. up to 10%–15% of the maximum test load), and the dynamic test by exciting short-term vibrations in structures to check the calculation systems of structures, reliability of test equipment and attachment of devices to the structure, and device readings and the alignment of the entire load-measuring system.

In engineering practice, in-situ field testing is usually performed when the entire bridge or its individual elements are being tested (decks, piers, etc.). These tests are carried out on new or reconstructed bridges after the completion of construction works and before the start of their operation (Fig. 12.2a). In some cases, single structures of bridges can be tested at production plants or construction sites before they are mounted (Fig. 12.2b). Newly manufactured structures or their models, as well as real-scale structures removed from the operating bridge, can be tested in a laboratory (Bačinskas et al., 2017).

12.1.2 Objectives of the tests

Depending on the nature and purpose of the tests, the objectives of the tests of footbridges may be (Lithuanian Road Administration, 2005) (a) to verify the quality of construction works and efficiency of newly built bridges, or bridges after repair or strengthening; (b) to determine the weak points or sections of the bridge; c) to improve theoretical calculation models of load effects and/or structural strength, or to identify uncertainties of the model parameters; (d) to determine the real resistance of single elements or the entire

Figure 12.2 Testing of bridge decks after construction (a) and before installation (b)

bridge, or load-bearing capacity reserve that cannot be reliably determined and evaluated analytically or by detailed inspection; or (e) to provide recommendations for the safe use of the bridge and its compliance with the requirements for the project and design standards.

The load-bearing capacity reserve is determined when there is demand for an increase in permanent load, for example, when it is decided to install engineering communications across the bridge which were not taken into account during the design stage, when the live load increased (the requirements of the design code changed), or upon extension of the design term of the service life of the bridge (Lithuanian Road Administration, 2005). There may be several goals when testing the bridge.

Static tests are carried out to determine (Lithuanian Road Administration, 2005) (a) the actual static systems of structures; (b) the impact of defects and damage on the behavior of structures in order to determine the need for intervention measures and possible efficiency; and (c) the compliance with the requirements for ultimate limit states of safety (when tested before decomposition) and serviceability provided in the design code, and in exceptional cases, to assess the adequacy of these requirements.

Dynamic tests of pedestrian bridges are carried out to check (Lithuanian Road Administration, 2005) (a) the modes and parameters of vibrations of bridges to determine their compliance with the standards and to ensure the safe and comfortable use of the bridge; (b) the causes of unacceptable vibrations of structures in order to reduce their values; and (c) real dynamic coefficients for the assessment of the load-carrying capacity of structures.

Information obtained during testing is also important in another aspect. It allows assessing the adequacy of the design strategy and theoretical models, taking into account specific aspects of the particular bridge type. Experimental data of each tested bridge complements both the country-specific and global databases of this field. Data obtained during testing can be used to improve the analyses of structures, design, construction, and monitoring systems and methods.

As a process, bridge tests can be divided into three stages (Lithuanian Road Administration, 2005): (a) preparation for testing; (b) conducting of testing; and (c) analysis of the results and assessment of the technical condition of the tested bridge. Each of these stages is discussed in detail in this chapter.

12.2 Preparation for testing

The chapter discusses the stages of preparation for tests of pedestrian bridges; presents the general aspects of preparation for tests; the necessity of the preliminary inspection of bridges prior to testing and describes its strategy; and provides information on the principles of establishing a test program and the loading of the bridge by static and dynamic loads. The information gathered and summarized at this stage is used in carrying out static and dynamic tests of pedestrian bridges.

12.2.1 General guidelines

Preparation for tests is performed at the first test stage. This is an important step in the process for collecting and processing information on the bridge under consideration. At this stage, documentation available on the bridge is thoroughly analyzed, including a bridge project with detailed drawings and calculations, reports on technical inspections

Field Testing of Pedestrian Bridges 255

and testing (if any), documentation of construction works carried out, and other related information. If the bridge is in service, it is also important to know the intensity of pedestrian traffic flow, as the bridge could have been designed for smaller loads than provided in valid design codes.

The inspection of the bridge structures is performed and the technical condition of the bridge is assessed at the initial phase of this stage. The theoretical numerical model of the bridge structures is developed based on available documentation and additional instrumental measurements (if required). The preliminary influence of defects identified at this stage and other deviations from the projects are taken into consideration. After performing the necessary calculations, it is determined which structures and critical sections of structures need to be tested, and the most dangerous limit states are identified. A test program, which determines the sequence and means by which the bridge will be loaded during testing and what results are expected to be obtained, is prepared based on the results of theoretical calculations. By selecting the loading method, an analysis of structural behavior is performed and the loading level is selected. The latter is determined by taking into account the response of bridge structures and the type of tests (see Section 12.1.1). The tests of the bridge must be planned and carried out in such a way as to provide detailed information on both the mechanical behavior of separate elements and the entire structure of the bridge, to reduce the uncertainties of calculation of the load-carrying capacity of the structures, and to increase the reliability and safety of the bridge. Each of the above phases is discussed in detail further in this chapter.

12.2.2 Preliminary inspection of the footbridge before the tests

The type and extent of the inspection of the bridge depends on the structural features of the bridge, its importance, technical condition, environmental conditions, and the level of detail and reliability of the existing technical documentation (Lithuanian Road Administration, 2005). Relevant normative documents or recommendations define the strategy and general requirements for bridge inspections. The main purpose of the inspection is to collect initial data on the condition of the bridge parts, structures, elements, the bridge environment, and conditions of use. If the data is lacking in the design documentation for proper and safe testing of the bridge, or the design documentation has not survived, the necessary information is collected at this particular stage.

The structures of pedestrian bridges are usually slender, so even minor defects in structures or inappropriate assumptions adopted at the design stage can have a significant effect on the load-bearing capacity and the serviceability of bridge structures. During the technical inspection, it is necessary to assess properly the conformity of structures with the design schemes adopted at the design stage, the stiffness of the joints and bearings, the overall interaction between structural elements, and other related effects.

The scope of the technical inspection of the bridge is highly dependent on the available design documentation. Verification of geometric parameters of the most important bridge structures (decks, piers, abutments, etc.) is carried out at the beginning of the inspection. If the main dimensions are not in accordance with the project, or the project of the bridge in service is not maintained, schemes or drawings of the bridge (front views, cross-sections, joints, etc.) must be prepared.

In the absence of precise information related to the properties of the materials or in the presence of material deterioration in structures, additional research on physical, mechanical,

256 Load Testing of Bridges

Figure 12.3 Application of destructive technique for the identification of the compressive strength of concrete

Figure 12.4 Application of nondestructive methods for the identification of mechanical properties of concrete (a) and steel (b)

and durability properties of materials are carried out. The latter may be subject to destructive (Fig. 12.3) or nondestructive (Fig. 12.4) methods. However, the best results are obtained using both methods at the same time. The following physical and mechanical properties are often investigated for concrete: density, compressive strength, and modulus of elasticity. If required, the level of carbonation of concrete, the quantity of chlorides, frost resistance, and other durability properties relevant to each individual case can be determined. If possible, the physical and mechanical properties of the reinforcement and the level of corrosion can be determined during the investigation on the reinforced concrete structures. The tensile strength and modulus of elasticity is usually determined for steel structures, and the condition of protective steel coatings and the level of corrosion damage is evaluated. The condition of the welded and bolted joints and the joining elements is also assessed. Timber pedestrian bridges are subject to a similar strategy.

After research on material properties, longitudinal and cross-sectional profiles of bridge decks are measured using geodetic methods. Deformations of load-bearing structures are measured as well, including deflections, settlement, displacements, and inclinations. The geodetic tools used for the bridge tests are discussed in more detail in Section 12.3.2.

(a) (b)

Figure 12.5 Crack pattern of the abutment (a) and inaccurate installation welded joint of truss elements (b)

Having drawn up the structural schemes of the bridge and assigned the appropriate material parameters to the separate elements, defects and damage (if any) in the structural elements and the reasons for their occurrence can be identified and the development trends and expected consequences can be explained. Schemes of resulting damage are drawn up for new bridges and basic parameters are identified, for example crack width and spacing (Fig. 12.5a), construction defects (Fig. 12.5b), and so forth. Damage recorded for the existing bridges is comparable to the data available from previous research. When examining a bridge, it is necessary to make sure that there are no hidden defects or damages such as reinforcement corrosion, reinforcement bar fractures, significant deviations from the design position, internal cavities and openings, fatigue cracking in steel, and so forth.

The analysis of the conditions of use of the bridge is carried out at the end of this stage, during which dead loads, intensity of traffic flow, possible past or present overloads, aggressive environmental conditions, and so forth are identified. Data accumulated in the above step is used to make the preliminary theoretical numerical model corresponding to the actual condition of the bridge structures at issue, which will be revised after carrying out static and/or dynamic tests. Having identified significant discrepancies between dimensions and material properties of structural elements, an assessment of limit states of test structures of the bridge is also performed.

The technical inspection of the bridge is summarized by identifying the most dangerous structural elements or their sections in regard of both the ultimate and the SLSs. The latter elements in most cases are tested, and their state and behavior are monitored in detail during the tests (Lithuanian Road Administration, 2005). The bridge cannot be tested in the following cases: the load-bearing structures are in a very bad condition and do not meet the requirements for limit states; there is no precise information on the strength of the foundations or supports; or there is a possibility of a sudden collapse of the load-bearing structures (Lithuanian Road Administration, 2005).

12.2.3 The test program

Establishment of the test program of the bridge is one of the essential phases in this process. It reflects the research objectives and the result to be achieved. The test program of the bridge consists of a test guide involving the owner of the bridge and the designer (if a new bridge or bridge after repair or reinforcement is being tested). The scope of the bridge test plan depends on the purpose of the test and the type of the bridge and its technical condition. An individual test program is usually established for each bridge. In the general case, the test plan specifies (Lithuanian Road Administration, 2005) (a) test objectives; (b) loading conditions; (c) test equipment and instruments; (d) measurement progress; and (e) expected results of the test.

Special attention at this stage is drawn to the description of the loading conditions. The program specifies load tools and load level, loading points, sections or areas, and load sequence. The loading stages and procedure during the test must comply with the use of structures under the most unfavorable conditions (Sétra, 2006). The higher the load level, the more reliable are the test results. On the other hand, testing becomes more dangerous due to the potential loss of load-bearing capacity. More detailed information on the methods of static and dynamic loads of the pedestrian bridges and the principles of distribution are provided in Section 12.2.

Equipment and instruments used for the tests must comply with the test type and must be suitable for setting of selected monitoring parameters within the appropriate limits of variation. Measurement parameters, measuring points, devices, their parameters and location schemes, methodology for accumulation of readings, and intensity must be described in the measurement chapter of the test program (Lithuanian Road Administration, 2005). The devices used for measurements and their location are briefly reviewed in Section 12.3.2.

The test strategy also depends on the financial capacity of the owner. On the one hand, the larger the scope of the measured parameters, the more information is obtained during the tests. On the other hand, the cost of testing increases. The level of the load and its distribution is provided in the test program, and the number of measuring points and locations depends on the static scheme of the bridge, the deck width, and the number of spans. Examples of test programs for different types of bridges are discussed in Sétra (2006).

The test program is summarized by presenting the preliminary theoretical results obtained in different phases of the test. The results of theoretical value that will be monitored during the bridge tests are indicated. Most often, these are displacements, deformations, vibration modes and frequencies, and so forth. Some of the received data (deformations, deflections) serve as reference values, which should not be exceeded during the tests.

The general manager of the tests signs the test program, which is also agreed with the owner of the bridge and in the above cases with the project manager. The latter must properly evaluate the adequacy and reliability of the solutions taken. The approved test program is the last step towards starting field testing of footbridges.

12.2.4 *Loading of the bridge*

This section provides brief information on loading of pedestrian bridges during static and dynamic tests, discusses the selection of loading tools and loading level during the static tests, and provides data on the excitation methods of self-vibration and forced vibrations of the bridge during the dynamic tests. The sequence of static and dynamic loads is discussed in more detail in Section 12.3.3.

12.2.4.1 *Static tests*

Static loading tests of the pedestrian bridge are important procedures for evaluating the stiffness, the load-carrying capacity and the construction quality of structures as well as ensuring safe further operation. The static loading test together with theoretical analysis can provide valuable information for the analysis and design of similar structures. The selection of static load depends on the type and geometric parameters of the bridge. When the bridge is not wide (width less than 2.5 m [8.2 ft]), the static test is most often performed by loading the deck of the bridge by precast concrete blocks (Figs. 12.6a and 12.6b), sandbags (Fig. 12.6c), containers with water (Fig. 12.6d), or by other means (van Nimmen et al., 2017, Bačinskas et al., 2014, Sétra, 2006). One of the advantages of this method is uniform distribution of loading without its concentration in the investigated sections. This is especially important for slender deck structures such as orthotropic steel decks, thin concrete slabs, and so forth. Another important aspect is that, in certain cases (e.g. using standard concrete blocks), the elements need not be weighed additionally, and the total weight of the load unit can be determined by the number of elements. However, this method is a time-consuming procedure since a crane or even several cranes are usually needed for deck loading. The placement of a single load unit onto the intended location often complicates a closed-type structure of the deck (e.g. arches, space trusses). In some cases, the lifting mechanisms and other tools such as mechanical loaders are used. Using a crane, a load unit can be positioned in the area near the bridge supports and transported by the loader to the intended load area. When the span of the deck is long, it may be necessary to use several cranes or to change the location of the crane. So depending on the length of the bridge, this process may take a long time.

As an alternative solution, vehicles loaded with various materials, construction or other machinery (e.g. road roller) can be used for static loading (Fig. 12.7). However, the use of this method is limited by the geometric and physical parameters of the bridge deck. First, it must be possible for a vehicle drive onto on a bridge. Second, the width of the deck must be sufficient for the vehicle (usually at least 2.5 m [8.2 ft]). In addition, the bridge deck must be able to withstand the concentrated load of the axle of the vehicle. This type of loading saves a lot of time, especially if several vehicles are used. In this case, the required load level is achieved quickly, and moving of the load to different positions does not cause any major difficulties.

Pedestrians (Byers et al., 2004), special bridge test machinery (Lantsoght et al., 2017b), hydraulic or screw jacks (Bagge et al., 2018), or other devices can also be used for loading of pedestrian bridges. The latter cases are more specific and apply only in exceptional cases. The pedestrian load is used when a low level of load is required. Special bridge testing machines are expensive. They are used only by specialized companies that supervise or test a large amount of bridges. On the other hand, these machines may not always be used due to restrictions on access to bridges. Hydraulic jacks are commonly used in carrying out failure field test of bridges.

Loads used in static tests must be easily placed and removed. Before testing, loads must be weighed and their dimensions measured. The load values are measured on the scales. It is recommended to set the load value with an accuracy of 3%–5% (Lithuanian Road Administration, 2005; Sétra, 2006). If vehicles or machinery are used for loading, the load of each axle and the total weight must be determined separately. If load weights are set in advance, it is necessary to protect them against possible weight increase (for example, due to rainwater). If the load is applied using hydraulic jacks, their calibration must be performed before the tests.

The level of the test load must be chosen in relation to the desired level of safety of the bridge in question and to ensure the safety of the bridge structures and the personnel conducting the test. If the test load is very high, the structure may be damaged or in critical

Figure 12.6 Different types of loading in static field tests of footbridges

Figure 12.7 Static loading of footbridge using loaded vehicles

cases lose its bearing capacity. If the test load is very small, it is not possible to reliably assess the real resistance of the structure, nor its safety and reliability. The static test load intensity can be calculated using the following equation (Lithuanian Road Administration, 2005):

$$\Delta = \frac{E_{stat}}{E_k(1+\phi)}, \tag{12.1}$$

where E_{stat} is the internal effect caused by the static loading (bending moment, shear force, flexural stress, etc.) in the section of the structure under consideration, E_k is the same effect from the characteristic live load assumed at the design stage, and $1 + \phi$ is the theoretical dynamic factor. If the dynamic factor is assessed within the representative values of the load model of the selected design code, the dynamic factor is $1 + \phi = 1$. Parameter E_{stat} is determined by performing a numerical analysis of the theoretical model of the bridge and assessing the distribution of the load between the structural elements, the physical and mechanical characteristics of the materials, defects, and other important parameters. Project data may be used for newly erected bridges if no significant deviations from design solutions were established during the initial inspection.

The owner of the bridge usually sets parameter Δ for the specific tests in accordance with the relevant normative documents or recommendations. Sétra (2006) recommends testing the bridge deck with loads that represent the action of traffic levels with a return period of between one week and one year. These recommendations also indicate that, in practice, the effects of the test loads must be between the effect of frequent traffic loads and three-quarters of the effects of the characteristic traffic loads defined in EN 1991–2 (2003).

The Lithuanian Road Administration (2005) set the following limits for parameter Δ. In regular tests, $0.7 \leq \Delta \leq 1.0$; in special tests, $1.0 \leq \Delta \leq n$, where $n = 1.3 - 0.001L$ for bridge spans $L < 100$ m [328 ft] and $n = 1.2$ for spans $L \geq 100$ m [328 ft].

The level of the test load depends on the type of the test. In the acceptance tests, the maximum test load must be close to the characteristic value of the accepted load effect. The test load may be lower (tests are more cautious) in the proof load tests since the real load-bearing capacity of the structures is not precisely known. When testing reinforced concrete structures for shear effects, it is recommended to accept lower load limits as the failure of elements in shear can be sudden, with no clear warning signs (cracks, large deflections, etc.).

12.2.4.2 Free vibration tests

Dynamic bridge tests can be divided into two essential stages. Modal analysis of free vibrations of the bridge is carried out during the first stage. At this stage, vibration modes and their corresponding parameters such as frequencies and damping ratios are determined experimentally. Having performed the measurements of free-vibration parameters, the compliance of the bridge to the comfort criteria provided in the selected design standards (see Section 12.5.5) and the critical values of the free-vibration parameters is verified. If the comfort criteria do not exceed the established critical limits, further research is not required unless the owner of the bridge requires it or when additional information is desired on forced vibrations of the bridges. If comfort criteria are not met, forced vibration tests are a must. The aspects of recent research are described in detail in Section 12.2.4.3. This section also discusses the excitation aspects of the free vibrations of the bridge.

The free vibrations of the bridge may be excited by the sudden removal of a weight attached to the bridge deck (Fig. 12.8a), by dropping a weight on the deck (Fig. 12.8b), or by using an impact hammer (Fig. 12.8c). The method of excitation strongly depends on the strength of the bridge and the required intensity of excitation. The most reliable method is the first one of the aforementioned as the free vibrations of the unloaded bridge is excited by the additional load. The second method can be used when the weight of the decreasing shock load is not significant compared to the weight of the bridge. This method is effective in exciting massive decks (e.g. reinforced concrete). The third method is applied both to measure the size of the shock load and to conduct an experimental modal analysis. It is necessary to excite more vibration modes in the investigated structures to obtain reliable results of dynamic analysis. Depending on the scope of the test, this is carried out by exciting the bridge with exterior effects in different sections or areas. More detailed information on the progress of dynamic measurements is provided in Section 12.3.3.2.

Figure 12.8 Different types of free-vibration excitation of pedestrian bridges

Figure 12.9 Excitation of forced vibrations of the bridge by walking or running of pedestrian groups

12.2.4.3 Forced and ambient vibration tests

When the comfort criteria for pedestrian bridges need to be checked, or additional information on bridge behavior and response to human induced actions is needed, forced and ambient vibration tests are performed. The main pedestrian comfort criterion is the amplitude of acceleration of pedestrian-induced vibrations, which depend on the resonant frequency of self-vibrations of the structure and the intensity and nature of the pedestrian load. Maximum acceleration of vertical, lateral, and torsional vibrations must not exceed the limit values summarized in Section 12.5.5.

In forced and ambient vibration tests, the bridge is excited by effects of varying intensity of a known size or by a random single person, a group of several people (Fig. 12.9), or by a crowd (Cremona, 2009). Forced vibrations are induced by walking, running, jumping, or otherwise moving pedestrians on the bridge. It is important to have the parameters of human loads used in the tests in carrying out comparative analysis of experimental and theoretical results. Static weight of pedestrian loads can be determined using scales. Pedestrian parameters can be determined using dynamometric force plates (Brownjohn et al., 2009) or platforms (Venuti and Bruno, 2011). Recently, Chen et al. (2019) proposed a power spectral-density approach to obtain samples for pedestrian walking load model parameters. A novel wireless insole pressure system was applied to measure the continuous walking-load time history. However, the technologies described above are more commonly used in research. In practical experiments, pedestrian parameters are usually determined by measuring walking distance, time, and the number of steps using simplified methods. The latter are used to determine the approximate frequencies induced by pedestrian effects.

Electromechanical shakers (Fig. 12.10) can be applied alternatively in a forced vibration test. The great advantage of this technology is that the bridge can be excited by the effects of both constant and variable frequencies. On the other hand, dynamic analysis of forced vibrations can be performed and the bridge response to different frequency effects can be obtained. This method is well suited for testing the resonance frequencies of the bridge experimentally.

Different types of signals are possible when a structure is tested with a shaker. Usually, a random or sine excitation signal is used (Data Physics, 2015). The random excitation

Figure 12.10 Electromechanical shakers for forced vibration tests

corresponds to the signal when the instantaneous spectrum amplitude and phase at a given frequency vary randomly from record to record. In the case of auto-spectrum, the random signal can be described by its amplitude probability density. The random signal amplitude probability is close to the characteristics of the Gaussian distribution. The random signal can be easily adjusted in a different frequency range seeking to obtain data on the response of the structure in a dynamic range of interest.

Sine excitation can be used in cases where high peaks can cause problems. This type of excitation can be applied to investigate the nonlinearity of the structure under test. Several measurements can be carried out for a certain frequency with different input levels. The sweeping technique is used to cover a wide frequency range. The frequency switching speed must be sufficiently low to allow the system to respond to changing frequencies.

When observing the bridges in service and seeking to obtain information on their dynamic characteristics, ambient vibrations tests can be performed (Omenzetter et al., 2013; O'Donnell et al., 2017). These tests can be used before strengthening or repair of the bridge in the event of instantaneous and intensive bridge loading (e.g. sports events) for identifying in scientific purposes the structural response to natural random excitation and other related cases. Since human walking has a stochastic character, it is expedient to conduct studies of vibrations for a long time using a permanent dynamic monitoring system. This strategy enables recording the maximum parameters occurring during random effects.

12.3 Organization of the tests

This chapter provides basic information on the organization of static and dynamic tests, briefly discusses the general requirements, the load sequence, the measuring equipment used and its positioning aspects, the course of measurement, and the frequency of data acquisition. The parameters of the bridge elements monitored during the tests are also discussed.

12.3.1 *General requirements*

Bridge tests can be carried out only after having reliably assessed the condition of the bridge and having agreed to the test plan. The bridge can be tested after sufficient time

Field Testing of Pedestrian Bridges 265

of completion of the construction works when the dead load is fully applied. When testing concrete structures, it is necessary to evaluate the history of the concrete compressive strength development. The bridge cannot be tested if concrete has not reached its design strength. During the test, the traffic must be limited or completely discontinued on the bridge and, if necessary, in its accesses and under it.

Before testing, distribution of the load must be marked on the bridge deck. The areas for mounting of measuring devices are marked on the surface of structures, where the appropriate measuring devices are mounted. The proof load test (Section 12.1.1) can be performed for the verification of operation of measuring devices.

When testing a bridge, a certain probability of damaging the structure always exists. Therefore, during testing it is necessary to follow the response of the structures to the load. Having noticed unpredictable deformations or in the event of unexpected damage, the test must be stopped. After the test, it is necessary to inspect the bridge and assess the condition of the bridge and to record defects occurred or additionally developed during the test. During the tests, it is necessary to observe and record the ambient conditions and especially significant changes in temperature, humidity, wind direction and speed, and so forth. In specific cases, when the structures are sensitive to these impacts, it is necessary to measure additionally the overall impact of the permanent, testing, and aforementioned loads on the load-bearing capacity of the bridge.

Loading of the bridge and its structures with the static or dynamic load of the same parameters, if possible, must be repeated several times to avoid measurement errors and to obtain reliable results. In these cases, the reliability of the results can be assessed by statistical methods.

12.3.2 Measuring techniques and equipment

Displacements at characteristic cross sections of the deck, relative strain of concrete, reinforcement or steel, stresses, settlement of the deck near supports, movements and/or inclinations of bearings, settlements and inclinations of bridge piers and abutments, buckling of compressive elements, prestressed forces, modes and frequencies of free and forced vibrations (longitudinal, transverse, torsional), amplitudes and acceleration of vibrations, and parameters of vibration damping are measured during static and dynamic tests. The type and extent of measured parameters mostly depend on the type of the bridge. Recommendations on monitored parameters for different types of bridge structures are presented in Sétra (2006). The number of devices must be sufficient to achieve the objectives of the tests and to assess the structural behavior. All devices should be reliable to work in harsh environmental conditions. Additional protective measures may be used, if necessary. Devices used to measure different parameters are discussed in more detail by Xu and He (2017). Brief information on the most commonly used devices is provided below.

Linear variable displacement transducers (LVDTs) are commonly used for measurement of displacements of bridge elements. LVDTs are arranged in characteristic cross sections on a fixed reference base. Sensors can be mounted directly on bridge structures by installing temporary supports (Fig. 12.11a) or by using a steel wire (Fig. 12.11b). LVDT parameters are selected depending on the variation limits of displacement and the desired measurement accuracy. The displacement transducer signal is transmitted to the receiver with further processing (Fig. 12.12). The processed data is transmitted to the computer, which shows displacements of the measured structures at all loading stages during the tests. However,

Figure 12.11 Measuring of bridge deck displacements using LVDT

Figure 12.12 Equipment for displacement data processing

the latter methods are suitable at low height of the bridge and the free space under the bridge. As the height rises, it becomes complicated to install the stable support. By joining the LVDT of the bridge and the deck with a steel wire, the results obtained due to temperature and wind effects can be insufficiently accurate. Moreover, these methods require a high installation effort.

Geodetic equipment, including leveling instruments (Fig. 12.13a) or total stations (Fig. 12.13b), can be used for testing of large span bridges when significant displacements due to the test load are expected. This technology makes it possible to obtain the displacement in a contactless way and does not require a high amount of time in the process of preparation for tests. On the other hand, measurement accuracy is usually lower than when using LVDTs. More advance wireless techniques based on application of smart radars, laser-based equipment, 3D scanners, and so forth can be used to measure displacements. An overview of the devices available for measuring bridge displacements is provided in Xu et al. (2019).

Different types of linear strain gages (Fig. 12.14) are used to measure the deformations of the elements. Depending on the measurement principle, deformations can be measured directly or calculated according to the recorded displacement. LVDT sensors are also

Figure 12.13 Measuring of bridge displacement using leveling instrument (a) and total station (b)

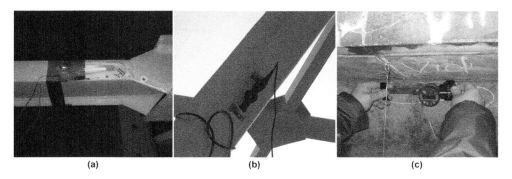

Figure 12.14 Different gages for strain measurement in bridge elements

suitable for this case. Knowing the mechanical characteristics and parameters of the materials, the stresses in the relevant area can be determined. The sensor data can be combined into a unified data recording system or the deformations can be read separately. Recently, fiber Bragg grating (Matta et al., 2008; Yau et al., 2013) or distributed optical fiber sensors (Bado et al., 2018) have been used for measuring deformation of building structures. However, these technologies are commonly applied for permanent monitoring systems of bridges. If such a system is installed on a bridge, its data can be used during the test.

For reinforced concrete bridges, measurement of the crack width is also performed during static testing. Cracks are usually measured using optical microscopes, special gages, digital crack meters or alternatively may be calculated from more advanced digital image correlation (DIC) techniques (Murray et al., 2014; Schmidt et al., 2014).

Dynamic measurement systems are used for identification of dynamic parameters of the bridge. Figure 12.15a presents the principle scheme of vibration measurement, data acquisition, and processing tools. The dynamic measurement system usually consists of high sensitivity accelerometers, portable equipment for collecting and processing measured results, and a personal computer. The accelerometers are fixed directly to the bridge

268 Load Testing of Bridges

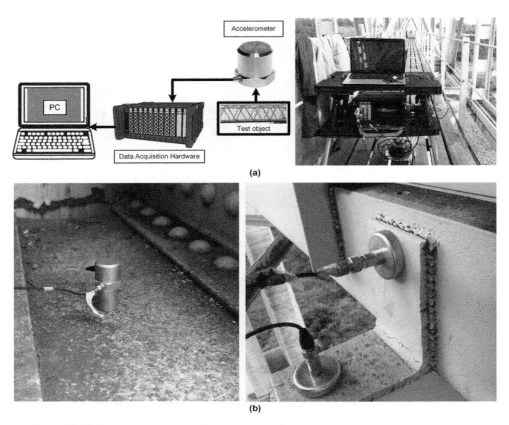

Figure 12.15 Dynamic equipment for measuring vibrations

surface or through the control pads attached to the surface of the bridge superstructure (Fig. 12.15b). Frequency contents can be obtained for all recordings by fast Fourier transform (FFT). Resonant frequencies are directly derived using, for example, peak picking or alternative methods (Omenzetter et al., 2013). The sampling frequency and frequency interval should be selected based on the data demands and the objective of the tests. Alternatively, laser doppler vibrometers or high-speed cameras also can be used to measure dynamic parameters.

The quantity and arrangement of measuring devices is very dependent on the amount of data to be obtained for the purpose of adequately evaluating the behavior of the structure. The more information is obtained, the more reliable conclusions of the test can be summarized. On the other hand, a larger quantity of equipment requires significantly higher time costs. The location of devices depends on the response of the structure. In cases where it is necessary to experimentally approximate the deformed bridge scheme, it is advisable to measure the displacements in the longitudinal direction at the supports, in the cross section with maximum displacement (usually in the middle of the span), and in intermediate cross sections (e.g. quarters of the span). It is advisable to perform appropriate measurements in the transverse direction in at least two cross sections depending on the type of the deck. An example of such an arrangement is shown in Figure 12.16a. A larger number of

Field Testing of Pedestrian Bridges 269

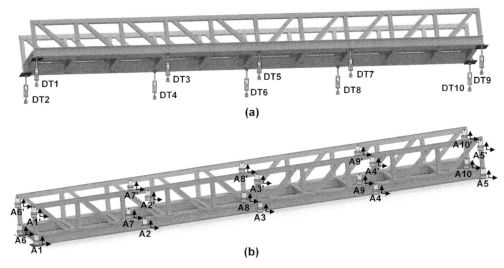

Figure 12.16 Locations of measurement points during static (a) and dynamic (b) tests

sensors have to be used to get a more accurate experimental deformed shape. If only the maximum value of displacement has to be obtained during the tests, measurements can be performed only in one cross section of the deck.

Similar trends are also valid for measurements of dynamic parameters. The number of accelerometers is taken based on experimental modes (vertical, horizontal, or rotary) and their parameters to be obtained. It is necessary to use more accelerometers to receive high order forms or to change their location during testing without changing the intensity of excitation. An example of the arrangement of the accelerometers is shown in Figure 12.16b. If parameter values of only first-order modes are to be recorded by dynamic tests, and the type of the mode itself is not important, locating the accelerometers in one of the characteristic cross sections will be sufficient. In this case, the direction of measurement depends on the type of mode the characteristics of which are to be measured.

12.3.3 Execution of the tests

12.3.3.1 Static tests

The main purpose of static tests is to identify the actual stiffness of the deck, taking into account the potential defects, structural integrity, and other appropriate effects on the bridge behavior. The test load is to be positioned on the deck along and across the bridge in the most unfavorable position with respect to the tested structure. The positioning of the test load depends on the static scheme of the bridge. In the general case, it is recommended to place the test load in the middle and quarters of the deck and above the supports. The load on the decks should be placed in adjacent spans in respect of continuous decks by causing the largest and smallest load effects. The test load on arch, suspension, and cable-stayed bridges must be placed asymmetrically with respect to the middle of the bridge span. The loading and unloading schemes must be chosen in such a way as to allow a proper assessment of the overall behavior of structures under different loading and unloading

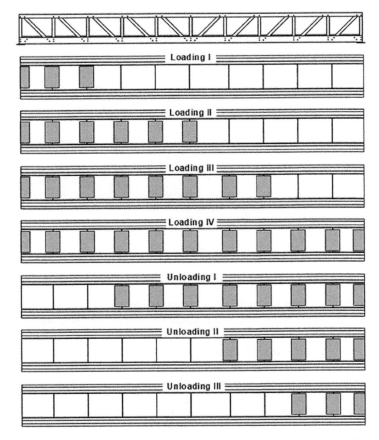

Figure 12.17 Example of loading and unloading stages during the static test of bridge deck

conditions, the response of symmetrically arranged elements in relation to similar load situations, and the recovery of structures after loading. An example of the deck loading and unloading of the pedestrian bridge is shown in Figure 12.17.

The test load must be increased gradually during loading and by uniform load steps, if possible. At least five load steps are recommended (Lithuanian Road Administration, 2005). Loading on the bridge must be carried out uniformly, without causing significant vibrations. The loading and unloading time should be as short as possible in order to avoid nonlinear deformations of the bridge elements. The load should be sustained at the final stage of the test. It is recommended to sustain the load for reinforced concrete decks for at least 30 min, and for steel decks for at least 15 min (Lithuanian Road Administration, 2005). The sustaining period of the load on the deck depends on the behavior of the bridge. If an increase in measured parameters is not recorded during the test, the load maintenance period may be shorter. Meanwhile, when the readouts indicated by devices increase, the load must be sustained until stabilization of the readings is achieved. The results of the static tests must be recorded at each loading stage before loading, immediately after loading, and every 5 to 10 min until the increase of values stabilizes (Lithuanian Road Administration, 2005). In the same aspects, it is also used to record changes in measured parameters at the time of

Field Testing of Pedestrian Bridges **271**

loading. The same aspects should be followed when recording changes in the measured parameters at the time of unloading. It is recommended to record the values indicated by all devices simultaneously and always in the same order, and the time of recording should be no longer than 5 min (Lithuanian Road Administration, 2005).

During the test it is necessary to observe the visual condition of the structures during loading, maintaining the load and removing it. It is required to record and measure the characteristics of existing and newly emerging defects, for example, crack widths. It is recommended to closely monitor the graphic or numerical values of elastic, residual, and total displacements during the test and compare them with the predicted theoretical values at each stage of loading. If the tests reveal significant disagreement between the preliminary results of the theoretical and experimental data, the tests must be stopped in order to identify the causes of these errors.

12.3.3.2 Dynamic tests

During dynamic testing, it is necessary to excite free and forced vibrations on the bridge deck. In order to excite as many forms of free vibrations as possible, the bridge on different sections is exposed to the same intensity of vertical or horizontal excitation effects (see Section 12.2.4.2). When testing a bridge with a shock load and seeking to protect the coating of the deck from damage, it is recommended to install a flexible deck (of sand, planks, rubber, etc.) under the load.

In the case of forced vibrations, the test is carried out when one pedestrian or a group of people walks, synchronously steps, jumps, or runs across the bridge. The frequency of vertical vibrations caused by normal walking is approximately 2 Hz. If possible, it is recommended to select the effects of pedestrians so that the vibrations they cause have a frequency close to the resonant one (for instance, when running). Other excitation cases when a group of several people walks or runs, sways rhythmically or walks synchronously, correspond to the accidental effects of people or acts of vandalism. Horizontal vibrations of the bridge may be excited when a person or group of people run along the bridge and suddenly synchronously stops, moves, or runs asymmetrically on the long axis of the bridge, and when a standing person or group of people swing simultaneously.

When forced vibrations are excited by the use of electromechanical shakers, they are positioned in characteristic cross sections to excite the modes of the expected frequency and shape. To excite horizontal modes, shakers are positioned asymmetrically with respect to the longitudinal axis of the bridge. Shakers generating horizontal forced vibrations can be used alternatively (Živanović et al., 2005). The aspects of analyzing the results of static and dynamic tests are discussed in Section 12.4.

12.4 Analysis of test results

This section gives a short overview of the methods used to identify static and dynamic parameters of footbridges. Static parameters of the tested bridge includes deformed shape of the structure, the developed strains and stresses in structural elements as well as development of crack width in case of RC pedestrian bridges. A more specific dynamic parameter covers the mode shapes, natural vibration frequencies, and damping characteristics of the bridge. In principle, static testing of footbridges is similar to other types of structures, where internal forces and deformed shape is of main concern. The

272 Load Testing of Bridges

identification of more specific dynamic parameters is discussed using the simple example of data transformation from time to frequency domains.

12.4.1 General guidelines

Important results for the assessment of safety and reliability of the structure are obtained at this stage. They help to reveal the real state of structures and to properly evaluate the interaction between different structural and nonstructural elements, the influence of material properties and defects on the behavior of the bridge, and other aspects of interest. In assessing test results, it is necessary to properly evaluate possible errors in measurements due to the accuracy of devices and equipment, the reliability of installation devices, inaccuracies in measuring the size of the test load and distribution on the real size structure, discrepancies in material properties, environmental conditions and changes, and other effects. If possible, it is recommended to assess the reliability of test results using statistical methods and identifying key statistical parameters.

It should be emphasized that the results obtained during the static and dynamic tests complement each other. The real stiffness of the bridge elements allowing reliable calibration of the theoretical model can be reliably identified by applying both tests. The calibrated model can continue to be used in the assessment of reliability and safety of the bridge by both deterministic and statistical methods. The information obtained is also an important starting point for further monitoring of possible changes in the stress and strain condition of the bridge.

12.4.2 Methods for identification of static and dynamic parameters of the bridge

12.4.2.1 Methods for identification of static parameters of the bridge

There are two main methods to identify static parameters of the bridge: (1) measurement of deformed shape of the structure and (2) measurement of the distribution of internal forces. In the former case, deflection of the bridge deck in longitudinal and transverse directions is measured. Displacements of the abutments or piers may be also measured if relevant (Lantsoght et al., 2017b). The displacements of decks under different loading conditions are measured in most cases. The displacements are determined by processing the LVDT data and by evaluating the displacements of the supporting parts (Fig. 12.18). If measurements are carried out several times, the average of different measurements is taken. The displacement results are important in determining the trends of deformation of bridge elements under different loading conditions. An example of the measurement of deformed shape of steel truss pedestrian bridge is shown in Figure 12.19.

The deformation of the bridge under a particular load serves as the main indicator of stiffness and integrity of the structure. The relationship of the applied load and measured deflection is also of primary importance: the nonlinear behavior of the bridge indicates the appearance of cracks, local plastic deformation in structural elements, and deformations in structural joints or bearings.

Another principal static parameter is the distribution of internal forces in structural elements. Particularly, this may be important in footbridges with advanced structural systems (e.g. cable-stay, suspension, or stress-ribbon bridges). As the internal force can hardly be

Field Testing of Pedestrian Bridges 273

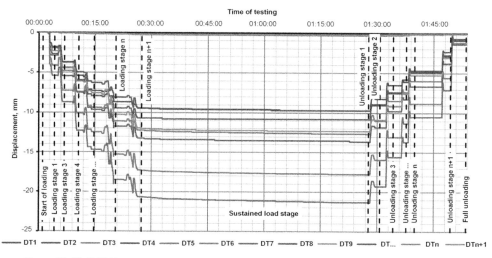

Figure 12.18 LVDT readings during the static test under different loading and unloading stages. Conversion: 1 mm = 0.04 in.

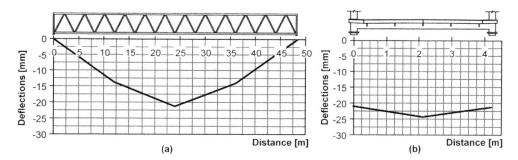

Figure 12.19 Experimental deformed shapes of the footbridge (Bačinskas et al., 2014): measured deflections of the bridge deck in longitudinal (a) and transverse directions (b). Conversions: 1 m = 3.3 ft, 1 mm = 0.04 in.

measured directly, strain gages are a common choice in such cases. The determination of deformations depends on the selected applied devices and technologies, the chosen measurement base, and other related aspects. In any case, deformations are determined either directly or according to measured parameters (displacement, resistance, etc.). From the measured deformations, it is further possible to calculate stresses and internal forces acting in a particular element. The latter depend on the material modulus of elasticity and real geometric characteristics.

The width of cracks of the reinforced concrete bridges is also measured during static tests. The changes of crack pattern during the test are observed. Monitoring of the width of the main cracks, distance between the cracks, and development trends of the cracks is documented at different load levels. The obtained data is used for the compilation of

274 Load Testing of Bridges

experimental schemes of development of cracks, which can be used to evaluate the nonlinear behavior of reinforced concrete elements.

12.4.2.2 Methods for identification of dynamic parameters of the bridge

Identification of dynamic parameters of footbridges is generally much more complex in comparison to measurement of static deflection or strain at particular points of the structure. The dynamic behavior of a structure in a given frequency range can be modeled as a set of individual modes of vibration. Each vibration mode can be described by three modal parameters: mode shape, natural frequency, and modal damping. Modal parameters represent the inherent properties of a structure which are independent to the excitation source. Principally, determination of modal parameters (which is commonly referred as modal analysis) is the main task in the dynamic analysis of the footbridges. Modal analysis may be accomplished either through analytical, numerical or experimental techniques.

In classical modal analysis (CMA), the modal parameters are found by fitting a model to frequency response functions (FRF) relating excitation forces to vibration responses. The FRF can be expresses as follows:

$$H(\omega) = \frac{Y(f)}{X(f)} = \frac{Output}{Input} = \frac{Motion}{Force} = \frac{Response}{Excitation} \tag{12.2}$$

where $H(\omega)$ is the frequency response function, $Y(f)$ is the output of the system, and $X(f)$ is the input of the system.

Extraction of the modal parameters from the FRF is based on the transformation of signals from time domain to frequency domain. To illustrate the principles of this transformation, let us consider a typical vertical force function in time of a normal walking pedestrian, shown in Figure 12.20a. The vertical axis of this function represents the ratio of measured force to static weight (G) of the pedestrian (dynamic coefficient). A normally walking pedestrian generates a continuous vertical force on the bridge, as shown in Figure 12.20b by the solid line. Let us assume that this vertical force was simplified by a function with period $T = 0.5$ s and frequency $fs = 2$ Hz, as shown in Figure 12.20c by the solid line.

So far, the variable on the horizontal axis was time. Such representation of the results is referred as analysis in the time domain. For identification of dynamic parameters, the transition from time domain to frequency domain is required. To transform signals from time domain to the frequency domain, a Fourier transform is performed. The idea of Fourier transform is based on the statement that every continuous wave can be approximated as the sum of sine and cosine waves. Hence, the force generated by a normally walking pedestrian can be approximated by the Fourier series (Štimac Grandić, 2015):

$$F(t) = G + G\sum_{i=1}^{n} \alpha_i \sin(2 \cdot \pi \cdot i \cdot f_s \cdot t - \phi_i) \tag{12.3}$$

where G is the static weight of the pedestrian, i is the harmonic number, f_s is the step frequency, φ_i is the phase shift of the ith harmonic, α_i is the Fourier coefficient for the ith harmonic, and n is the total number of contributing harmonics.

Using Equation 12.3, the function $F(t)$ was decomposed in three harmonic parts with different frequencies, amplitudes, and phases as shown in Figure 12.20d by black (1st

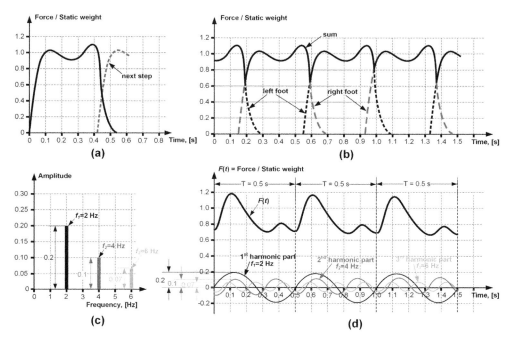

Figure 12.20 Basis of footbridge dynamics: (a) and (b) typical vertical force function for normal walking (adopted from Živanović et al. [2005]); (c) presentation of harmonic functions in frequency domain; and (d) approximation of vertical force by harmonic functions (Fourier transform)

harmonic part), gray (2nd harmonic part) and light gray (3rd harmonic part) lines. These three harmonic components can also be shown in the frequency domain representation (Fig. 12.20c). Note that the diagram in Figure 12.20c shows exactly the same harmonic functions, but the presentation of the results is different.

In the discussed example, the vertical force function $F(t)$ was decomposed into three harmonic components. The recorded signals during dynamic bridge testing are generally much more complex. Figure 12.21 shows the schematic example of signal processing in a steel truss pedestrian bridge and extraction of FRF.

The footbridge shown in Figure 12.21 is excited with a hammer impact. As may be expected, impact of the hammer generates a single force impulse and decaying acceleration curve in time domain, as may be seen from diagrams shown in the frequency domain. The impact of the hammer may be considered as the input whereas the measured acceleration of the bridge deck is the output of the system. It should be noted that the measured acceleration curve is dependent both on the excitation signal (hammer impact) and the dynamic properties of the structure under test.

Using the FFT method, the recorded force impulse of the hammer and acceleration of the bridge deck are transformed from the time to the frequency domain. FRF is then obtained by dividing the output of the system (accelerations in frequency domain) by the input (force of hammer impact in frequency domain) as shown in Figure 12.21. The FRF reflects the dynamic properties of the structure itself, rather than properties of the external environment

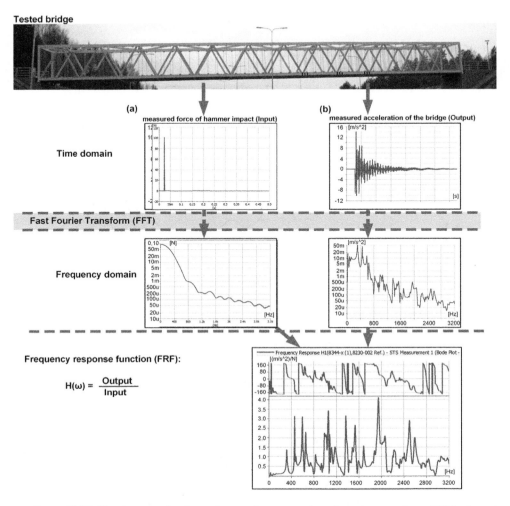

Figure 12.21 Signal processing from time to frequency domain and extraction of FRF of steel truss pedestrian bridge. Conversion: 1 (m/s^2)/N = 0.454 1/lb

or means of excitation. From the FRF it is possible to extract modal characteristics of the structure:

- Peaks of frequencies indicate the natural frequencies of the structure under test (Fig. 12.22).
- The width of the peaks indicates the damping characteristics of the structure: the wider the peak, the heavier the damping. Damping ratio can be calculated directly from the FRF (Omenzetter et al., 2013):

$$\xi = \frac{f_{2a} - f_{2b}}{f_{2a} + f_{2b}} \tag{12.4}$$

where frequencies f_{2a} and f_{2b} are schematically shown in Figure 12.22.

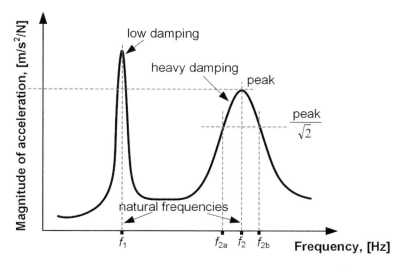

Figure 12.22 Modal characteristics obtained from FRF. Conversion: 1 (m/s²)/N = 0.454 1/lb

Using the amplitude and phase of multiple FRF, it is possible to determine the mode shapes of the structure. To accurately model the associated mode shape, frequency response measurements must be made over a sufficient number of degrees of freedom (DOF) to ensure enough detailed coverage of the structure under test.

The damping ratio of the structure may be also estimated using a method of logarithmic decrement, which is perhaps the most popular time-response method used to measure damping. The logarithmic decrement represents the rate at which the amplitude of a free damped vibration decreases. It is defined as the natural logarithm of the ratio of any two successive amplitudes, as schematically shown in Figure 12.23. The damping ratio is expressed as:

$$\varsigma = \frac{1}{2\pi p} \ln \frac{u_n}{u_{n+p}} \qquad (12.5)$$

where u_n is the amplitude of vibration at time t_n, u_{n+p} is the amplitude of vibration at time t_{n+p}, and p is the time period between the measured peaks.

If the input to the system is unknown (for example in case of an ambient vibration survey), operational modal analysis (OMA), is performed, which is based on the vibration responses only. OMA is used instead of CMA for accurate modal identification under actual operating conditions and in situations where it is difficult or impossible to control an artificial excitation of the structure. In this case, different identification techniques for modal parameters are used, such as frequency domain decomposition (FDD), enhanced frequency domain decomposition (EFDD), and stochastic subspace identification (Gentile and Gallino, 2008; van Nimmen et al., 2017; Živanović et al., 2005).

Another important dynamic characteristic is the acceleration of the footbridge deck. Acceleration is limited to assure the comfort of pedestrians (see Section 12.5.5). Different types of accelerometers are used for structural applications: capacitive, piezoelectric, strain

278 Load Testing of Bridges

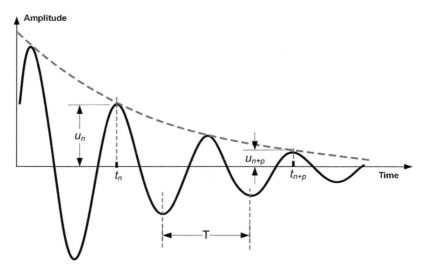

Figure 12.23 Schematic representation of algorithmic decrement

gage, fiber grating, micro-electro-mechanical systems (MEMS), or servo accelerometers (Omenzetter et al., 2013).

12.4.3 Presentation of results

After testing of the bridge, the gathered experimental data should be sufficient to evaluate the structural condition and make a decision on further operation. Generally, the following data should be presented (Lithuanian Road Administration, 2005; Sétra, 2006):

- Structural drawings and technical parameters of the bridge, notation and dimensions of the main structural elements;
- Material characteristics, determined from laboratory or field tests;
- Deterioration or damage of the load-bearing elements;
- Theoretical model of the bridge, describing the assumed simplifications, modeling techniques, and material properties;
- Test loads on the bridge: type, duration, and distribution of loads;
- Details on testing and measurement of bridge response: test procedure, types and distribution of measurement devices, and weather conditions;
- Analysis of the experimental data, comparison with the theoretically obtained results, and evaluation of footbridge condition based on test results.

12.5 Theoretical modeling of tested bridge

This section gives some insights to the theoretical modeling techniques used for bridges and criteria to evaluate the structural condition of the tested structure. A precise numerical model may be a powerful tool to study the structural behavior of the bridge. Moreover, it may help to evaluate the damage of the structure and plan the further maintenance and inspection. In this respect, the model updating procedure is briefly discussed, emphasizing the most common modeling errors. Finally, code requirements for serviceability of footbridges are summarized, focusing on the dynamic characteristics of the structures.

12.5.1 Introduction

Theoretical modeling of the bridge is essential in the interpretation of the experimental results and evaluation of structural condition. Significant discrepancies between experimentally obtained data and theoretical modeling indicates possible damage of structure, flaws in design, construction, or maintenance. With modern computational tools it is possible to generate refined numerical bridge models, taking into account nonlinear material properties, stiffness of the supports or influence of nonstructural elements (pavement, railings, stairs, etc.). Such refined models allow not only to assess the condition of the structure but also to plan the test procedure itself (distribution of static loads, positions of accelerometers during dynamic testing).

12.5.2 Modeling techniques

In recent decades the finite element (FE) analysis method has emerged as the most powerful and versatile tool for structural analysis. The FE models are also broadly used in the modeling of static and dynamic characteristics of footbridges (van Nimmen et al., 2014; Dulińska et al., 2016). The main advantage of the FE models is the ability to represent complex geometry of the particular structure.

The finite element models, generated from the detailed structural drawings and field observations, serve as powerful tools for the static (determining internal forces, stresses, and strains in the structural elements, support reactions, deformations of the structure) and dynamic (calculation of natural frequencies and mode shapes) analysis of the bridge. With the increasing computational capabilities, it is not necessary to simplify the structure assuming idealized boundary conditions, using beam elements or simulating plane structure. Moreover, refined FE models enable a deeper insight in the structure, and help to interpret the experimental results, particularly regarding the mode shapes. Once the FE model is generated and properly tuned, it may be employed in condition assessment of the bridge and for asset management reasons (Votsis et al., 2017).

As footbridges are generally slender structures, the contribution of nonstructural elements and the actual support stiffness may significantly influence the modeling results, especially in the assessment of dynamic characteristics of the structure. In the study of Votsis et al. (2017) it was found that the dynamic response of a fiber reinforced polymer (FRP) suspension footbridge was sensitive to the performance of the parapet system (parapet posts, handrails, and footrails). In the numerical and experimental analysis of a steel girder footbridge performed by van Nimmen et al. (2017), it was determined that staircases and bike paths installed on the bridge deck influenced the natural frequency of bending modes. In the same study it was also found that the natural frequencies of the bridge were also sensitive to the assumed support conditions. Živanović et al. (2006) introduced longitudinal springs at the supports for a more realistic representation of the vibration modes. There also exists uncertainty in the actual mechanical material

280 Load Testing of Bridges

characteristics, stiffness of joints, and damage level of structural elements (Gentile and Gallino, 2008).

The discussed modeling uncertainties may be determined or tuned only if the experimental data of the modeled bridge are present. Model updating procedure will be further discussed in Section 12.5.4.

12.5.3 Comparison of experimental and theoretical results

Comparison of theoretical and experimental results are usually performed in two stages. In the first stage, theoretical models are created using the project data or alternatively data obtained from the inspections prior the tests. In the second stage the theoretical model is updated taking into consideration the results obtained during the static and dynamic tests. Aspects of model updating are briefly discussed in Section 12.5.4.

Numerical model of the bridge may help for better insight to the structure, especially when experimental data is presented. Comparing simulated and experimentally determined behavior of the structure, it is possible to assess the structural condition of the bridge and possibly damaged members. The structural condition of the bridge may be represented by the coefficient K, expressed as:

$$K = \frac{S_{exp}}{S_{calc}} \tag{12.6}$$

where S_{exp} and S_{calc} are the experimental and calculated structural effect (stress, deflection, etc.), respectively.

The structural coefficient K should be in the limits of 0.6–1.1, depending on the type of the bridge: for steel bridges, $0.8 < K \leq 1.05$; for reinforced concrete bridges, $0.6 < K \leq 1.1$; and for prestressed concrete bridges, $0.7 < K \leq 1.05$ (Lithuanian Road Administration, 2005).

Low values of the factor K indicate that the structure potentially has additional sources of load-bearing capacity. This situation possibly may arise due to conservative design methods, differences in real and idealized structural system (stiffness of joints, bearings, material properties, etc.) or the influence of nonstructural members (additional layers on the bridge deck, hand railings, etc.). On the other hand, a structural factor significantly exceeding 1 indicates flaws in the design or construction; it also may indicate that the bridge is damaged. Static tests in this case should be stopped and a detailed investigation of the bridge should be performed.

Another important parameter to assess the structural condition of the footbridge from the static tests is the ratio of residual to maximal deflection, α, expressed as:

$$\alpha = \frac{y_{pl}}{y_{tot}} \tag{12.7}$$

where y_{pl} is the residual deflection determined from the first unloading cycle; y_{tot} is the total measured deflection of the bridge.

The total deflection of the bridge is determined after the stabilization of deformation (the growth of displacements in 5 min should not exceed 5%). For new steel bridges the allowed value of parameter α should not exceed 0.1 and for RC bridges 0.3 (Lithuanian Road Administration, 2005). High values of residual deflection indicate cracking, damage, or local deformations in the structure. An example of a comparison of theoretical and experimental displacements is presented in Figure 12.24.

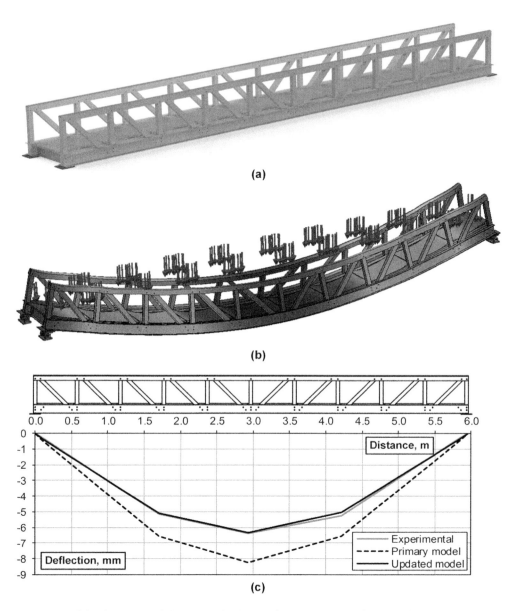

Figure 12.24 FE model of the tested footbridge (a), theoretical deformed scheme due to static loading (b), and comparison of theoretical and experimental results (c). Conversions: 1 m = 3.3 ft, 1 mm = 0.04 in.

Regarding the modeling of the dynamic behavior of the bridge, the common outcomes are the mode shapes and natural frequencies of the structure. The obtained natural frequencies can be directly compared with those experimentally measured (Fig. 12.25). However, evaluating the accuracy in modeling of mode shapes is a more complex issue. One of the

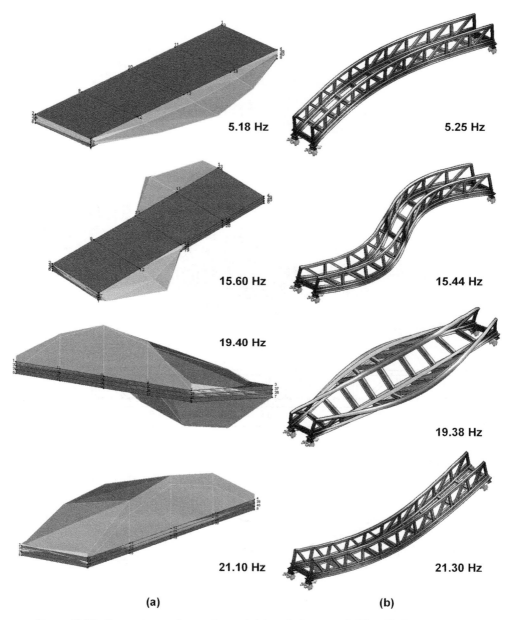

Figure 12.25 Comparison of experimental (a) and theoretical (b) self-vibration modes and frequencies

possible ways is to compare experimentally and theoretically obtained modal vectors, reflecting the deformed shape of the structure. In this approach, the modal assurance criterion (MAC) is used to evaluate the adequacy between theoretical and experimental modal vectors. The modal assurance criterion is defined as a scalar constant relating the degree of

Field Testing of Pedestrian Bridges 283

consistency between two modal vectors as follows (Pastor et al., 2012):

$$MAC = \frac{|\{\phi_A\}_r^T\{\phi_X\}_q|^2}{(\{\phi_A\}_r^T\{\phi_A\}_r)(\{\phi_X\}_q^T\{\phi_X\}_q)} \tag{12.8}$$

where $\{\varphi_X\}_q$ is the test modal vector, mode q; and $\{\varphi_A\}_r$ is the compatible analytical modal vector, mode r.

The modal assurance criterion varies from 0 (representing no consistent correspondence) to 1 (representing fully consistent correspondence). In practice, values larger than 0.9 indicate good consistency of the simulated mode shape. Accurate modeling of mode shapes is more difficult in comparison to finding natural frequencies and commonly requires a model updating procedure (van Nimmen et al., 2017), which is discussed in the next section.

12.5.4 Model updating

The FE model of the tested bridge assists in the interpretation of the obtained results and evaluation of the structural condition. Generally, the initial FE model of the bridge is generated before testing, based on the project documentation, field observations, and measurements. Such initial models are particularly valuable in the preliminary analysis of dynamic characteristics of the footbridge, determining expected natural frequencies and mode shapes. The simulated shapes give a general understanding about the dynamic behavior of the bridge and help to determine the required number and exact positions of accelerometers.

However, even the most refined initial FE models generated with the best engineering judgment often lead to modeling errors of about 30%–40% in comparison to experimentally obtained results (Živanović et al., 2005). Such discrepancies are particularly characteristic to the prediction of dynamic characteristics (natural frequencies and the mode shapes) of the footbridge. Modeling errors generally may be attributed to one of the following groups (Omenzetter et al., 2013):

1 Material characteristics. Elastic and shear modulus and density of the materials may differ from those assumed in the initial design. Such uncertainties are especially characteristic to bridges where timber is used as the main structural material.
2 Modeling errors. Such errors are related to the choice of finite element type, meshing parameters, incompatibility, or inadequate contact conditions between meshes.
3 Support conditions of the bridge. In the initial FE model, ideal support conditions (it is a common practice in the initial model to allow full horizontal translation of one of the supports) are assumed. In real bridges, additional restraint at the support may arise due to friction of the bearings and the bridge deck (van Nimmen et al., 2017). Expansion joints may also exhibit particular stiffness (Živanović et al., 2005), which affects the dynamic characteristics of the footbridge.
4 Structural uncertainties. The stiffness of the bridge may increase from the additional non-load-bearing elements (hand railings, bicycle routes, etc.). Stiffness is also affected by the rigidity of joints and damaged parts of the bridge (Gentile and Gallino, 2008). Moreover, in heritage and historical bridges, it is common that not all design (or repair) documentation is available. During the service life, such bridges may be repaired and

284 Load Testing of Bridges

strengthened several times, and information about strengthening is not initially available.

In practice, it is difficult to take all these possible uncertainties into account when developing the initial FE model. After the field tests, it becomes possible to update the initial FE model, tuning it as close as possible to the real bridge behavior.

The FE updating (or tuning) procedure may be performed manually or automatically. In the former case, the designer varies the material parameters, stiffness of structural or nonstructural elements, support conditions, or material properties to match the simulated and experimental parameters of the bridge. Manual model tuning is essential in case some structural elements were missed in the initial model (e.g. the thickness of piers or bridge deck was increased after strengthening) (Živanović et al., 2005). Manual tuning may be more efficient if a sensitivity analysis is performed prior to model tuning. Another method for FE model updating is automatic model tuning. In this case, candidate parameters for model updating are selected (such as material properties, geometrical parameters, or boundary conditions) and grouped in a vector. Similarly, output of the system (natural frequencies, mode shapes) also form a respective vector. The objective of the analysis is to minimize the difference or error between these two vectors, adjusting the selected parameters of the bridge and improving correlation between the experimental and analytical model (Omenzetter et al., 2013). Automatic model updating gives a precise and optimal combination of bridge parameters which best suits the desired properties (natural frequencies, mode shapes) of the bridge.

An example of model updating procedure is given in Figure 12.24c. The analysis of experimental and theoretical results of the bridge model carried out by Bačinskas et al. (2017) is presented in the figure. The primary model corresponds to the FE modeling results of the bridge performed at the initial stage before the tests. After carrying out experimental tests, the model is adjusted according to the actual characteristics of the materials as well as the stiffness of the joints and support parts. As a result, good agreement between theoretical and experimental displacements is obtained for the updated model.

12.5.5 *Code requirements for serviceability of footbridges*

For footbridges, three main serviceability criteria may be delineated: (1) maximal deflection, (2) allowable crack width (for reinforced concrete bridges), and (3) comfort due to vibrations.

The serviceability criteria is generally specified for each project and agreed with the client. The recommended values by different codes and guidelines are briefly discussed below.

Footbridges are generally much more slender structures than railway and road bridges. This causes a common problem of assurance of the comfort criteria regarding vibrations of the bridge deck. Human perception of vibration is frequency dependent. In the frequency range of 1–10 Hz, the perception is proportional to acceleration, whereas in the range of 10–100 Hz velocity becomes the major factor of vibration perception (Dulińska et al., 2016). For footbridges, comfort criteria become critical at low vibration frequencies, generally below 5 Hz. Such frequencies are related to specific dynamic loads generated by pedestrians. The normal walking pedestrian generates periodic vertical, longitudinal, and lateral load components on the bridge deck which are related to step frequency. For normal walking steps, the frequency falls between 1.7–2.2 Hz, for fast walking 2.2–2.4 Hz, and for running 2.7–3.3 Hz (Štimac Grandić, 2015). Obviously, it may be expected that vertical and longitudinal load components will have the same frequency, whereas the frequency of lateral component will be divided by two. This may be explained by the nature of walking or running: left and right feet generate equivalent forces in longitudinal and vertical directions. Lateral load is generated from moving the center of mass from left to right foot and equals half of the step frequency (van Nimmen et al., 2014).

If the natural frequency of the bridge falls in the interval of pedestrian-induced loads, the risk of resonance occurs. Due to specific loads induced by the pedestrians on the bridge, generally only modes with a natural frequency less than 3–5 Hz have to be taken into account for vibration serviceability assessment (EN 1990, 2002; EN 1991–2, 2003; Schlaich et al., 2005). According to EN 1991–2 (2003), the verification of comfort criteria should be performed if the natural frequency of the deck is less than 5 Hz for vertical vibrations and 2.5 Hz for horizontal (lateral) and torsional vibrations. International design recommendations (Schlaich et al., 2005) also adopt the same limit value for vertical and horizontal frequencies: 5 and 2.5 Hz, respectively.

As was previously mentioned, the perception of vibration in the low frequency range is related to the acceleration. In the European design code, the allowable acceleration of a footbridge is limited to 0.7 m/s^2 (2.3 ft/s^2) for vertical vibrations and 0.2 m/s^2 (0.7 ft/s^2) for horizontal vibrations. Such limitations are related to the human perception of acceleration: 0.1 m/s^2 (0.3 ft/s^2) is slightly perceivable; 0.55 m/s^2 (1.8 ft/s^2) is unpleasant; and 1.8 m/s^2 (5.9 ft/s^2) is intolerable (Dulińska et al., 2016).

Design codes also provide the recommended values to evaluate structural damping. In general, damping characteristics of a footbridge depend on the material type and the structural system. The damping ratio is also sensitive to the vibration amplitude: higher amplitudes result in higher damping ratios (Sétra, 2006). The exact values of damping ratios for the particular bridge can only be determined by experimental measurements. Approximate values for different types of bridges are presented in Table 12.1.

Regarding the maximal allowable deflection, a common practice is to relate maximal deflection with the span of the footbridge. The deflection due to the service pedestrian live load should not exceed 1/360 (AASHTO, 2009) of the length of the span. In general practice, the limitation of 1/400 of the span length may be adopted.

Table 12.1 Approximate values of logarithmic decrement of structural damping (data taken from EN 1991-1-4)

Bridge type	Structural damping*, δ_s
Steel (welded)	0.02
Steel (high resistance bolts)	0.03
Steel (ordinary bolts)	0.05
Steel and concrete composite	0.04
Concrete (prestressed without cracks)	0.04
Concrete (with cracks)	0.1
Timber	0.06–0.12
Aluminum alloys	0.02
Fiber reinforced plastic	0.04–0.08

*Note: The relationship between the structural damping ratio ζ and the logarithmic decrement due to structural damping ζ_s is $\zeta = \delta_s/2\pi$

The limitation of crack width is related to the corrosion prevention and durability of the structure. The maximal allowable crack width depends on the exposure class of the structure and type of prestressing, and varies from 0.2 to 0.4 mm (0.008 to 0.016 in.) (EN 1992-1-1, 2004).

12.5.6 *Evaluation of footbridge condition based on test results*

After field tests and theoretical modeling of the bridge, the real structural condition may be evaluated. Depending on the gathered experimental data and obtained modeling results, one of the following decisions may be made:

- The footbridge fulfills the current requirements of the design codes and may be normally used.
- The footbridge can be used only with limited loads (limiting the possible crowd loading). Rehabilitation is needed to assure normal operation.
- The footbridge does not fulfill the requirements of the current design code and cannot be used. Reconstruction, strengthening, or special measures to increase the load-bearing capacity or limit the excessive vibrations should be done.

12.6 Concluding remarks

Transport infrastructure plays a vital role in the economy and society of any country. The development and expansion of various industries is accompanied by the intensity and growth of transport flows. Cyclist and pedestrian bridges form an important part of the transport infrastructure in cities, on motorways, and on railways. They are often built for the development of recreational and entertainment areas. From the structural point of view, pedestrian bridges are exceptional structures. Nowadays innovative footbridges most often are designed as slender structural systems that are particularly sensitive to human-induced effects such as vibrations, kinematic displacements, and so forth.

The behavior of both new and existing bridges can be analyzed using modern computer software programs based on finite element technique. Despite advanced modern design technology, numerical models of pedestrian bridges in most cases have a fair number of uncertainties. The latter can be identified by field load testing of footbridges. The data obtained during the tests show the real structural behavior under static or dynamic loads.

Static testing of footbridges is similar to other types of structures, where internal forces and deformed shape are of main concern. The main objective of static tests is to identify the actual stiffness of the deck, taking into account the potential defects, structural integrity, and other appropriate effects on the bridge behavior. The loading and unloading schemes must be chosen in such a way as to allow a proper assessment of the overall behavior of structures under different loading and unloading conditions, the response of symmetrically arranged elements in relation to similar load situations, and the recovery of structures after loading. The relationship of the applied load and measured deflection is also of primary importance: the nonlinear behavior of the bridge indicates the appearance of cracks, local plastic deformation in structural elements, and deformations in structural joints or bearings.

The identification of specific dynamic parameters (mode shape, natural frequency, and modal damping) of footbridges are generally much more complex. A special technique, based on the data extraction from frequency response functions, is commonly used to obtain dynamic parameters of the footbridges. Despite the developed advanced numerical modeling techniques, even the most refined initial FE models often lead to modeling errors of about 30%–40% in comparison to experimentally obtained results. Such discrepancies are particularly characteristic to the prediction of dynamic characteristics of the footbridge. In this respect, model updating procedures could be efficiently used to calibrate the numerical model with the measured outcomes of the tested bridge. A precise numerical model may be a powerful tool to study the structural behavior of the bridge. Moreover, it may help to evaluate the damage of the structure and plan further maintenance and inspection.

The field test of the bridge is considered as a reliable tool for the assessment of the behavior and technical condition of the newly constructed, strengthened, or repaired bridge, or the bridge operated without additional intervention measures when its structures are subjected to static and/or dynamic loading. Information obtained during testing allows assessing the real behavior of the bridge under consideration, as well as the adequacy of the design and maintenance strategy and theoretical models, taking into account specific aspects of a particular bridge type. Experimental data of each tested bridge complements both the country-specific and global database in the field. Data obtained during testing can be used to improve the analyses of structures, design, construction, and monitoring systems and methods.

Acknowledgments

The first and the second authors gratefully acknowledge the financial support provided by the European Social Fund according to the activity "Improvement of researchers" qualification by implementing world-class research and development projects of Measure No. 09.3.3-LMT-K-712 (Project No. 09.3.3-LMT-K-712–01–0145). The authors also wish to thank Domas Čėsna and Vilius Karieta, TEC Infrastructure JSC, for kindly providing photos of constructed and tested bridges. The authors would like to express their deepest gratitude to Deividas Rumšys for technical support during the preparation of this chapter.

Note

1 Editor's note: The reader should note that these definitions differ from the AASHTO definitions of diagnostic and proof load testing.

References

AASHTO (2009) *LRFD Guide Specifications for the Design of Pedestrian Bridges*. American Association of State Highway and Transportation Officials, Washington.

Bačinskas, D., Kamaitis, Z., Jatulis, D., Kilikevičius, A., Gudonis, E., Danielius, G., Tamulėnas, V. & Rumšys, D. (2014) Field load testing and structural evaluation of steel truss footbridge. In: *Environmental Engineering: Proceedings of the 9th International Conference on Environmental Engineering, May 22–23, 2014, Vilnius, Lithuania.*, Technika, Vilnius. p. 6.

Bačinskas, D., Rimkus, A., Rumšys, D., Meškėnas, A., Bielinis, S., Sokolov, A. & Merkevičius, T. (2017) Structural analysis of GFRP truss bridge model. *Procedia Engineering*, 172, 68–74.

Bado, M., Casas, J. & Barrias, A. (2018) Performance of Rayleigh-based distributed optical fiber sensors bonded to reinforcing bars in bending. *Sensors*, 18(9), 3125.

Bagge, N., Popescu, C. & Elfgren, L. (2018) Failure tests on concrete bridges: Have we learnt the lessons? *Structure and Infrastructure Engineering*, 14(3), 292–319.

Brownjohn, J., Živanović, S. & Pavic, A. (2009) Crowd dynamic loading on footbridges. In: Caetano, E., Cunha, A., Hoorpah, W. & Raoul, J. (eds.) *Footbridge Vibration Design*. CRC Press, London. pp. 135–166.

Burgoyne, C. J. & Head, P. R. (1993) *Aberfeldy Bridge: An Advanced Textile Reinforced Footbridge*. [Lecture] Techtextil Symposium, Frankfurt, Germany, June, Lecture no.: 418. pp. 1–9.

Byers, D. D., Stoyanoff, S. & Boschert, J. P. (2004) Dynamic testing of the I-235 Pedestrian Bridge for human induced vibration. *IBC Pittsburgh*, 34, 8.

Cai, H., Abudayyeh, O., Abdel-Qader, I., Attanayake, U., Barbera, J. & Almaita, E. (2012) Bridge deck load testing using sensors and optical survey equipment. *Advances in Civil Engineering*, 11.

Chen, J., Wang, J. & Brownjohn, J. M. (2019) Power spectral-density model for pedestrian walking load. *Journal of Structural Engineering*, 145(2). https://doi.org/10.1061/(ASCE)ST.1943-541X.0002248

Cremona, C. (2009) Dynamic investigations of the Solferino footbridge. In: *Proceedings of the 3rd International Operational Modal Analysis Conference (IOMAC 2009)*, Portonovo, Italy, 4–6 May, pp. 72–81.

Data Physics (2015) *Modal Testing with Shaker Excitation*. [Online] Available from: http://blog.dataphysics.com/modal-testing-with-shaker-excitation/ [accessed 21 November 2018].

Dulińska, J., Murzyn, I., Kondrat, K. & Trybek, S. (2016) Analysis of the dynamic characteristics and vibrational comfort of selected footbridges over the S7 national road. *Czasopismo Techniczne*, (Budownictwo Zeszyt 2-B 2016), 47–59.

Dymond, B. Z., Roberts-Wollmann, C. L., Wright, W. J., Cousins, T. E. & Bapat, A. V. (2013) Pedestrian bridge collapse and failure analysis in Giles County, Virginia. *Journal of Performance of Constructed Facilities*, 28(4), 8.

EN 1990 (2002) Eurocode. *Basis of Structural Design*. European Committee for Standardization, Brussels.

EN 1991–2 (2003) Eurocode 1. *Actions on Structures, Part 2: Traffic Loads on Bridges*. European Committee for Standardization, Brussels.

EN 1992–1-1 (2004) Eurocode 2. *Design of Concrete Structures, Part 1–1: General Rules and Rules for Buildings*. Brussels, European Committee for Standardization.

Eriksson, P. (2013) *Vibration Response of Lightweight Pedestrian Bridges*. MSc thesis, Chalmers University of Technology, Sweden.

European Commission (2017) *EC: Europe on the Move: An Agenda for a Socially Fair Transition towards Clean, Competitive and Connected Mobility for All*. [Online] Available from: https://ec.europa.eu/transport/sites/transport/files/com20170283-europe-on-the-move.pdf [accessed 31 October 2018].

Eurostat (2018a) *Road Accident Fatalities: Statistics by Type of Vehicle*. [Online] Available from: https://ec.europa.eu/eurostat/statistics-explained/index.php/Road_accident_fatalities_-_statistics_by_type_of_vehicle#Pedestrians_particularly_at_risk_in_Baltic_countries_and_in_Romania [accessed 31 October 2018].

Eurostat (2018b) *Rail Accident Fatalities in the EU*. [Online] Available from: https://ec.europa.eu/eurostat/statistics-explained/index.php/Rail_accident_fatalities_in_the_EU [accessed 31 October 2018].

Faber, M. H., Val, D. V. & Stewart, M. G. (2000) Proof load testing for bridge assessment and upgrading. *Engineering Structures*, 22(12), 1677–1689.

Gentile, C. & Gallino, N. (2008) Condition assessment and dynamic system identification of a historic suspension footbridge. *Structural Control and Health Monitoring*, 15(3), 369–388.

Humar, G. (ed.) (2014) *Footbridges: Small Is Beautiful*. European Council of Civil Engineers (ECCE). Grafika Soča d. o. o, Nova Gorica.

Idelberger, K. (2011) *The World of Footbridges: From the Utilitarian to the Spectacular*. Ernst & Sohn, Berlin.

Kamaitis, Z. (1995) *Condition State and Assessment of Reinforced Concrete Bridges*. Monograph.: Technika, Vilnius. (in Lithuanian).

Keil, A. (2013) *Pedestrian Bridges: Ramps, Walkways, Structures*. Munich, Walter de Gruyter.

Lantsoght, E. O., van der Veen, C., de Boer, A. & Hordijk, D. A. (2017a) Collapse test and moment capacity of the Ruytenschildt reinforced concrete slab bridge. *Structure and Infrastructure Engineering*, 13(9), 1130–1145.

Lantsoght, E. O., van der Veen, C., de Boer, A. & Hordijk, D. A. (2017b) State-of-the-art on load testing of concrete bridges. *Engineering Structures*, 150, 231–241.

Lithuanian Road Administration (2005) ST 188710638.10:2005. Field load testing of roadway bridges. LRA, Vilnius (in Lithuanian).

Matta, F., Bastianini, F., Galati, N., Casadei, P. & Nanni, A. (2008) Distributed strain measurement in steel bridge with fiber optic sensors: Validation through diagnostic load test. *Journal of Performance of Constructed Facilities*, 22(4), 264–273.

Murray, C., Hoag, A., Hoult, N. A. & Take, W. A. (2014) Field monitoring of a bridge using digital image correlation. *Proceedings of the Institution of Civil Engineers-Bridge Engineering*, 168(1), 3–12.

O'Donnell, D., Wright, R., O'Byrne, M., Sadhu, A., Edwards Murphy, F., Cahill, P. & Popovici, E. (2017) Modelling and testing of a historic steel suspension footbridge in Ireland. *Proceedings of the Institution of Civil Engineers-Bridge Engineering*, 170(2), 116–132.

Omenzetter, P., Beskhyroun, S., Shabbir, F., Chen, G. W., Chen, X., Wang, S. & Zha, W. (2013) *Forced and Ambient Vibration Testing of Full Scale Bridges*. Research report of the Project No. UNI/578. Earthquake Commission, Wellington, New Zealand.

Pastor, M., Binda, M. & Harčarik, T. (2012) Modal assurance criterion. *Procedia Engineering*, 48, 543–548.

Roos, I. (2009) *Human Induced Vibrations on Footbridges: Application & Comparison of Pedestrian Load Models*. MSc thesis, Delft University of Technology, The Netherlands.

Sánchez-Silva, M., Frangopol, D. M., Padgett, J. & Soliman, M. (2016) Maintenance and operation of infrastructure systems. *Journal of Structural Engineering*, 142(9), 1–17.

Schlaich, M., Brownlie, K., Conzett, J., Sobrino, J., Strasky, J. & Takenouchi, K. (2005) *Guidelines for the Design of Footbridges: Guide to Good Practice*. The International Federation for Structural Concrete (*fib*), Technical Report Prepared by Task Group 1.2, *fib* bulletin number: 32.

Schmidt, J. W., Hansen, S. G., Barbosa, R. A. & Henriksen, A. (2014) Novel shear capacity testing of ASR damaged full scale concrete bridge. *Engineering Structures*, 79, 365–374.

Sétra (2006) *Loading Tests on Road Bridges and Footbridges*. Technical Department for Transport, Roads and Bridges Engineering and Road Safety, Paris.

Štimac Grandić, I. (2015) Serviceability verification of pedestrian bridges under pedestrian loading. *Tehnički vjesnik*, 22(2), 527–537.

Stratford, T. (2012) The condition of the Aberfeldy Footbridge after 20 years in service. In: *Proceedings of the International Conference Structural Faults and Repair 2012, 3–5 July 2012*, Edinburgh, United Kingdom, p. 11.

van Nimmen, K., Lombaert, G., De Roeck, G. & van den Broeck, P. (2014) Vibration serviceability of footbridges: Evaluation of the current codes of practice. *Engineering Structures*, 59, 448–461.

van Nimmen, K., Van den Broeck, P., Verbeke, P., Schauvliege, C., Mallié, M., Ney, L. & De Roeck, G. (2017) Numerical and experimental analysis of the vibration serviceability of the Bears' Cage footbridge. *Structure and Infrastructure Engineering*, 13(3), 390–400.

Venuti, F. & Bruno, L. (2011) Footbridge lateral vibrations induced by synchronised pedestrians: An overview on modelling strategies. In: Papadrakakis, M., Fragiadakis, M. & Plevris, V. (eds.) *COMPDYN 2011: Proceedings of the 3rd ECCOMAS Thematic Conference on Computational Methods in Structural Dynamics and Earthquake Engineering*. Corfu, Greece, 25–28 May 2011. p. 36.

Votsis, R. A., Stratford, T. J., Chryssanthopoulos, M. K. & Tantele, E. A. (2017) Dynamic assessment of a FRP suspension footbridge through field testing and finite element modelling. *Steel and Composite Structures*, 23(2), 205–215.

Xu, Y. L. & He, J. (2017) *Smart Civil Structures*. CRC Press, Boca Raton.

Xu, Y. L., Brownjohn, J. M. & Huseynov, F. (2019) Accurate deformation monitoring on bridge structures using a cost-effective sensing system combined with a camera and accelerometers: Case study. *Journal of Bridge Engineering*, 24(1). https://doi.org/10.1061/(ASCE)ST.1943-541X.0002248

Yau, M. H., Chan, T. H., Thambiratnam, D. P. & Tam, H. Y. (2013) Static vertical displacement measurement of bridges using fiber Bragg grating (FBG) sensors. *Advances in Structural Engineering*, 16(1), 165–176.

Živanović, S., Pavic, A. & Reynolds, P. (2005) Vibration serviceability of footbridges under human-induced excitation: A literature review. *Journal of Sound and Vibration*, 279(1–2), 1–74.

Živanović, S., Pavic, A. & Reynolds, P. (2006) Modal testing and FE model tuning of a lively footbridge structure. *Engineering Structures*, 28(6), 857–868.

Author Index

Note: Page numbers in italics indicate figures; page numbers in bold indicate tables.

Anay, R. *24*

Bačinskas, Darius 249, 253, 259, *273*, 284
Bandhauer, Christian G. H. *14*
Barker, M. G. 24, 79, 84, 147, 155, 167, 176, *179*
Bernoulli, Daniel 10
Bliss, John 20
Bolle, Guido 9, 10, 12, 18, 80
Bonifaz, Jonathan 130, 155, 157
Burr, Theodore 20, *21*

Camino, S. *116, 117*
Casas, Joan R. 79, 217, 218, 219
Copper, Theodore 22

Denkhahn *18*
Diaz Arancibia, Mauricio 201, 203, 206, 208, 213

ElBatanouny, Mohamed K. 9
Euler, Leonhard 10

Fennis, S. *103, 120*
Frýba, L. 54, **55, 57**, 79, 132, 222

Hale, Enoch (Colonel) 20
Harris, Devin K. 155, 156, 166
Hernandez, Eli S. 23, 24, 155, 161, 168, 181, 184–185, **186, 191**, 193–194
Hordijk, D. *118, 131, 139*
Howe, William 21

Jakubovskis, Ronaldus 249

Kilikevičius, Arturas 249
Koekkoek, R. T. *103*
Kulibin, Ivan Petrovich 9–11

Lantsoght, Eva O. L. 3, 29, 32, 73, 78–80, 82, *83*, 84, 97, *102*, 105–107, *108*, 110–111, *121*, 129, 141, 148, 155, 171, 199, 253, 259, 272
Long, Stephen 21

Mörsch, E. *17*
Moulton, Stephen 21
Myers, John J. 79, 155–156, 161, 168, 181, 184–185, **186, 191**, 193–194, 199

Nowak, A. S. *81*

Okumus, Pinar 201, 203, 206, 208, 213
Olaszek, Piotr 78, 217, 220, 223, *224, 225, 227*, 228, *231, 232, 237, 238, 239, 240, 244, 245*

Palmer, Timothy 20, *20*
Pirner, M. 54, **55, 57**, 79, 132
Pratt, Caleb 21
Pratt, Thomas 21

Rider, Nathaniel 21
Roossen, M. *114*

Sanchez, Telmo A. 78, *132*, 155, 157
Schacht, Gregor 9, 19, 24,67, 84
Schmidt, Jacob W. 73, 82, 97, 129, 133, 141, 147, 267
Stamm **16**

Tharmabala, T. *81*

Wernwag, Louis 20, *21, 22*

Yang, Y *123*

DOI: https://doi.org/10.1201/9780429265426

Subject Index

Note: Page numbers in italics indicate figures; page numbers in bold indicate tables.

Aberfeldy Footbridge 251
Allowable Stress Rating (ASR) 23
American Association of State Highway and Transportation Officials (AASHTO) 23; AASHTO *Manual for Bridge Evaluation* 37–38, 104, 106–107, 148, 167, 171, 181, 197; AASHTO *Manual for Condition Evaluation of Bridges* 23; AASHTO *Manual for Maintenance Inspection of Bridges* 23; LRFD approach 196–197; LRFD *Bridge Design Specifications* 182, 197, 201, 202, 213, 215; LRFD girder distribution factors **196**
American Association of State Highway Officials (AASHO) 23
American Railroad Bridges (Copper) 22
arch bridges, masonry 86–87
ARCHES Project 220
assessment calculations, finite element model 106–108
assessment process, safety and 220–223
automated total station (ATS) *185*, 185–186

B-40-870 (bridge in Wisconsin) 202; bending strain influence and moment distribution *213*, 213–214; characteristics of 202; construction photograph *202*; coordination of load test 204–205; cross section of *202*; data acquisition for load test 209; deck strains during short-term loading 214, *214*; goals of load testing 202–203; instrumentation plan for testing 205–209; load configurations and locations *210*, **210**, 210–211; loading 209–211; load type and magnitude 209, *210*; planning and scheduling testing 211; preliminary analytical model 203–204;

recommendations 214–216; redundancy and repeatability 211; results 212–214; sensor locations *208*, 208–209, *209*; sensor types and application methods 205–207, *206*, *207*; shear strain influence lines and shear distribution *212*, 212–213
Bavaria 15
BELFA-loading vehicle: proof load testing *19*, 19–20
"Blaues Wunder" bridge, opening ceremony of *13*
bridge(s): assessment of 148–149; British guidelines 34–36; current development for improving 67; Czech Republic and Slovakia guidelines 54–56; estimation of behavior during load test 108–109; French guidelines 53–54; German guidelines 30–34; Hungarian guidelines 65–66; inspection of 99–103; Irish guidelines 36–37; Italian guidelines 64; load testing of 130–132; load testing practices 67; Polish guidelines 65; Spanish guidelines 57–64; Switzerland 64–65; United States guidelines 37–43; *see also* B-40-870 (bridge in Wisconsin); load test execution; pedestrian bridge(s); proof load testing; static load testing
Bridge A7957, 181; analytical and experimental evaluation results **198**; automated total station (ATS) *185*, 185–186; cross section of *183*; data acquisition 184–186; description of 182, *182*, 182–183, *183*; design parameters **196**; diagnostic load test of 182, 199; diagnostic load test program 186–187; dynamic amplification factor (DAF) 194; dynamic and quasi-static vertical deflection *194*; dynamic load allowance (DLA)

294 Subject Index

193–195, **195**; dynamic load test 187, *190*; dynamic load test results 193–195; elevation *182*; girder distribution factors 195–197; girders' longitudinal strain 191–193, **192**; H20 dump truck dimensions *186*; installation of embedded sensors 183–184; instrumentation plan 183–186; lateral distribution factor 191, **192**, 193, **193**; load rating by field load testing 197–198; load rating data **198**; non-contact remote data acquisition systems *185*; plan view of *182*; recommendations for 199; remote sensing vibrometer (RSV-150) *185*, 186; static load test 187, *188–189*; static load test results 187, 190–193; truck weights for load test **186**; vertical deflection in static test 187, 190, **191**; vibrating wire strain gages (VWSG) 183–184, *184*

Bridge Inspector's Reference Manual 99

bridge load testing: BELFA-loading vehicle *19*, 19–20; in Europe 11–20; in North America 20–24; potential for existing structures 24, **25**; *see also* diagnostic load testing; dynamic load testing; proof load testing; static load testing

bridge type(s): masonry 86–87; plastic composites 5, 80, 101, 165, **285**; prestressed concrete 85–86; reinforced concrete 84, *85*; steel 84; timber 87

British guidelines: load testing 34–35, 74; nomenclature 34; *see also* United Kingdom

Broughton Suspension Bridge, collapse of 14–15

building(s) (US): ACI 318–14 code for new 48–49; ACI 437.1, 43–48; ACI 437.2M–13 code for existing 49–52; cyclic loading protocol from ACI 437.1R–07 45, *45*; load-deflection curve for six cycles *47*; load-deflection curve for two cycles *47*; loading protocol for monotonic load test procedure *51*; load testing 4–5; testing guidelines in United States 43–52; test load magnitude (TLM) 44–45, 50

cable-stayed bridge: analysis of speed of vertical displacements stabilization **229**; lateral view of tested bridges *225*, *226*

civil engineering 9; Rensselaer School 20

collapses: bridge during load tests **16**; Broughton Suspension Bridge 14–15

Colossus Bridge Philadelphia 21, *22*

communication: diagnostic load testing 163–164; load test execution 137, *137*, 211, 215

concrete: destructive technique for compressive strength of *256*; nondestructive technique for mechanical properties of *256*; shear failure 110

concrete bridge(s) 77; analysis of speed of vertical displacements stabilization **229**; distribution of wheel print to center of cross section *106*; linear final element model 105; prestressed 85–86; reinforced 84, *85*; *see also* B-40-870 (bridge in Wisconsin); Bridge A7957

contact sensors: linear variable differential transducers (LVDTs) *85*, 86

Czech Republic 6, 29, 68; acceptance criteria 54–56; dynamic load tests 56, **57**; guidelines 54–56

data acquisition and visualization: load testing 133–135; real-time system 133, *135*; software for 134; weatherproof boxes for equipment 134, *136*

deflection measurement: evaluating 16, 18–19; proof load test 12, *13*

diagnostic load testing: background 217–223; on bridges after strengthening 156; calibration of analytical model for existing bridges 167–168; classification **25**; definition 217–218; dynamic load testing examples 232–245; dynamic tests 222–223; evaluation of results for new bridges 168–171; examples of 223–245; for existing bridge 161–162; goal of 30, 155–156, 172; goals for new and existing bridge **99**; improved assessment for existing bridges 171–172; investigation range 219; loading methods 162–163; measurement methods 219–220; measuring techniques 246; monitoring bridge behavior during 163–164; for new bridge 157–161; objectives of 218; on-site validation and test data review 164–166; planning and execution 218–220; post-processing of 79, 141; preparation for 157–162; procedures for execution of 162–164; processing and reporting test data 166; processing results of 164–168; program for Bridge A7957 186–187; recommendations for practice 246–247; results and safety assessment 220–223; static load testing examples 223–232; static tests 221–222; structural behavior in 155–156, 172; test(s) 3–4, 78; transverse flexural distribution 79; types 78–79; updating finite element model with measurements 147–148; verification of responses for new bridges 166–167; *see also* dynamic load testing; static load testing

Dornie Bridge 34

Dutch Guidelines for the Assessment of Bridges RBK 77

dynamic load testing 222–223; cutoff frequencies as filtration parameters 242–243, **244**; examples of 236–245; extrapolation of values for quasi-static load 237–243; extrapolation of values for quasi-static speed 233–235; extrapolation of values under higher speed 236, 243–245; extreme displacement amplitude extrapolation *244*, 244–245, *245*; measured time histories of vertical displacements for train passages 237, *239*, *240*, *241*; quasi-static value estimation using FIR filters 238–240, *241*, *242*, **243**; recommendations for practice 246–247; train passage results using FIR filters 238–240, *241*, *242*; view of tested bridges *237*, *238*; *see also* diagnostic load testing of bridges

Ecuador: Spanish loading recommendations for bridges 158; *see also* Los Pajaros Bridge; Villorita Bridge

elastic deformation 12

elastomeric bearings, support deformations 143–144, *144*

electrical resistance strain gages 205, *206*, *208*, 215

Europe 6, 7; bridge assessment in 148–149; bridge load testing in 11–20; transport infrastructure 249

European Union (EU), transport infrastructure 249–250

execution of load test *see* load test execution

Extended Strip Model 110

failure test(s) 74, 78, 81–82; application of load for *82*; classification **25**

field testing: load rating of Bridge A7957 by 197–198; methods of 4; recommendations 199; *see also* Bridge A7957

finite element model 23; assessment calculations 106–108; assessment of existing bridge 167–168; comparing results with measured strains during test *133*;

development of 104–106; distribution of shear stresses in transverse direction *108*; estimation of bridge behavior during load test 108–109; live-load model 105; load testing and 133; pedestrian bridges 279–280; preliminary 203–204, 215; self-weight 105; shear capacity considerations 110–111; superimposed dead load 105; typical loads in 105; updating for tested bridge 283–284; updating with measurement data 147–148

footbridge(s): basis of dynamics of 274–275, *275*; code requirements for serviceability of 284–286; comparing self-vibration modes and frequencies 281–283, *282*; different types of *250*; evaluation of condition 286; experimental deformed shapes of 272–273, *273*; finite element model of tested 280, *281*; identification of dynamic parameters 287; modal assurance criterion (MAC) 282–283; preliminary inspection before tests 255–257; static loading using loaded vehicles *261*; types of loading in static field tests of 259, *260*; *see also* pedestrian bridge(s)

forced and ambient vibration tests pedestrian bridges *263*, 263–264, *264*

France 6, 29, 68; bridge collapses during load tests **16**; evaluation of load test 54; guidelines 53–54; recommendations for load application 53–54

free vibration tests pedestrian bridges 262, *262*

frozen bearing 100; example of *100*

Germany 6, 29, 68; bridge collapses during load tests **16**; guidelines 30–34; load testing guidelines 30–31; load testing of bridge in Alexisbad *18*; load testing practices 67; requirements for crack width **33**; safety philosophy 31–32, *32*; stop criteria 32–34; suspension bridge over Saale River 14, *14*; target proof load 31–32

girder distribution factors Bridge A7957 195–197

Highway Structures Information System 202

Hudson River (Waterford, NY) 20, *21*

humidity, correcting for influence of 144, *145*

Hungary 6, 29, 68; guidelines 65–66; limitations to deviation between measured and calculated deformations **67**

hydraulic jack(s) *see* load application

296   Subject Index

inaugural celebrations, loading test at 12, *13*
inspection: bridge 99–103; frozen bearing 100; limitations of testing site 101–103; map of bridge damages 97–98, *102*
Ireland 6, 29, 68; evaluation of load test 37; load testing guidelines 36–37; recommendations for applied loading 37
Italy 6, 29, 68; guidelines 64

Kosovo: bridge collapses during load tests **16**; steel truss bridge over Morava River 12

lateral distribution factor: deflection measurements 191, *192*; strain measurements 193, **193**
Latvia 249; bridge collapse during load tests **16**
linear variable differential transducers (LVDTs) 193, 207, *207*; example load test 207; loading and unloading stages of pedestrian bridges 272, *273*; masonry bridges 86; monitoring concrete slab bridge *85*; pedestrian bridges 265–266, *266*; sensor plan *121, 124*
Lithuanian Road Administration 252, 254, 257, 261, 270–271
Load and Resistance Factor Rating (LRFR) 23
load application: counterweights *119*; dead weight with cement bags *117*; dead weight with water *116*; dump trucks *115*; equipment for 130–132; hydraulic jacks 114, *115, 116–117, 119,* 131, 142; load testing vehicle *118*; methods *115, 116*; sensor plan 119–121, *121, 124,* 124–127; steel spreader beam *119*
Load Factor Rating (LFR) 23
load test(s): acoustic emissions sensors *123*; Alexisbad, Germany *18*; book outline 6–8; book structure *6*; bridge collapses during **16**; bridge structure or element 74; considerations 73–78; counterweights usage *119*; dead weight application with cement bags *117*; dead weight application with water *116*; determination for bridge 82–84; determining objectives for 98–99, 126; diagnostic 78–79; estimation of bridge behavior during 108–109; exterior and global measurements 122–123; failure tests 81–82; finite element model 104–111; flowchart for determination of *75*; formulating recommendations for maintenance or operation 149; interior measurements 123–125; levels of approximation *83*, 104;

limitations of testing site 101–103; load application methods *115*; load application with load testing vehicle *118*; loading requirements 113–117; local measurements 122; map of damages 97–98, *102*; measurement engineers *114*; measurement goals for 145–146; measurements and sensor plan 119–125; personnel requirements 113; planning 111–113, 126–127; potential for evaluating existing structures 24, **25**; preparation for 97–98; proof 79–81; questions for 90–91; recommendations for reporting 149–150; safety requirements for 87–90; sensor plan example *121, 124*; single hydraulic jack *119*; steel spreader beam *119*; structural safety 90; time schedule for planning *112*; traffic control and safety 117–119; traffic overview during *120*; types and goals 73–74; *see also* diagnostic load testing; proof load testing; types of 78–82; *see also* B-40-870 (bridge in Wisconsin)
load test execution: communication in 137, *137*; data acquisition and visualization equipment 133–135; interpreting measurements of 136–137; loaded area and measured deformations *134*; loading equipment for 130–132, *131*; measurement equipment for 132–137; measurement requirements 132–133; real-time data visualization *135*; safety during 129–130, 137–138; sensors for 135–136; weatherproof boxes for data acquisition system *136*; *see also* post-processing measurement data
Los Pajaros Bridge: configuration of trucks for load test *160*; cross section of *157*; design code 158–159; diagnostic load test with loading vehicles on *162*; evaluating results of 169, *170*; load cases for *157, 158, 159,* 159–160; new bridge diagnostic testing 157–161; predicted deflections for *161*; review of load data *165*
LRFR (load and resistance factor rating), rating factor for 38

maintenance, formulating recommendations for 149, 151
Manual for Bridge Evaluation 37–38, 104, 106–107, 148, 167, 171, 181, 197
Manual for Bridge Rating through Load Testing 167; bridge guidelines in United

Subject Index   297

States 37–43; determination of K_{b1}, K_{b2}, and K_{b3} **42**; determination of rating factor after diagnostic load test 41–42; determination of rating factor after proof load test 42–43; execution of load tests 40; personnel requirements 113; preparation of load tests 39–40; test termination 38–39; *see also* United States

Manual for Condition Evaluation and Load and Resistance Factor Rating (LRFR) of Highway Bridges (MCE) 197

masonry bridges, structure considerations 86–87

measurement(s): application of acoustic emissions sensors *123*; data acquisition and visualization equipment 133–135; exterior and global 122–123; interior 123–125; interpretation during load test 136–137; local, 122; methods 219–220; real-time data visualization *135*; reporting of 144–147; requirements for load testing 132–133; sensor plan and 119–125; sensors for 135–136; updating finite element model with data, 147–148; *see also* post-processing of measurement data

Middle Ages 9

Milwaukee County Department of Transportation 204, 209

Missouri *see* Bridge A7957

Missouri Department of Transportation (MoDOT) 182

Mohawk Bridge (Schenectady, NY) 21, *21*

Morava River, steel truss bridge 12

Münchensteiner Bridge, collapse of 15, *17*

National Bridge Inspection Standards (NBIS) 23

Neva River, design of bridge over *10*

New Mexico, photograph of load test *24*

North America 6, 7; assessment for existing bridges 171; bridge load testing in 20–24

operation, formulating recommendations for 149

pedestrian bridge(s) 5; Aberfeldy Footbridge 251; analysis of test results 271–278; dynamic measurement systems 267–268; dynamic tests 271; execution of tests 269–271; forced and ambient vibration tests *263*, 263–264, *264*; free vibration tests 262, *262*; loading and unloading stages during static test of bridge deck *270*; loading

equipment for 131; loading of bridge 258–264; locations of measurement points *269*; measuring bridge displacement 266, *267*; measuring techniques and equipment 265–269; methods of identification of dynamic parameters of 274–278; methods of identification of static parameters of 272–274; objectives of tests 253–254; organization of tests 264–271; pedestrian loads 259; preliminary inspection of footbridge before tests 255–257; preparation for testing 254–264; presentation of results 278; serviceability limit states (SLSs) 250; static loading using loaded vehicles 260, *261*; static tests 258–259, *260*, 261–262, 269–271; strain measurement 266–267, *267*; test program 257–258; theoretical modeling of tested bridge 278–286; transport infrastructure 249–251, 286–287; types of footbridges *250*; types of tests 252–253; vibrations using dynamic equipment *268*

personnel: measurement engineers *114*; requirements: load tests 113

Poland 6, 29, 68; guidelines 65; Road and Bridge Research Institute 223

post-processing of measurement data 141–142, 150–151; applied load determination 142, *143*; correction for support deformations 143–144; correction for temperature and humidity influence 144, *145*; elastomeric bridge bearings 143, *144*; longitudinal deflection profiles for load levels 146, *146*; recommendations for reporting load tests 149–150; reporting measurements 144–147; verification of data 142–143

prestressed concrete bridge(s): analysis of speed of vertical displacements stabilization **229**; analysis of vertical displacements stabilization *232*; Bridge A7957 in Missouri 182; lateral view of tested bridges *227*, *228*; structure considerations 85–86; *see also* Bridge A7957

proof load test(s) 3–4, 29, 78; BELFA-loading vehicle for *19*, 19–20; bridge in Rotterdam 12; classification **25**; deflection measurement 12, *13*; field measurements to update finite element model 148; finite element models 111; goals for new and existing bridge **99**; identifying uncertainties 80; loading equipment for 130–131; pedestrian bridges

252–253; post-processing of 141; reporting measurements 146–147; time-dependent damage 80; truncation of probability density function of resistance after *81*; types 79–81; using enormous masses and loads *17*; viaduct with open traffic lane *103*

punching shear capacity 110

railroad/railway bridges 5; dynamic load tests of 56; dynamics of 217, 222–223, 236–237, *237*, **243**; filtering method of 233; guideline for load tests 19; loading of 35, 54; Münchenstein collapse 15, *17*; quasi-static value estimating displacement 246; Spanish guidelines for 57; transport safety of 249–250; Tübingen, Germany 12

Rating Factors 106–108, 162; load testing of existing bridge 162, 171

reinforced concrete bridges: analysis of speed of vertical displacements stabilization **229**; analysis of vertical displacements stabilization *231*; lateral view of tested bridges *225*, *226*; structure considerations 84, *85*

reliability index 8, 43, 221

remote sensing vibrometer (RSV-150) *185*, 186

Rosmosevej Bridge, application of load for collapse test *82*

Rotterdam (Netherlands), proof load testing of bridge 12

Ruytenschildt Bridge, collapse test 82, *82*

Saale River, suspension bridge over 14, *14*

safety 3, 4, 7; assessment 220–223; load test execution 129–130, 137–138; personnel and traveling public 88–90, *89*; protective clothing for load test *139*; requirements during load testing 87–90; signposting for detour *139*; structural 90

safety engineer 91; certified 113; personnel and traveling public 88–89; traffic control 117–118

Schuylkill River, "Permanent Bridge" at Philadelphia 20

sensor(s): displacement transducers 207, *207*, *209*; electrical resistance strain gages 205, *206*, *208*, 215; instrumentation of bridge B-40-870 205–209; load testing of bridges 135–136; vibrating wire strain gages (VWSG) in Bridge A7957 183–184, *184*; VWSG in bridge B-40-870, 205, *206*, *208*

sensor plans: application of acoustic emissions sensors *123*; examples *121*, *124*; measurements and 119–125

Slovakia 6, 29, 68; acceptance criteria 54–56; dynamic load tests 56, **57**; guidelines 54–56

soft load test 217–218

Spain 6, 29, 68; acceptance criteria for dynamic load tests 62–64; cycle of loading and unloading *61*; diagnostic load test guidelines 156; guidelines 57–64; highway bridges 57; loading requirements 58–59; pedestrian bridges 57; railway bridges 57; recommendations for execution of dynamic load tests **63**; stabilization of measurements *60*; stop and acceptance criteria for static load tests 59–62

static load testing: analysis of speed of vertical displacements stabilization **229**; dead loads for 130; displacements in time function *224*; estimation of elastic and permanent values 223–224; examples of application to different bridge types 224–232; lateral view of tested bridges *225*, *226*, *227*; recommendations for practice 246; *see also* diagnostic load testing of bridges

steel, nondestructive technique for mechanical properties of *256*

steel bridges 77; analysis of speed of vertical displacements stabilization **229**; analysis of vertical displacements stabilization *230*; lateral view of tested bridges *225*; structure considerations 84

suspension bridge, collapse of Broughton Suspension Bridge 14–15

Switzerland 6, 29, 68; bridge collapses during load tests **16**; bridge test over Thur River 12; collapse of Münchensteiner Bridge 15, *17*; guidelines 64–65; road bridge near Salez 14

temperature, correcting for influence of 144, *145*

The Netherlands *82*, 88, 118, 171

theoretical modeling of tested bridge 278–286; code requirements for serviceability of footbridges 284–286; comparing experimental and theoretical results 280–283; evaluation of footbridge condition 286; modeling techniques 279–280; structural damping **285**; updating model 283–284

thrust line method 10

Thur River, test of bridge 12

timber bridges, structure considerations 87
time schedule, planning load tests *112*
transport infrastructure 249–250, 286–287;
  see also pedestrian bridge(s)

ultimate limit state (ULS): bridge assessment
  149; diagnostic load test 79, 156; estimating
  36; load testing of bridges 19; static
  tests 254
United Kingdom 6, 29, 68; British guidelines
  34–36; evaluation of load test 36; load
  testing guidelines 34–35; preparation and
  application of loading 35
United States 6, 68; bridges using *Manual
  for Bridge Rating through Load Testing*
  37–43; tests on buildings in 43–52; *see also*

B-40-870 (bridge in Wisconsin); Bridge
  A7957; building(s) (US)
Unity Check (load testing) 75, 83, 91, 106–108,
  162, 171

vibrating wire strain gages (VWSG): bridge
  B-40-870, 205, *206*; installation in Bridge
  A7957 183–184, *184*
Villorita Bridge 157; load testing of *132*;
  symmetry of load cases 164, *165*

weigh-in-motion (WIM) system 218
Whipple Squire 21
Wisconsin Department of Transportation
  (DOT) 202, 204
Wisconsin Highway Research Program 204

Structures and Infrastructures Series

Book Series Editor: Dan M. Frangopol

ISSN: 1747–7735

Publisher: CRC Press/Balkema, Taylor & Francis Group

1. Structural Design Optimization Considering Uncertainties
   Editors: Yiannis Tsompanakis, Nikos D. Lagaros & Manolis Papadrakakis
   ISBN: 978-0-415-45260-1 (Hb)

2. Computational Structural Dynamics and Earthquake Engineering
   Editors: Manolis Papadrakakis, Dimos C. Charmpis,
   Nikos D. Lagaros & Yiannis Tsompanakis
   ISBN: 978-0-415-45261-8 (Hb)

3. Computational Analysis of Randomness in Structural Mechanics
   Christian Bucher
   ISBN: 978-0-415-40354-2 (Hb)

4. Frontier Technologies for Infrastructures Engineering
   Editors: Shi-Shuenn Chen & Alfredo H-S. Ang
   ISBN: 978-0-415-49875-3 (Hb)

5. Damage Models and Algorithms for Assessment of Structures
   under Operating Conditions
   Siu-Seong Law & Xin-Qun Zhu
   ISBN: 978-0-415-42195-9 (Hb)

6. Structural Identification and Damage Detection using Genetic Algorithms
   Chan Ghee Koh & Michael John Perry
   ISBN: 978-0-415-46102-3 (Hb)

7. Design Decisions under Uncertainty with Limited Information
   Efstratios Nikolaidis, Zissimos P. Mourelatos & Vijitashwa Pandey
   ISBN: 978-0-415-49247-8 (Hb)

8. Moving Loads – Dynamic Analysis and Identification Techniques
   Siu-Seong Law & Xin-Qun Zhu
   ISBN: 978-0-415-87877-7 (Hb)

9. Seismic Performance of Concrete Buildings
   Liviu Crainic & Mihai Munteanu
   ISBN: 978-0-415-63186-0 (Hb)

DOI: https://doi.org/10.1201/9780429265426

10. Maintenance and Safety of Aging Infrastructure
    Dan M. Frangopol & Yiannis Tsompanakis
    ISBN: 978-0-415-65942-0 (Hb)

11. Non-Destructive Techniques for the Evaluation of Structures
    and Infrastructure
    Belén Riveiro & Mercedes Solla
    ISBN: 978-0-138-02810-4 (Hb)

12. Load Testing of Bridges – Current Practice and Diagnostic Load Testing
    Editor: Eva O.L. Lantsoght
    ISBN: 978-0-367-21082-3 (Hb)

13. Load Testing of Bridges – Proof Load Testing and the Future of Load Testing
    Editor: Eva O.L. Lantsoght
    ISBN: 978-0-367-21083-0 (Hb)